T0348379

VOLUME ONE HUNDRED AND THIRTEEN

Progress in
# MOLECULAR BIOLOGY
# AND TRANSLATIONAL
# SCIENCE

Fluorescence-based Biosensors

From Concepts to Applications

VOLUME ONE HUNDRED AND THIRTEEN

# PROGRESS IN MOLECULAR BIOLOGY AND TRANSLATIONAL SCIENCE

## Fluorescence-based Biosensors
### From Concepts to Applications

Edited by

**MAY C. MORRIS**
*CRBM-CNRS-UMR5237,*
*1919 Route de Mende,*
*34293 Montpellier, France*

ELSEVIER

AMSTERDAM • BOSTON • HEIDELBERG • LONDON
NEW YORK • OXFORD • PARIS • SAN DIEGO
SAN FRANCISCO • SINGAPORE • SYDNEY • TOKYO
Academic Press is an imprint of Elsevier

Academic Press is an imprint of Elsevier
The Boulevard, Langford Lane, Kidlington, Oxford, OX51GB, UK
32, Jamestown Road, London NW1 7BY, UK
Radarweg 29, PO Box 211, 1000 AE Amsterdam, The Netherlands
225 Wyman Street, Waltham, MA 02451, USA
525 B Street, Suite 1900, San Diego, CA 92101-4495, USA

First edition 2013

**Library of Congress Cataloging-in-Publication Data**
A catalog record for this book is available from the Library of Congress

**British Library Cataloguing in Publication Data**
A catalogue record for this book is available from the British Library

ISBN: 978-0-12-386932-6
ISSN: 1877-1173

For information on all Academic Press publications
visit our website at store.elsevier.com

Printed and bound by CPI Group (UK) Ltd, Croydon, CR0 4YY
Transferred to digital print 2012

# CONTENTS

# CONTRIBUTORS

**Jean-François Bodart**
Laboratoire de Régulation des Signaux de division EA4479 University Lille1, and
Groupement de Recherche Microscopie Fonctionnelle du vivant, GDR2588-CNRS,
Villeneuve d'Ascq, France

**Alexei A. Bogdanov**
Laboratory of Molecular Imaging Probes, Department of Radiology, University of
Massachusetts Medical School, Worcester, Massachusetts, USA

**Sebastien Deshayes**
Centre de Recherches de Biochimie Macromoléculaire, Department of Chemical Biology
and Nanotechnology for Therapeutics, CRBM-CNRS, UMR-5237, UM1-UM2,
University of Montpellier, 1919 Route de Mende, Montpellier, France

**Gilles Divita**
Centre de Recherches de Biochimie Macromoléculaire, Department of Chemical Biology
and Nanotechnology for Therapeutics, CRBM-CNRS, UMR-5237, UM1-UM2,
University of Montpellier, 1919 Route de Mende, Montpellier, France

**Yoichiro Fujioka**
Laboratory of Pathophysiology and Signal Transduction, Hokkaido University Graduate
School of Medicine, N15W7, Kita-ku, Sapporo, Japan

**Jonathan B. Grimm**
Janelia Farm Research Campus, The Howard Hughes Medical Institute, Ashburn, Virginia,
USA

**Laurel M. Heckman**
Janelia Farm Research Campus, The Howard Hughes Medical Institute, Ashburn, Virginia,
USA

**Laurent Héliot**
Biophotonic Team (BCF), IRI CNRS, University Lille1–University Lille2 USR 3078, and
Groupement de Recherche Microscopie Fonctionnelle du vivant, GDR2588-CNRS,
Villeneuve d'Ascq, France

**Robert M. Hoffman**
AntiCancer, Inc., and Department of Surgery, University of California, San Diego,
California, USA

**Julie Kniazeff**
CNRS, UMR-5203, Institut de Génomique Fonctionnelle; INSEM U661, and Universités
de Montpellier 1&2, UMR-5203, Montpellier, F-34094, France

**Andrey S. Klymchenko**
Laboratoire de Biophotonique et Pharmacologie, UMR 7213 CNRS, Université de
Strasbourg, Faculté de Pharmacie, Illkirch Cedex, France

**Luke D. Lavis**
Janelia Farm Research Campus, The Howard Hughes Medical Institute, Ashburn, Virginia, USA

**Aymeric Leray**
Biophotonic Team (BCF), IRI CNRS, University Lille1–University Lille2 USR 3078, France, and Groupement de Recherche Microscopie Fonctionnelle du vivant, GDR2588-CNRS, France

**Mary L. Mazzanti**
Laboratory of Molecular Imaging Probes, Department of Radiology, University of Massachusetts Medical School, Worcester, Massachusetts, USA

**Yves Mely**
Laboratoire de Biophotonique et Pharmacologie, UMR 7213 CNRS, Université de Strasbourg, Faculté de Pharmacie, Illkirch Cedex, France

**May C. Morris**
CRBM-CNRS-UMR 5237, Chemical Biology and Nanotechnology for Therapeutics, Montpellier, France

**Shigeyuki Nakada**
Mitsui Engineering and Shipbuilding Co. Ltd., Tamano Technology Center, 3-16-1 Tamahara, Tamano, Okayama, Japan

**Yusuke Ohba**
Laboratory of Pathophysiology and Signal Transduction, Hokkaido University Graduate School of Medicine, N15W7, Kita-ku, Sapporo, Japan

**Jean-Philippe Pin**
CNRS, UMR-5203, Institut de Génomique Fonctionnelle; INSEM U661, and Universités de Montpellier 1&2, UMR-5203, Montpellier, F-34094, France

**Laurent Prézeau**
CNRS, UMR-5203, Institut de Génomique Fonctionnelle; INSEM U661, and Universités de Montpellier 1&2, UMR-5203, Montpellier, F-34094, France

**Marcin Ptaszek**
Department of Chemistry and Biochemistry, University of Maryland Baltimore County, Baltimore, Maryland, USA

**Philippe Rondard**
CNRS, UMR-5203, Institut de Génomique Fonctionnelle; INSEM U661, and Universités de Montpellier 1&2, UMR-5203, Montpellier, F-34094, France

**Franck B. Riquet**
Laboratoire de Régulation des Signaux de division EA4479 University Lille1, and Groupement de Recherche Microscopie Fonctionnelle du vivant, GDR2588-CNRS, Villeneuve d'Ascq, France

**Pauline Scholler**
CNRS, UMR-5203, Institut de Génomique Fonctionnelle; INSEM U661; Universités de Montpellier 1&2, UMR-5203, Montpellier, F-34094, and Cisbio Bioassays, Parc Marcel Boiteux, BP 84175, Codolet, France

**François Sipieter**
Biophotonic Team (BCF), IRI CNRS, University Lille1–University Lille2 USR 3078;
Laboratoire de Régulation des Signaux de division EA4479 University Lille1, and
Groupement de Recherche Microscopie Fonctionnelle du vivant, GDR2588-CNRS,
Villeneuve d'Ascq, France

**Corentin Spriet**
Biophotonic Team (BCF), IRI CNRS, University Lille1–University Lille2 USR 3078, and
Groupement de Recherche Microscopie Fonctionnelle du vivant, GDR2588-CNRS,
Villeneuve d'Ascq, France

**Dave Trinel**
Biophotonic Team (BCF), IRI CNRS, University Lille1–University Lille2 USR 3078, and
Groupement de Recherche Microscopie Fonctionnelle du vivant, GDR2588-CNRS,
Villeneuve d'Ascq, France

**Eric Trinquet**
Cisbio Bioassays, Parc Marcel Boiteux, BP 84175, Codolet, France

**Masumi Tsuda**
Laboratory of Pathophysiology and Signal Transduction, Hokkaido University Graduate
School of Medicine, N15W7, Kita-ku, Sapporo, Japan

**Thi Nhu Ngoc Van**
CRBM-CNRS-UMR 5237, Chemical Biology and Nanotechnology for Therapeutics,
Montpellier, France

**Pauline Vandame**
Laboratoire de Régulation des Signaux de division EA4479 University Lille1, and
Groupement de Recherche Microscopie Fonctionnelle du vivant, GDR2588-CNRS,
Villeneuve d'Ascq, France

**Pierre Vincent**
Groupement de Recherche Microscopie Fonctionnelle du vivant, GDR2588-CNRS,
Villeneuve d'Ascq, and Neurobiologie des processus adaptatifs—UMR 7102, UPMC, Paris,
France

**Jurriaan M. Zwier**
Cisbio Bioassays, Parc Marcel Boiteux, BP 84175, Codolet, France

# PREFACE

One of the major challenges of modern biology and medicine consists in visualizing biomolecules in their natural environment with the greatest level of accuracy so as to gain insight into their behavior in physiological and pathological settings. Together with the discovery of the green fluorescent protein and the development of genetically encoded autofluorescent variants of GFP, recent advances in the synthesis of small-molecule fluorescent probes, and the explosion of fluorescence-based imaging technologies, a new generation of tools has emerged, providing a means of probing molecular targets and biomarkers in a sensitive and noninvasive fashion. Fluorescent biosensors constitute attractive tools for probing protein/protein interactions, conformational changes, and posttranslational modifications in complex biological samples and in living cells, offering a means of studying the biological activity and dynamics of biomolecular targets in real time, with high temporal and spatial resolution. Whether genetically encoded or based on peptide or protein scaffolds, biosensors find useful applications for fundamental purposes as well as in biomedical and drug discovery programs, allowing to address issues that previously remained unresolved due to technological constraints.

This volume will attempt to convey the many exciting developments the field of fluorescent biosensors and reporters has witnessed over the recent years, from concepts to applications, including chapters on the chemistry of fluorescent probes and on technologies for monitoring protein/protein interactions and imaging fluorescent biosensors in cultured cells and *in vivo*. Other chapters are devoted to genetically encoded reporters or to protein and peptide biosensors, together with examples illustrating their application to fundamental studies and to biomedical and drug discovery developments, *in vitro*, *in cellulo*, and *in vivo*.

I am most grateful to the authors, without whom this book would not have been possible. The contributors have written authoritative chapters that will be useful to a wide audience of scientists, whether newcomers or specialists in the field. I would also like to thank the staff of Elsevier, in particular, Sarah Latham, for her help and patience throughout the preparation of the volume. I hope that the readers of this volume will appreciate its content and find it both useful and scientifically stimulating.

MAY C. MORRIS

# INTRODUCTION

In the postgenomic and postproteomic era, one of the major challenges in biological sciences consists in probing molecules in their natural environment so as to monitor changes in their spatiotemporal dynamics, relative abundance, and biological activity in response to specific stimuli or drugs, in a physiological context as well as in a pathological setting. Since biological processes are intrinsically dynamic, it has become essential to develop tools that report on biological phenomena in a suitable fashion, conveying an overall sense of their dynamics and reversibility with high spatial and temporal resolution.

Biosensors are commonly defined as "analytical devices" that combine a biological component or a biomimetic material (e.g., tissue, microorganism, receptor, enzyme, antibody, nucleic acid, etc.), responsible for sensing a specific analyte or target, with a physicochemical transducer (optical, electrochemical, thermometric, piezoelectric, magnetic, or mechanical), which converts the process of recognition into a detectable and measurable signal.[1] The discovery of the green fluorescent protein (GFP), the development of genetically encoded autofluorescent protein variants of GFP,[2,3] together with significant advances in the synthesis of small fluorescent dyes,[4,5] and the concomitant explosion of fluorescent imaging technologies over the past decade prompted the development of an attractive class of biosensors, known as fluorescence-based biosensors, that report on protein/protein interactions, conformational changes, enzymatic activities, and posttranslational modifications through changes in fluorescence.[6–9] Fluorescence detection is currently one of the most widely used approaches in biomolecular imaging due to its high intrinsic sensitivity and selectivity. Indeed, fluorescence lends itself to nondestructive imaging, thereby preserving the sample and the molecules of interest within. Moreover, imaging of fluorescent signals provides a high degree of temporal and spatial resolution and allows for real-time tracking of a molecule in motion in a complex solution or environment. As such, fluorescent biosensors constitute extremely useful tools for the detection of biomolecules for monitoring dynamic molecular events and visualizing dynamic processes in complex biological samples in living cells and in animal models.

Two large classes of kinase biosensors have been developed: genetically encoded single-chain FRET reporters and fluorescent peptide and protein biosensors. The development of GFP-based reporters led to the development of genetically encoded FRET reporters, which have been successfully applied to study the behavior of biomolecules in living cells with high spatial and temporal resolution.[2,3,6,7,9] In parallel, combined efforts in fluorescence chemistry and in chemical biology have led to the design of an entirely different family of biosensors, based on peptide, polypeptide, or polymeric scaffolds, onto which small synthetic probes with particular photophysical properties, such as environmental sensitivity, are incorporated.[4,5] These nongenetic biosensors allow to monitor molecular activities *in vitro* and in more complex biological samples, including cell extracts and living cells, with an equally successful outcome.[7–9]

Today, these tools offer a wide panel/range of opportunities for fundamental research and biomedical applications, allowing to address issues which could not be resolved previously, due to a lack of appropriate technologies. Fluorescent biosensors not only provide a means of studying the behavior of enzymes in their natural environment, they are also very useful for comparative studies of biological processes in healthy and pathological conditions and may further be applied to diagnostic approaches to assess the status of a specific target, to highlight molecular and cellular alterations associated with pathological disorders, and to monitor disease progression and response to therapeutics. Moreover, fluorescent biosensors constitute attractive and powerful tools for drug discovery programs, from high-throughput screening assays to postscreen characterization of hits and preclinical evaluation of candidate drugs.

This volume will attempt to provide a general overview of the state-of-the-art in the field of fluorescent biosensors – from general concepts on fluorescence and on the chemistry of fluorescent probes, to the design and engineering of fluorescent biosensors, and their application *in vitro*, in living cells and in animal models.

A first set of chapters will describe the chemistry behind fluorescent probes and provide several examples of small-molecule fluorogenic probes, environment-sensitive probes, and probes suited for *in vivo* biological applications. In Chapter 1, Lavis *et al.* describe the chemistry of fluorogenicity and the different classes of small-molecule fluorophores available for biological applications. Their chapter highlights both general strategies and unique approaches that are employed to control fluorescence and modulate fluorescent properties using chemistry. Klymchenko and Mely, in Chapter 2, provide an overview of fluorescent-sensitive dyes that undergo changes in their color and brightness in response to changes in their environment. They further describe their use as reporters of biomolecular interactions in biosensing applications, in particular, for monitoring protein conformational changes and interactions with nucleic acids and with lipids. Ptaszek, in Chapter 3, addresses the requirements of fluorescent dyes for *in vivo* applications. He describes the optical, chemical, and biological properties of red and near-infrared dyes and discusses the challenges in design and development of novel dyes which are suitable for *in vivo* imaging, with special focus on self-illuminating fluorophores and fluorophores for multicolor *in vivo* imaging.

A second set of chapters addresses fluorescence principles and technologies. Chapter 4, by Deshayes and Divita, discusses fluorescence approaches for monitoring interactions between biological molecules *in vitro* and explain the subtleties and challenges for monitoring steady-state reactions and dynamic conformational changes. They describe several examples of fluorescence studies of protein/protein, protein/nucleic acids, protein/small molecules, and biomembrane/biomolecule interactions. Riquet *et al.*, in Chapter 5, provide an overview of fluorescence imaging approaches and practical methodology for FRET imaging of genetically encoded fluorescence reporters in living cells. In particular, they provide some well-documented examples of FRET biosensors for imaging protein kinase activity. They further discuss ongoing efforts to improve biophotonic techniques and genetically encoded probes for qualitative and quantitative studies of dynamic molecular processes in living cells.

In Chapter 6, Van and Morris provide an overview on protein kinase biosensors from genetically encoded to peptide/protein-based biosensors to study protein kinases, one of the major classes of enzymes underlying biological signaling pathways. These tools provide means of probing protein kinase activity and conformation, thereby allowing to investigate protein kinase behavior, dynamics, and regulation in living cells in real time, with high spatial and temporal resolution. We discuss developments in fluorescent biosensor technology related to protein kinase sensing and the different strategies employed to monitor kinase activity, regulation, and subcellular dynamics in living cells. Moreover, we discuss their application in imaging disease and monitoring response to therapeutics, to high-throughput screening and drug discovery programs. Chapter 7, by Scholler *et al.*, describes the application of fluorescent sensing strategies to identify inhibitors of G protein-coupled receptors in high-throughput screening assays based on time-resolved Förster resonance energy transfer technology (TR-FRET and HTRF®). Ohba *et al.*, in Chapter 8, review the molecular bases of genetically encoded fluorescent proteins and fluorescent protein biosensors and describe strategies to apply these biosensors for biomedical purposes in the clinic.

Finally, two chapters are dedicated to the application of fluorescent reporters to *in vivo* imaging studies. In Chapter 9, Bogdanov and Mazzanti describe fluorescent macromolecular sensors to probe enzymatic activities, based on polymeric scaffolds that are cleaved by proteases in diseased tissues, thereby releasing optically silent fluorophores, which allow imaging disease *in vivo*. In addition, they review the application of these probes in various areas of experimental medicine and provide an overview of currently available techniques for animal imaging using visible and near-infrared light. Chapter 10, by R. Hoffmann, describes the use of genetically encoded fluorescent proteins as *in vivo* sensors, with particular focus on strategies for imaging different aspects of cancer in living animals, including tumor cell mobility, invasion, metastasis, and angiogenesis. In his chapter, he further describes whole-body imaging technology to monitor the dynamics of metastatic cancer and determine the efficacy of antitumor and antimetastatic drugs in mouse models.

## REFERENCES

1. Turner APF, Karube I, Wilson GS. *Biosensors: fundamentals and applications.* Oxford: Oxford University Press; 1987.
2. Zhang J, Campbell RE, Ting AY, Tsien RY. Creating new fluorescent probes for cell biology. *Nat Rev Mol Cell Biol* 2002;**3**:906–18.
3. Chudakov DM, Matz MV, Lukyanov S, Lukyanov KA. Fluorescent proteins and their applications in imaging living cells and tissues. *Physiol Rev* 2010;**90**:1103–63.
4. Lavis LD, Raines RT. Bright ideas for chemical biology. *ACS Chem Biol* 2008;**3**:142–55.
5. Loving GS, Sainlos M, Imperiali B. Monitoring protein interactions and dynamics with solvatochromic fluorophores. *Trends Biotechnol* 2010;**28**:73–83.
6. Ibraheem A, Campbell RE. Designs and applications of fluorescent protein-based biosensors. *Curr Opin Chem Biol* 2010;**14**:30–6.
7. Wang H, Nakata E, Hamachi I. Recent progress in strategies for the creation of protein-based fluorescent biosensors. *Chembiochem* 2009;**10**:2560–77.
8. Pazos E, Vázquez O, Mascareñas JL, Vázquez ME. Peptide-based fluorescent biosensors. *Chem Soc Rev* 2009;**38**:3348.
9. Morris MC. Fluorescent biosensors of intracellular targets from genetically encoded reporters to modular polypeptide probes. *Cell Biochem Biophys* 2010;**56**:19–37.

CHAPTER ONE

# The Chemistry of Small-Molecule Fluorogenic Probes

**Jonathan B. Grimm, Laurel M. Heckman, Luke D. Lavis**
Janelia Farm Research Campus, The Howard Hughes Medical Institute, Ashburn, Virginia, USA

## Contents

*Progress in Molecular Biology and Translational Science*, Volume 113
ISSN 1877-1173
http://dx.doi.org/10.1016/B978-0-12-386932-6.00001-6

## Abstract

Chemical fluorophores find wide use in biology to detect and visualize different phenomena. A key advantage of small-molecule dyes is the ability to construct compounds where fluorescence is activated by chemical or biochemical processes. Fluorogenic molecules, in which fluorescence is activated by enzymatic activity, light, or environmental changes, enable advanced bioassays and sophisticated imaging experiments. Here, we detail the collection of fluorophores and highlight both general strategies and unique approaches that are employed to control fluorescence using chemistry.

# 1. INTRODUCTION

Small-molecule fluorophores are indispensible tools for modern scientific research. Such compounds find use as biomolecule labels, cellular stains, ion indicators, and enzyme substrates, enabling a wide variety of biochemical and biological experiments.[1–9] A benefit of small-molecule fluorophores is their ability to use organic chemistry to fine-tune the structure of the dye.[5] The entire lexicon of organic synthesis can be leveraged to prepare molecules with properties that are tailored for a specific application. Chemistry can also be used to control the fluorescence of a dye. This ability to regulate the fluorescence of small-molecule fluorophores forms the basis for a large number of useful fluorogenic probes. In this chapter, we detail the collection of different fluorophore structures and the chemistry of fluorogenicity. Many of the strategies to modulate fluorescence behavior are generalizable, applying across many classes of small-molecule fluorophores. However, each small-molecule fluorophore scaffold displays unique properties that can be exploited to give useful fluorogenic molecules. An understanding of the nuanced properties of each dye class will enable the effective use of existing probes and the development of new fluorogenic molecules.

## 2. FLUORESCENCE AND FLUOROPHORES

Fluorescence is the emission of a photon by an excited-state molecule following the absorption of light. The phenomenon of fluorescence from the small molecule quinine was first described in 1845 by Herschel[10] and further elucidated in 1852 by Stokes.[11] Fluorescence remained a novelty for nearly a century until the first commercial fluorometers[12] and fluorescent microscopes[13] appeared in the 1950s. Since that time, fluorescence has evolved from a curiosity to a crucial scientific tool. Fluorescent measurements are highly sensitive, allowing measurement and visualization of fluorescent compounds against a background of a myriad of nonfluorescent molecules. Fluorescent materials enable biological imaging, high-throughput screening, genome sequencing, and many other useful technologies.[1–9]

Key properties of fluorophores include the absorption maximum ($\lambda_{max}$), the emission maximum ($\lambda_{em}$), the extinction coefficient ($\varepsilon$), and the fluorescence quantum yield ($\Phi$). The difference between $\lambda_{max}$ and $\lambda_{em}$ is termed the "Stokes shift" in homage to Stokes.[11] The extinction coefficient, or molar absorptivity, is a measure of the probability of light absorption by the dye. The quantum yield, or quantum efficiency, is the ratio of the number of photons emitted to the number of photons absorbed. The relative brightness of fluorophores can be determined by comparing values of $\varepsilon \times \Phi$, which takes into account both the photons absorbed and the efficiency of the fluorescence process.[5] For use in biological experiments, other properties of fluorophores become important, such as solubility, tendency for aggregation, photobleaching rates, and sensitivity to environments. In this chapter, we move from the blue to the red region of the electromagnetic spectrum, detailing the $\lambda_{max}$, $\lambda_{em}$, $\varepsilon$, and $\Phi$ of the major classes of dyes, and discussing the chemical strategies used for controlling fluorescence.

## 3. MODES OF FLUORESCENCE MODULATION

Chemistry provides diverse methods to control the fluorescence intensity of small-molecule dyes. Figure 1.1 presents schematic representation of these different modes of modulation and corresponding examples of fluorogenic compounds. The most straightforward method to control fluorescence is to install a blocking group onto the dye that suppresses or

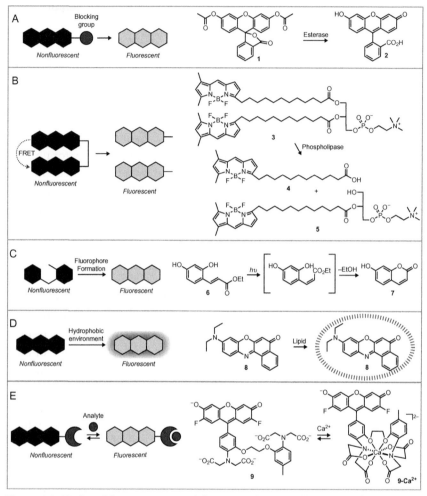

**Figure 1.1** Modes of fluorescence modulation involving small molecule fluorophores.

eliminates fluorescence. Fluorescence is restored by removal of this group through an enzyme-catalyzed reaction, photolysis, or another covalent bond cleavage. A classic example is fluorescein diacetate (**1**) shown in Fig. 1.1A. Acetylation of the phenolic oxygens of fluorescein forces the molecule to adopt a nonfluorescent, "closed" lactone form. Hydrolysis of the acetate esters by chemical or enzymatic means yields the highly fluorescent fluorescein in the "open" form (**2**; $\lambda_{max}/\lambda_{em} = 490/514$ nm, $\varepsilon = 9.3 \times 10^4$ $M^{-1}$ $cm^{-1}$, and $\Phi = 0.95$).[3,4] Control of the open–closed equilibrium in fluoresceins and rhodamines is a versatile method for constructing fluorogenic biological probes (see Sections 6 and 7).

A second approach to modulate fluorescence is to use Förster resonance energy transfer (FRET).[14,15] FRET entails energy transfer from the excited state of one "donor" dye to another "acceptor" dye. In most cases, FRET occurs between two structurally distinct dyes, where the emission spectrum from the donor fluorophore overlaps the absorption spectrum of the acceptor dye.[15] In dyes with small Stokes shifts, however, FRET between two dyes with the same structure (i.e., homo-FRET) is favorable. A useful application of this energy transfer uses boron dipyrromethene (BODIPY) dyes (Section 5), which are environmentally insensitive and show small Stokes shifts of $<20$ nm.[3] An example of a fluorogenic compound exploiting homo-FRET between two BODIPY dyes is phospholipase substrate **3** (Fig. 1.1B), which is relatively nonfluorescent because of energy transfer between the two fluorophore moieties incorporated into the lipid chains. Enzyme-catalyzed hydrolysis of the ester bond yields fatty acid **4** and phospholipid **5**, both of which are highly fluorescent ($\lambda_{max}/\lambda_{em} = 505/514$ nm, $\varepsilon = 9.1 \times 10^4$ M$^{-1}$ cm$^{-1}$, and $\Phi = 0.94$).[2,3,16]

A third method to modulate the fluorescence of fluorophores is to modify the core structure of the dye using chemistry. An example of this strategy is the *trans*-cinnamic acid derivative **6** shown in Fig. 1.1C. Illumination of such molecules with UV light causes *trans* → *cis* isomerization. The *cis* form undergoes rapid lactonization with cleavage of the ester bond, generating a fluorescent coumarin **7** (see also Section 4).[17,18] Changes in environment such as polarity can also elicit increases in fluorescence, resulting in fluorogenic dyes. One example of this phenomenon is the phenoxazine dye Nile Red (**8**; Fig. 1.1D; Section 8). Compound **8** absorbs at 591 nm, emits at 657 nm, and is relatively nonfluorescent in aqueous solution. In nonpolar media, such as xylene, Nile Red undergoes a dramatic hypsochromic shift ($\lambda_{max}/\lambda_{em} = 523/565$ nm) and an increase in quantum yield.[19] This allows fluorescent staining of hydrophobic regions, such as lipid droplets, in living cells.[20]

Finally, changes in the electronic structure of the dye or its appendages can cause changes in fluorescence.[1] An example of a probe type utilizing this process are the fluorescent Ca$^{2+}$ indicators[21] such as Fluo-4 (**9**; Fig. 1.1E). In the *apo* state, the lone pairs of electrons on the aniline moieties of the 1,2-bis(*o*-aminophenoxy)ethane-*N*,*N*,*N'*,*N'*-tetraacetic acid (BAPTA) chelation motif[22] quench fluorescence because of photoinduced electron transfer (PeT). Ca$^{2+}$ chelation changes the energy of these lone pairs of electrons, making PeT less efficient and leading to a large increase in fluorescence.[21,23]

Collectively, these disparate modes of modulation enable the construction of fluorogenic molecules from almost any fluorophore scaffold.

## 4. COUMARINS

### 4.1. Overview

Figure 1.2 shows examples of coumarin-based fluorophores and fluorogenic compounds. The coumarin framework is a privileged structure in organic chemistry and is found in many natural products and pharmacological agents. Molecules bearing a heteroatom at the 7-position of coumarins are widely used fluorophores. 7-Hydroxycoumarin (i.e., umbelliferone) is a natural product found in the *Umbelliferae* family (e.g., carrots). To improve chemical stability, a methyl group is often added to yield 4-methylumbelliferone (4-MU, **10**). The phenolate form of **10** is highly fluorescent, exhibiting strong absorption in the UV ($\lambda_{abs} = 360$ nm, $\varepsilon = 1.7 \times 10^4$ M$^{-1}$ cm$^{-1}$), emission in the blue ($\lambda_{em} = 450$ nm), and good quantum yields ($\Phi = 0.63$).[24] The p$K_a$ of 4-MU is around 7.8, making it largely protonated (and less fluorescent) at neutral pH. Thus, the p$K_a$ of 7-hydroxycoumarins is often modified by attachment of halogen substituents to improve performance at physiologically relevant pH values.[24]

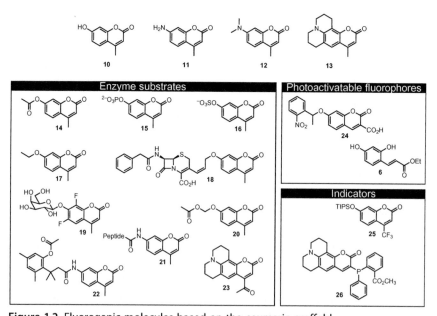

**Figure 1.2** Fluorogenic molecules based on the coumarin scaffold.

The analogous 7-aminocoumarins such as **11–13** are also important fluorophores. Compound **11** shares similar spectral properties to hydroxycoumarin **10** ($\lambda_{max}/\lambda_{em} = 351/430$ nm, $\varepsilon = 1.8 \times 10^4$ M$^{-1}$ cm$^{-1}$, $\Phi = 0.75$), but the fluorescence of this dye is insensitive to pH levels above pH 5.[2,3] Moreover, alkylation of the nitrogen substituent can yield significant bathochromic shifts, allowing access to molecules with longer wavelengths such as 7-(diethylamino)-4-methylcoumarin **12** ($\lambda_{max}/\lambda_{em} = 380/456$ nm) and julolidine derivative **13** ($\lambda_{max}/\lambda_{em} = 396/489$ nm, $\Phi = 0.66$).[25]

The fluorescence of coumarins can be modulated by structural modifications using several strategies. For hydroxycoumarins, attachment of acyl or alkyl substituents on the oxygen elicits a hypsochromic shift and lowers fluorescence quantum yield.[26] Likewise, attachment of electron–withdrawing acyl groups on the nitrogen of aminocoumarins decreases fluorescence.[27] As detailed below, this property has been exploited in the design of numerous useful probes. In addition, chemical reactions that construct or modulate the coumarin core can be used to create fluorogenic compounds.

## 4.2. Fluorogenic enzyme substrates

The strategies detailed above have been used to generate numerous fluorogenic enzyme substrates based on the flexible coumarin core. Attachment of acyl groups on 4-MU can yield useful substrates for esterases and lipases such as acetate **14**.[28,29] The phosphate ester **15** and sulfate ester **16** also serve as probes for phosphatases[26] and sulfatases,[30] respectively. Simple alkoxy derivatives, such as ethoxycoumarin **17**, are valuable substrates for cytochrome P450 enzymes where oxidative dealkylation releases 4-MU.[31] Alkylation with more exotic moieties can enable measurement of other enzyme activities. Hydrolysis of the β-lactam amide bond in ether **18** by β-lactamase yields an intermediate that decomposes to release fluorescent **10**.[32] Substrates for the widely used β-galactosidase can also be built from coumarins. Compound **19** is a galactosidase substrate that bears two fluorine substituents, which lower the p$K_a$ and increase the fluorescence of the resulting coumarin dye at physiological pH.[26] Other glycosylated coumarins have been used as enzyme substrates.[33] Alkylated hydroxycoumarins can be employed as esterase substrates by appending acyloxymethyl ethers to the phenolic oxygen to yield compounds such as **20**.[34] These highly stable esterase substrates can enable careful screening of esterase activity[35] and delivery of labels inside living cells.[36]

Aminocoumarins serve as another type of scaffold for enzyme substrates. Amidation of aminocoumarins with peptide chains, as in compound **21**, can yield valuable protease substrates.[37] This strategy has been expanded to perform screening of protease activities against substrate libraries.[38] The low pH sensitivity of aminocoumarins makes it desirable to use this scaffold for other enzymes. Installation of a self-immolative linker between the substrate moiety and the fluorophore increases the diversity of enzyme substrates that can be built upon a given scaffold. For example, esterase substrate **22** is prepared by inclusion of a "trimethyl lock" unit[39] between an acetate and the 7-aminocoumarin fluorophore **11**.[27] Hydrolysis of the acetate yields an intermediate that undergoes a rapid $N \rightarrow O$ acyl shift, releasing the fluorophore. Attachment of blocking groups at other positions can also modulate fluorescence. Compound **23** exhibits low fluorescence ($\Phi < 0.1$), presumably due to the electron-withdrawing nature of the ketone. Dehydrogenase-catalyzed reduction of the ketone yields a highly fluorescent coumarin alcohol.[40]

## 4.3. Photoactivatable probes

In addition to enzyme activity, light can be used to turn on fluorescent species. Analogous to the coumarin-based enzyme substrates, attachment of photolabile groups to the hydroxyl groups of hydroxycoumarins quenches fluorescence, yielding "caged" fluorophores. Compounds such as **24** bear an *ortho*-nitrobenzyl cage as a blocking group on the phenol. These caged coumarins exhibit unexpected and desirable properties including high two-photon cross sections and extraordinary photochemical quantum yields.[41] These unexpected properties are likely caused by the quenched coumarin moiety acting as an antenna for the cage. Absorption of light by the coumarin moiety, followed by energy transfer to the cage, causes the desirable properties observed. These properties have enabled sophisticated experiments to trace gap junctions connectivity in developing animals.[42]

As mentioned in Section 3, another method to modulate fluorescence using light is to prepare *trans*-cinnamic acid derivatives such as compound **6** (see also Fig. 1.1C). Illumination of such molecules with UV light causes *trans* $\rightarrow$ *cis* isomerization.[17,18] The cis form undergoes rapid lactonization with cleavage of the ester bond. This strategy is unique to the coumarins and has the added benefit of being able to release a molecule, such as an alcohol, concomitantly with photoactivation. This raises the possibility of releasing two fluorophores with a single photon.

## 4.4. Indicators

In addition to being sensitive to enzymes and light, coumarins can serve as environmental indicators. Attachment of groups that are removed by specific ions constitutes a valuable and general strategy for constructing such sensors. For example, compound **25** is formed by attachment of a triisopropylsilyl (TIPS) group to a hydroxycoumarin. Removal of this blocking group with fluoride ion occurs in a stoichiometric fashion and allows the detection of F$^-$ in both aqueous and organic media.[43] Another example of a coumarin-based sensor is compound **26**. Here, the phosphine moiety at the 3-position quenches fluorescence until activation with an azide in the Staudinger ligation reaction. This fluorescent coumarin becomes covalently attached in this reaction, constituting a bioorthogonal fluorogenic ligation to azide-containing molecules.[44]

# 5. BODIPY DYES

## 5.1. Overview

The BODIPY dyes, shown in Fig. 1.3, are a unique and highly flexible dye class.[45–47] A simple BODIPY, such as compound **27**, exhibits $\lambda_{abs}/\lambda_{em} = 505/511$ nm, $\varepsilon = 9.1 \times 10^4$ M$^{-1}$ cm$^{-1}$, and $\Phi = 0.94$.[2,3] The spectral properties of these dyes can be red-shifted by extending the conjugated system through alkenyl or aryl substitution. For example, BODIPY **28** exhibits $\lambda_{max}/\lambda_{em} = 544/570$ nm and dye **29** displays even longer wavelengths

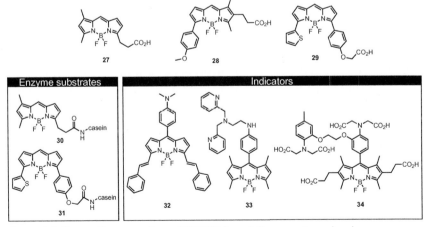

**Figure 1.3** Boron dipyrromethene (BODIPY)-based fluorogenic molecules.

$(\lambda_{max}/\lambda_{em} = 588/616$ nm, $\varepsilon = 6.8 \times 10^4$ M$^{-1}$ cm$^{-1}$, and $\Phi = 0.84$).[2,3] Hallmarks of these dyes are their insensitivity to environment and relatively small Stokes shifts ($<20$ nm). The main utility of BODIPY dyes has been as biomolecule labels. Unlike many other fluorophore classes, the BODIPY core structure has no obvious points of attachment for enzyme- or photolabile functionality. This limits the production of fluorogenic molecules based on the BODIPY dyes. Still, there are a few examples of fluorogenic molecules, especially as sensors based on PeT. Moreover, the utility of BODIPY dyes continues to grow with the development of advanced molecules, suggesting that this versatile dye scaffold is far from fully explored.[45,46]

## 5.2. Enzyme substrates

Although it is difficult to install blocking groups on the BODIPY structure, the small Stokes shift of these dyes makes them susceptible to quenching by homo-FRET. As demonstrated in phospholipase substrate **3** (Fig. 1.1B, Section 3), labeling a biomolecule with multiple BODIPY dyes yields a quenched substrate; hydrolysis of this labeled molecule elicits a large increase in fluorescence.[16] The environmental insensitivity of these dyes makes it possible to prepare fluorogenic protease, glycosylase, and lipase substrates based on this strategy.[3] For example, overlabeling of casein with BODIPYs **27** and **29** give conjugates **30** and **31**, which are useful fluorogenic substrates for a variety of proteases.[48]

## 5.3. Indicators

BODIPY dyes are modulated efficiently by PeT, allowing construction of a variety of fluorogenic sensors based on this scaffold.[47] A prototypical example is compound **32**, which incorporates a dimethylaniline moiety into a red-shifted BODIPY (protonated form: $\lambda_{max}/\lambda_{em} = 634/652$ nm, $\varepsilon = 10.4 \times 10^4$ M$^{-1}$ cm$^{-1}$, and $\Phi = 0.75$) to give a fluorogenic pH sensor with a p$K_a$ of 2.3.[49] Further elaboration by addition of various chelating groups to the BODIPY core gives valuable fluorescent ion indicators useful in biological applications. Compound **33** bears a $N,N$-bis(2-pyridylmethyl) ethylenediamine chelation motif and shows an 20-fold increase in fluorescence upon binding to Zn$^{2+}$ ion with $\lambda_{max}/\lambda_{em} = 499/509$ nm.[50] Compound **34** contains a BAPTA moiety to recognize calcium ion and is an exciting calcium indicator exhibiting a large 250-fold increase in fluorescence upon $Ca^{2+}$ binding ($\lambda_{max}/\lambda_{em}=522/536$). Indicator **34** can be further derivatized with a SNAP-Tag ligand to allow labeling of specific proteins in living cells for spatially restricted calcium imaging.[51]

## 6. FLUORESCEINS

### 6.1. Overview

Fluorescein (**2**) was first synthesized in 1871 by Baeyer[52] and has played a pivotal role in fluorescence technology. Fluorescein was one of the first labels used in immunofluorescence microscopy experiments[13] and is still widely employed in many different biological and biochemical applications.[3] This persistence can be attributed to its ease of synthesis, the excellent fluorescence properties of the dye ($\lambda_{max}/\lambda_{em}=490/514$ nm, $\varepsilon=9.3\times10^4$ $M^{-1}$ $cm^{-1}$, and $\Phi=0.95$), and the many modes of modulation available with this dye scaffold.[3,4] Fluorescein is most fluorescent as the phenolate and bears a phenolic $pK_a$ of 6.4.[53] The proximity of this $pK_a$ to physiologically relevant pH values has led to the development of halogenated derivatives of fluorescein such as $2',7'$-difluorofluorescein (Oregon Green; **36**). Oregon Green demonstrates similar optical properties to fluorescein ($\lambda_{max}/\lambda_{em}=492/516$ nm, $\varepsilon=8.6\times10^4$ $M^{-1}$ $cm^{-1}$, and $\Phi=0.92$) but displays a lower phenolic $pK_a$ of 4.6.[54] Likewise, $2',7'$-dichlorofluorescein (**36**) exhibits a drop in $pK_a$ value to 5.0 and shows a modest bathochromic shift with $\lambda_{max}/\lambda_{em}=502/525$ nm, $\varepsilon=10.1\times10^4$ $M^{-1}$ $cm^{-1}$, and $\Phi=0.88$.[55] Halogenation with bromine or iodine causes even larger bathochromic shifts but severely reduces the quantum yield.[56]

As mentioned in Section 3, a key property of fluoresceins is the equilibrium between an "open" fluorescent quinoid form and a "closed" non-fluorescent lactone. The structures of these molecular forms are shown in Fig. 1.4 (**2-open** and **2-closed**). While the open fluorescent form predominates in aqueous solution, acylation or alkylation of the phenolic oxygens can lock the molecule in the nonfluorescent lactone. Thus, many of the blocking group strategies used for hydroxycoumarins can be applied to fluoresceins. Fluorogenic compounds based on fluorescein are superior to the coumarin dyes, however, as the open–closed equilibrium completely interrupts the extended conjugation of the dye and thus endows fluorogenic fluoresceins with extremely high contrast. Fluorescein is also efficiently modulated by PeT; attachment of various groups on the pendant phenyl ring of fluorescein can yield useful sensor molecules and ion indicators.

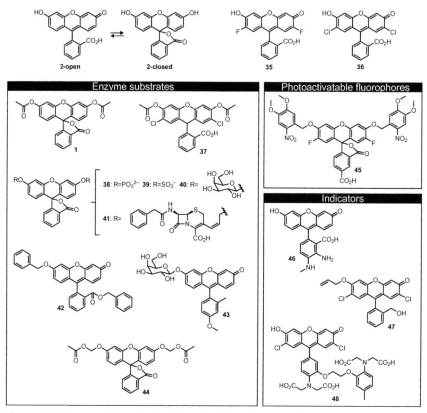

**Figure 1.4** Fluorogenic molecules based on fluorescein.

## 6.2. Enzyme substrates

Fluorescein diacetate (**1**; Fig. 1.4) is the archetypal enzyme substrate. Rotman and Papermaster showed that this lipophilic compound could cross biological membranes and be hydrolyzed by esterases inside living cells.[57] Diacylfluoresceins are widely used as esterase and lipase substrates and to deliver fluorescein–containing molecules to cells.[3,58] Fluoresceins can also be reduced to form leuco-fluorescein (i.e., fluorescin) derivatives, which are useful probes for reactive oxygen species (ROS). Compound **37** is a reduced dichlorofluorescein that is activated by esterases and oxidation.[59] Still other fluorogenic fluorescein diesters can be prepared, including diphosphate **38** and disulfate **39**.[60–62]

Alkylation of the phenolic oxygens on fluorescein also forces the molecule into the nonfluorescent lactone form. This strategy has yielded useful substrates for glycosylases, such as the β-galactosidase substrate **40**.[60,63] Substrates for β-lactamase, such as **41**, can be synthesized based on this scaffold.[64] Despite the utility of fluorescein diether substrates, the synthesis of such compounds can be difficult. Treatment of fluoresceins with alkylating agents typically gives an ether–ester mixture, such as observed with dibenzylfluorescein **42**.[65] While still useful for some assays—compound **42** is a cytochrome P450 substrate with modest utility[66]—the ether–ester cannot adopt the colorless lactone form and displays a much higher fluorescence background than a diether substrate. To overcome this synthetic issue, fluorescein diethers such as **41** can be prepared using a reduced fluorescein intermediate, which allows efficient alkylation of both phenolic oxygens.[64,67] Another strategy to circumvent the diether problem is to tune the electronics of the dye such that only one masking group is required to quench fluorescence. Fluorescein derivative **43** bears an electron-rich pendant ring that quenches the fluorescence of the dye when it is alkylated. Glycolysis catalyzed by β-galactosidase yields the bright Tokyo Green fluorophore.[68] Finally, fluorescein diethers can be used to probe esterase activity. Fluorescein diacetate **1** is notoriously unstable in aqueous solution, making it of limited utility for the sensitive measurement of esterase activity in complex biological environments. Insertion of an oxygen–methylene moiety into the ester bond yields fluorescein di(acetoxymethyl ether) substrates such as **44**. These compounds are chemically stable, allowing whole-cell analysis of endogenous esterase activity and discovery of selective esterase–ester pairs.[58,69]

## 6.3. Photoactivatable probes

Like the dialkylfluorescein enzyme substrates, attachment of photolabile groups to the oxygens of fluorescein can yield "caged" fluorescein. These photoactivatable dyes have proved instrumental in the elucidation of different dynamic cellular processes.[70] As mentioned above, dialkylation of fluorescein is hampered by the competing reactivity of the carboxylate on the bottom ring. Use of reduced fluorescein intermediates has enabled the synthesis of new photoactivatable dyes such as the caged difluorofluorescein **45** bearing a carboxyl group for bioconjugation.[67] This general strategy should allow the synthesis of many caged fluorescein derivatives with diverse photolabile groups.

## 6.4. Indicators

Attachment of disparate groups to fluorescein can endow this dye with sensitivity to environmental changes; numerous indicator molecules have been built from this scaffold. Three main strategies of sensing are exemplified by the following examples. Diaminofluorescein **46** is relatively nonfluorescent ($\Phi = 0.004$) as a result of quenching by PeT. Irreversible reaction with nitric oxide (NO) yields a triazine, which shows a significant increase in fluorescence intensity ($\Phi = 0.78$), allowing sensing of NO in a biological context.[71] Reactive groups on the phenolic oxygens can also quench fluorescence. Allyl ether **47** is relatively nonfluorescent but can be activated through deallylation with Pd. This could allow sensitive detection of catalyst contaminants in pharmaceuticals.[72] Perhaps the most important fluorescein-based indicators are the BAPTA-based "Fluo" calcium ion sensors originally developed by Tsien and coworkers. The dichlorofluorescein derivative **48** (Fluo-3) shows $>100$-fold increase in fluorescence intensity upon binding to $Ca^{2+}$.[3,21] Importantly, this and related molecules can be delivered to the interior of cells and have been instrumental for high-throughput screening in living cells[73] and functional imaging in intact brain tissue.[74]

# 7. RHODAMINES

## 7.1. Overview

The amino isologs of fluorescein, namely, the rhodamines, were first described in the 1880s.[75–77] The simplest rhodamine, rhodamine 110 (**49**; Fig. 1.5), has similar optical properties to fluorescein ($\lambda_{max}/\lambda_{em} = 496/517$ nm, $\varepsilon = 7.4 \times 10^4$ $M^{-1}$ $cm^{-1}$, and $\Phi = 0.92$) but, like aminocoumarins, is pH insensitive.[3,78] Also analogous to fluorescein, acylation of the rhodamine nitrogens in dye **49** locks the molecule into a closed lactone form that is colorless and nonfluorescent. This enables a variety of fluorogenic compounds to be built on $N,N'$-diacylrhodamines.[67,78,79] However, unlike fluoresceins, alkylation of the aniline nitrogens does not enforce the closed form but instead maintains high fluorescence and elicits a bathochromic shift. For example, tetramethylrhodamine (**50**) displays $\lambda_{max}/\lambda_{em} = 540/565$ nm and $\varepsilon = 9.5 \times 10^4$ $M^{-1}$ $cm^{-1}$ albeit with a lower quantum yield of 0.68.[2] Fully alkylated rhodamines with more rigid structures impose even larger spectral shifts, as in the julolidine derivative rhodamine 101 (**51**; $\lambda_{max}/\lambda_{em} = 575/596$ nm). Rhodamine 101 also displays high quantum efficiency approaching unity due to the rigid structure of the dye.[80] Although this

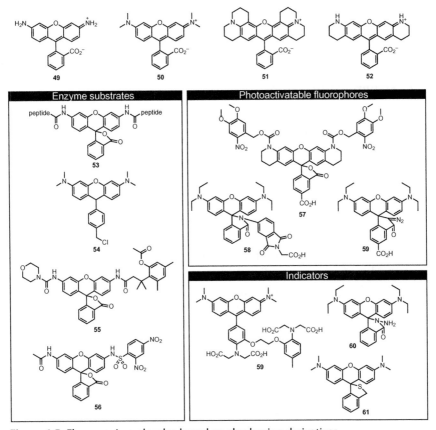

**Figure 1.5** Fluorogenic molecules based on rhodamine derivatives.

modification causes a useful shift in photophysical properties, full alkylation precludes attachment of blocking groups by acylation. Compound **52** (Q-rhodamine) balances these two effects, as partial alkylation with a rigid tetrahydroquinoline moiety yields the desired shift in wavelength ($\lambda_{max}$/ $\lambda_{em} = 537/556$ nm), but still allows acylation to produce fluorogenic compounds.[67,81]

In addition to acylation of the nitrogen substituents, the zwitterionic nature of the rhodamine structure in aqueous solution allows a mode of fluorescence modulation that is unique to the rhodamine and isologous dyes. Amidation of the *ortho*-carboxyl groups on the pendant phenyl rings prompts the formation of nonfluorescent rhodamine lactams, which can serve as photoactivatable dyes[82,83] and irreversible fluorescent sensors.[9,84,85] Finally, like the related fluoresceins, addition of different functionality on the bottom ring of rhodamines can also be used to modulate fluorescence, usually through PeT.

## 7.2. Enzyme substrates

Similar to fluorescein, acylation of rhodamine nitrogens locks the molecule into a closed lactone form. This allows the construction of a variety of fluorogenic substrates based on $N,N'$-diacylrhodamines. The classic rhodamine-based enzyme substrates are dipeptidyl rhodamines (e.g., **53**), which serve as important protease substrates for many enzymes[86] including the caspases.[87] Rhodamines can also be reduced to nonfluorescent leuco forms. Compound **54** (i.e., Mitotracker Orange) is oxidized within the mitochondria; the electrophilic chloromethyl group serves to retain the compound in that organelle for imaging.[3] Other enzyme activities can be probed with rhodamine-based probes. As with the aminocoumarins, insertion of a self-immolative linker between the substrate moiety and the dye can improve stability and expand the substrate scope of the dye scaffold. Asymmetric rhodamine **55** bears a trimethyl lock acetate on one of the nitrogen substituents and a urea on the other. The trimethyl lock acetate makes this compound a highly stable esterase substrate; the nonhydrolyzable urea functionality preserves significant fluorescence while dramatically improving enzyme kinetics.[78] The trimethyl lock strategy has been expanded to create rhodamine-based substrates for phosphatases,[88] reductases,[89] and cytochrome P450 enzymes.[90] Rhodamine sulfonamides, such as compound **56**, have also been used to assay glutathione transferase activity. Enzyme-mediated attack of glutathione on the dinitrophenyl ring results in extrusion of $SO_2$ and release of a fluorescent rhodamine species.[91]

## 7.3. Photoactivatable probes

Rhodamines can serve as valuable photoactivatable dyes. The most versatile method to cage rhodamines is via acylation of the nitrogens, as in caged Q-rhodamine **57**, which locks the molecule in the nonfluorescent lactone form.[67,70,81] Unfortunately, the installation of caging groups is hampered by the poor nucleophilicity of the rhodamine nitrogens. Like fluoresceins, the use of reduced leuco-rhodamine derivatives as intermediates can significantly improve the synthesis of caged rhodamine dyes.[67] Alternatively, caged rhodamines can be accessed directly from fluorescein derivatives using Pd-catalyzed cross-coupling.[79] Compound **57** has been used for photoactivatable localization microscopy (PALM) imaging of cellular DNA.[67]

Because of the problems encountered with the synthesis of N-acyl caged rhodamines, other design strategies have been developed for photolabile rhodamines. As mentioned above, amidation of the *ortho*-carboxyl group of rhodamines yields derivatives that adopt a nonfluorescent, UV-absorbing lactam form under certain conditions. Illumination with short-wavelength light can cause the molecule to open, transiently generating a fluorescent species. Photochromic rhodamines such as compound **58** have been used to perform super-resolution imaging.[92] Another strategy to generate caged rhodamine dyes is exemplified in compound **59**. Here, the molecule is locked in a nonfluorescent form through formation of a spiro-carbocycle, which remains closed regardless of nitrogen substitution. Illumination causes a photochemical rearrangement of the diazo moiety, ultimately yielding a fluorescent rhodamine species.[93]

## 7.4. Indicators

Given the tunability of wavelengths, the rhodamines serve as useful fluorogenic sensors across a large spectral range. Like the fluoresceins, the addition of an aniline-containing chelation motif can yield fluorogenic sensors that are governed by PeT. An example is the calcium indicator Rhod-2 (**59**). One caveat with rhodamine-based sensors is their lower sensitivity than the fluorescein-based Fluo dyes **9** and **48**.[21] This difference stems from the efficiency of PeT being based on the $\lambda_{max}$ and redox potential of the dye.[94] The PeT process is less favorable with dyes exhibiting high $\lambda_{max}$ values. Ion indicators built from newer rhodamine isologs that exhibit optimal redox potentials show higher sensitivity while retaining relatively long wavelengths.[95] In addition to PeT-modulated probes, the open–closed equilibrium of rhodamines can be used to generate sensors. Rhodamine lactams find wide use as irreversible "turn-on" probes.[9,85] For example, compound **60** remains nonfluorescent until interaction with Cu.[96] Likewise, rhodamine **61** adopts a nonfluorescent form because of the nucleophilicity of the thiol group. Oxidation of the thiol to a sulfonate by hypochlorous acid yields a species that exists in the open fluorescent state, allowing sensing of this ROS.[97]

## 8. PHENOXAZINE DYES

### 8.1. Overview

The use of a phenoxazine core yields small fluorophores with significant bathochromic shifts in comparison to analogous fluorescein and rhodamine dyes. Examples of this dye class are shown in Fig. 1.6. The most widely used oxazine dye is resorufin **62**, which exhibits $\lambda_{max}/\lambda_{em} = 572/585$ nm, $\varepsilon = 5.6 \times 10^4$ M$^{-1}$ cm$^{-1}$, and $\Phi = 0.74$ as the anion.[98] Like fluoresceins, resorufin is most fluorescent as a phenolate (p$K_a = 5.8$), making alkylation or acylation a useful strategy to suppress fluorescence. However, owing to the structure of the resorufins, only one blocking group can be attached; the resulting O-alkyl or O-acylresorufins do not display the complete color change of the xanthene dyes and still absorb visible light.[99] This complicates multicolor experiments using resorufin-based fluorogenic compounds. Another strategy to control fluorescence is through oxidation or reduction chemistry, with both transformations causing a significant decrease in fluorescence.[3,100]

The amine-containing phenoxazine dyes such as Nile Red (**8**) and Cresyl Violet (**63**), although less utilized as fluorogenic probes, exhibit useful

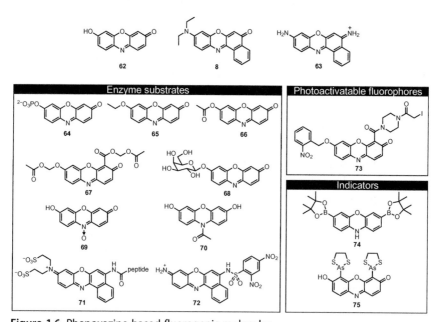

**Figure 1.6** Phenoxazine-based fluorogenic molecules.

fluorescence properties.[101] As mentioned in Section 3, the optical properties of Nile Red show high sensitivity to the polarity of the environment. This allows fluorescent staining of hydrophobic regions, such as lipid droplets, in living cells. In water, Nile Red displays low fluorescence and relatively long absorption and emission maxima with $\lambda_{max}/\lambda_{em} = 591/657$ nm. In organic solvents, however, compound **8** exhibits $\lambda_{max}/\lambda_{em} = 523/565$ nm and a large increase in quantum yield.[19] This dramatic change has been exploited to prepare fluorogenic stains for cellular membranes and lipid droplets.[20] The environmental sensitivity of phenoxazine dyes can also be used to read out protein conformation.[102] Cresyl Violet (**63**) exhibits a $\lambda_{max}$ that is ~100 nm longer than the analogous rhodamine 110 (**49**), with $\lambda_{max}/\lambda_{em} = 591/654$ nm, $\varepsilon = 8.3 \times 10^4$ M$^{-1}$ cm$^{-1}$, and $\Phi = 0.54$.[103] Like rhodamines, the fluorescence of Cresyl Violet and similar dyes can be modulated by acylation of the aniline nitrogens. This modification dramatically shifts the $\lambda_{max}/\lambda_{em}$ to shorter wavelengths and decreases the quantum yield, making this dye a useful scaffold for fluorogenic molecules.[27,104]

## 8.2. Enzyme substrates

The relatively long absorption and emission spectra of resorufin are advantageous for bioassays because of reduced interference from cellular autofluorescence. Analogous to the hydroxycoumarin and fluorescein-based enzyme substrates, phenol substitution of resorufin causes a decrease in fluorescence. Thus, resorufin has been used to construct a variety of red-shifted enzyme substrates. These include the phosphatase substrate **64**[105] and the cytochrome P450 substrate **65**.[106] Resorufin acetate **66** is a fluorogenic dye that can report on the activity of a variety of enzymes including esterases and proteases.[107,108] The instability of the resorufin–ester bond and the propensity of acyl resorufins toward nucleophilic attack[98] hamper the utility of acyl resorufins, especially in living cells. Substitution of the acetate ester with an acetoxymethyl ether group as in compound **67** improves stability; the extra carboxyl group increases cellular retention of the molecule.[58] Glycosylated resorufins, such as β-galactosidase substrate **68**, find wide use for many applications because of the red-shifted properties of the released resorufin.[109]

In addition, the oxidation state of the dye can be modulated to create fluorogenic molecules. For example, resazurin (**69**) is a strongly absorbing ($\lambda_{max} = 602$ nm, $\varepsilon = 4.2 \times 10^4$ M$^{-1}$ cm$^{-1}$) but nonfluorescent molecule.[110] Reduction of this compound by endogenous oxidoreductases, such as diaphorase, inside living cells yields the fluorescent resorufin (**62**), making this molecule a valuable reagent for the determination of cell viability.[111] Resorufin can also be reduced to a nonfluorescent form as in compound **70**. This molecule is oxidized by $H_2O_2$ in a horseradish peroxidase-catalyzed reaction to yield resorufin and is widely used in the enzyme-linked immunosorbent assay (ELISA).[3,100]

The amine-containing phenoxazoles are also useful scaffolds for enzyme assays. Analogous to the rhodamine-based protease substrates, acylation of Cresyl Violet and its derivatives with amino acids gives valuable fluorogenic compounds.[27,104,112,113] One example is compound **71**, which releases a far-red fluorophore ($\lambda_{max}/\lambda_{em}=633/675$ nm, $\varepsilon=3.6\times10^4$ $M^{-1}$ $cm^{-1}$, and $\Phi=0.10$) upon cleavage of the peptide.[114] In a similar vein, the Cresyl Violet sulfonamide **72** has been described as a fluorogenic glutathione transferase substrate.[91]

## 8.3. Photoactivatable probes

Photolabile "caged" resorufin was developed by Mitchison for use in imaging assays to determine the dynamic behavior of actin filaments.[99] A benefit of this dye is the relatively long wavelengths ($\lambda_{max}/\lambda_{em}=572/585$ nm) when compared to the caged fluoresceins and rhodamines. However, like the other alkyl resorufins, the quenching of the fluorescence is incomplete, limiting the utility of this dye in experiments that require high-contrast dyes.

## 8.4. Indicators

The long wavelength of resorufin makes it an attractive platform for sensors. One example is the reduced boronic ester **74**, which reacts spontaneously with $H_2O_2$ to yield the fluorescent resorufin. Other phenolic fluorophores, such as fluorescein, can be masked in a similar fashion.[115] Resorufin can also be appended with arsenical substituents to give the resorufin arsenical hairpin binder (ReAsH; **75**). This compound is nonfluorescent in solution but becomes fluorescent upon binding to an engineered tetracysteine motif on a protein.[116] Yet another sensor molecule based on the phenoxazines is Nile Red (**8**). As highlighted in Section 3, this compound becomes fluorescent in a lipophilic environment. In addition to staining lipid droplets in cells,[20] this fluorogenic phenomenon has also been used to perform super-resolution microscopy on cellular membranes.[117]

## 9. ACRIDINONES

### 9.1. Overview

Substitution of the oxygen in the phenoxazine core with a quaternary carbon to yield 9$H$-dialkylacridinones elicits a further bathochromic shift in spectral properties. The most widely used acridinone is 7-hydroxy-9$H$-(1,3-dichloro-9,9-dimethylacridin-2-one) (DDAO; **76**; Fig. 1.7), which displays $\lambda_{max}/\lambda_{em} = 646/659$ nm and $\varepsilon = 5.6 \times 10^4$ M$^{-1}$ cm$^{-1}$.[3,118] DDAO enjoys this utility because of its straightforward synthesis and because the two chloro substituents ensure a low phenolic p$K_a$ value of 5.0.[119] Like the other phenolic dyes, the phenolate is the most fluorescent form; alkylation or acylation diminishes fluorescence, providing a means to fluorogenic compounds. One caveat with DDAO is the relatively high fluorescence of the $O$-alkylated or $O$-acylated form, although this is mitigated by the dramatic (>200 nm) hypsochromic shift in absorbance when derivatized.

### 9.2. Fluorogenic enzyme substrates and photoactivatable probes

Fluorogenic molecules based on DDAO are surprisingly limited, especially given the red-shifted spectra of the released fluorophore, which minimize interference from endogenous or exogenous fluorescent molecules.[118] The acridinone scaffold has been used to create a few useful substrates, including phosphatase substrate **77**[120] and β-galactosidase substrate **78**.[118] The sulfatase substrate **79** has also been reported and found useful to assay aryl sulfatases, an important enzyme for bacterial infection.[121] Attachment of

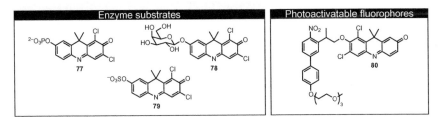

**Figure 1.7** Fluorogenic acridinones.

photolabile group to the phenolic oxygen allows synthesis of caged acridinone **80**. The unique biphenyl cage in this compound can be photolyzed with either one- or two-photon illumination inside living cells.[119]

## 10. CYANINES

### 10.1. Overview

The cyanine dyes are strongly absorbing molecules with the general structure $R_2N—(CH=CH)_n—CH=N^+R_2$ as shown in Fig. 1.8. The lipophilic nature of these dyes limited their utility to cellular membrane probes[122] and DNA stains[123] until the development of sulfonated indocarbocyanine derivatives such as **81–83**.[124] These "CyDyes" are useful labels for proteins and nucleic acids and remain some of the most widely used fluorophores in biological and biochemical experiments.[125] Cyanine dye labels are named according to the number of carbons between the indoline moieties. Cy3 (**81**) exhibits $\lambda_{max}/\lambda_{em} = 554/568$ nm, $\varepsilon = 1.3 \times 10^5$ M$^{-1}$ cm$^{-1}$, and $\Phi = 0.14$; these spectral properties are comparable to those of tetramethylrhodamine **50** or Q-rhodamine **52**. Cy5 (**82**) shows longer wavelengths at $\lambda_{max}/\lambda_{em} = 652/672$ nm, $\varepsilon = 2.0 \times 10^5$ M$^{-1}$ cm$^{-1}$, and $\Phi = 0.18$. Still longer wavelengths can be accessed by lengthening the polymethine chain further as in Cy7 (**83**; $\lambda_{max}/\lambda_{em} = 755/778$ nm).[126]

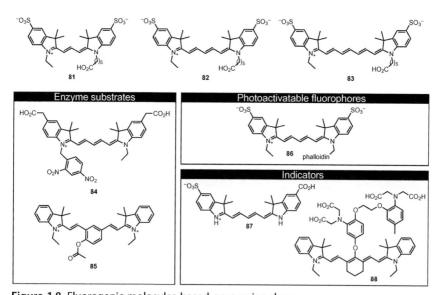

**Figure 1.8** Fluorogenic molecules based on cyanine dyes.

Modulation of the cyanines is possible through several strategies. First, carefully chosen blocking groups can be attached to the cyanine nitrogens to quench fluorescence.[127] An alternative strategy is to modify the polymethine chain with heteroatoms and attach blocking groups at those positions.[6,128] In addition, many fluorogenic cyanine constructs are based on FRET between two different cyanine dyes (e.g., Cy3 and Cy5).[129] Finally, modulation of the fluorescence intensity by PeT is possible, but inefficient because of the long absorption of this dye class.[130]

## 10.2. Enzyme substrates

In addition to several FRET-based probes, unimolecular cyanine-based fluorogenic probes have been developed. An example is compound **84**, which is an effective fluorogenic substrate for *Escherichia coli* nitroreductase. Enzyme-catalyzed reduction of a nitro group on the dinitrophenyl ring leads to bond cleavage and a concomitant increase in fluorescence.[127] Modulation at the polymethine chain is also possible. Fusing a phenol into a Cy7-like structure allows modulation of cyanine fluorescence by attaching blocking groups onto the phenolic oxygen. For example, acetate **85** exhibits low fluorescence. Cleavage of the ester group to liberate the phenol elicits a large shift in spectral properties ($\lambda_{max}/\lambda_{em} = 570/715$ nm) and a significant increase in quantum yield.[131]

## 10.3. Photoactivatable probes

Linking two cyanine dyes via a photocleavable linker can produce useful fluorogenic compounds where the fluorescence is modulated by FRET. Cy3–Cy5 dimers have been conjugated to proteins on the surface of living cells.[129] In addition, the cyanine dyes exhibit photoswitching under certain redox conditions that can be harnessed for super-resolution fluorescence microcopy. When illuminated at $\lambda_{max}$ in the presence of thiol reducing agents, cyanines conjugates such as the actin-binding Cy5-phalloidin (**86**) can adopt a nonfluorescent state that can be switched to a fluorescent state by illumination with shorter wavelength light. This "dark state" could be a reduced radical[132] or a thiol adduct of the dye.[133] Regardless, this property has been used to perform stochastic optical reconstruction microscopy (STORM) on a variety of samples.[134]

## 10.4. Indicators

Like the cyanine-based enzyme substrates, substitution on the cyanine nitrogens can elicit large changes in fluorescence. Unsubstituted cyanine dyes such as **87** (CypHer-5) are fluorescent at low pH values ($pK_a = 6.1$) and thus useful for imaging vesicle acidification.[135] Linking a cyanine dye that is susceptible to oxidation to a more stable dye can be used to prepare fluorogenic sensors for oxidative stress based on FRET.[136] Finally, like many other classes of dyes, appending chelating moieties such as BAPTA can yield fluorogenic ion sensors based on PeT.[137] Although compounds such as **88** have been prepared, the long wavelength of the cyanine dye ($\lambda_{max}/\lambda_{em} = 766/782$ nm) decreases the efficiency of the PeT process.[130] Thus, molecule **88** exhibits high basal fluorescence and only a modest threefold increase in fluorescence upon calcium ion binding.[137]

## 11. OTHER FLUOROGENIC SCAFFOLDS

### 11.1. Overview

As shown in Fig. 1.9, several other fluorophore types have been used to construct fluorogenic molecules. Derivatives of polycyclic aromatic compounds such as the pyrene-based 8-hydroxypyrene-1,3,6-trisulfonic acid (HPTS, **89**) and the amino–naphthalimide Lucifer Yellow (**90**) are widely used

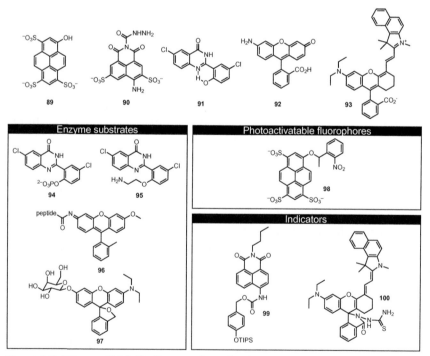

**Figure 1.9** Other scaffolds for fluorogenic molecules.

fluorophores.[138,139] Appropriate structural modification of these dyes through the techniques discussed previously can yield fluorogenic derivatives. Another unique dye scaffold is 2-(2′-hydroxy-5′-chlorophenyl)-6-chloro-4($^3H$)-quinazolinone (HPQ, **91**). This compound is insoluble in water and highly fluorescent in the solid state ($\lambda_{max}/\lambda_{em} = 375/510$ nm). Derivatization of the phenolic oxygen blocks the formation of the internal hydrogen bond and gives soluble nonfluorescent substrates.[140] Finally, hybrid structures between two dye classes can yield fluorescent molecules with nuanced properties. These include the "rhodols" such as compound **92**, which share characteristics of both fluorescein and rhodamine dyes.[141] Another fusion of dye classes are the rhodamine–cyanine hybrids such as compound **93**, which absorb at long wavelengths ($\lambda_{max}/\lambda_{em} = 720/750$ nm), but can be modulated using similar strategies to fluorogenic rhodamines.[142]

## 11.2. Fluorogenic enzyme substrates

Like the more common fluorophores, attachment of enzyme-labile moieties to these fluorescent compounds can yield substrates for various enzymes. In particular, the HPQ scaffold has been used to prepare a variety of fluorogenic substrates by derivatization of the phenolic oxygen.[140] The unique precipitating HPQ product links the fluorescence signal to the position of the enzyme, which is useful for immunohistochemistry, mRNA *in situ* hybridization, and microarray experiments. This nondiffusible substrate can also be used to map enzyme activity and location in living cells. The most widely used HPQ substrate is phosphate **94**, which is highly soluble and an excellent substrate for alkaline phosphatase.[143,144] Another example is amine **95**, which has been used to illuminate the activity of cellular monoamine oxygenase.[145]

The rhodols can also serve as useful scaffolds for enzyme substrates. Analogous to the Tokyo Green substrate **43** (Fig. 1.4), compound **96** is nonfluorescent when acylated. Removal of the peptide by protease activity affords a large increase in fluorescence; the oxygen substituent in **96** can be used to anchor substrates to a solid support.[146] Another rhodol-based substrate is β-galactoside **97**. Here, the open–closed equilibrium of the rhodol is modified by reduction of the *ortho*-carboxyl group to the alcohol. Upon alkylation of the rhodol phenol, the compound adopts a closed, non-fluorescent form around pH 7. Glycolysis elicits a large increase in fluorescence; this compound can be used to measure β-galactosidase activity in living cells.[147]

## 11.3. Photoactivatable probes

HPTS (**89**) is most fluorescent in the phenolate form and has found use as a pH sensor with a $pK_a \approx 7.3.$[139] Thus, like other phenolic fluorophores, alkylation of the phenol is a useful strategy to control fluorescence. This has been used to prepare highly soluble caged fluorophores such as compound **98**.[148] Caged HPTS has been used to measure diffusion in neuronal spines[149] and for calibration during photo-uncaging experiments.[150]

## 11.4. Indicators

In addition to being useful labels and fixable polar tracers,[3,138] the aminonapthalimides are modulated by attachment of labile moieties to the aniline nitrogen, making them useful scaffolds for fluorogenic molecules. Acylation of this group causes a sharp decrease in fluorescence intensity. Compound **99** is a sensor for fluoride ion, where $F^-$ causes cleavage of the TIPS group and extrusion of quinone methide and $CO_2$, leading to the release of the free amino–naphthalimide ($\lambda_{max}/\lambda_{em} = 421/508$ nm).[151] Similar naphthalimide sensors have been built for other analytes.[152,153] Finally, the rhodamine–cyanine hybrids can serve as turn-on fluorescent sensors. Because this dye system is imitative of rhodamine dyes, these compounds exist in equilibrium between a fluorescent, open form and a nonfluorescent, closed form. Akin to the rhodamine dyes, lactam derivatives, including molecule **100**, adopt the closed, nonfluorescent form. Oxidation of the thiosemicarbazide by hypochlorate recapitulates the fluorescent structure.[142]

## 12. CONCLUSIONS AND FUTURE DIRECTIONS

In this chapter, we described the major classes of fluorescent dyes and the chemistry involved in controlling the optical properties of these dyes. We note some general design principles. Oxygen-containing fluorophores, such as methylumbelliferone, fluorescein, resorufin, and DDAO, are typically most fluorescent in the deprotonated state. Thus, O–alkylation or O-acylation of these fluorophores can be used to suppress fluorescence. Likewise, N-acylation of nitrogen-containing fluorophores, such as rhodamines, significantly decreases fluorescence intensity. Fluorogenic molecules based on fluoresceins, rhodamines, and some rhodols have the added benefit of adopting a colorless form upon derivatization, yielding fluorogenic compounds with extremely high contrast. Rhodamines can also be masked by amidation of the *ortho*-carboxyl group to yield colorless nonfluorescent lactams. This can be used to prepare reversible or irreversible fluorogenic molecules.

FRET and PeT are alternative methods to modulate fluorescence intensity. These strategies are especially useful for fluorophores that lack obvious points of attachment for blocking groups, such as the BODIPY dyes and some cyanine derivatives. In particular, sensors based on PeT have enabled important biological experiments to visualize changes in ionic concentration. Finally, dyes that exhibit intrinsic sensitivity to the environment, such as phenoxazines, can serve as valuable fluorogenic compounds without the need for attachment of any moiety.

The design principles of fluorogenic molecules are well established. Future advances in small-molecule fluorogenic compounds will depend on improved synthetic methods to efficiently build finely tuned derivatives for specific biological experiments. Efficient generation of libraries of dyes will allow the discovery of additional probes with intrinsic sensitivity and affinity for different cellular environments. Moreover, advances in the chemistry of fluorescent probes will allow the synthesis of new hybrid structures that combine advantages of multiple dyes to create fluorogenic probes of ever-increasing sophistication.

## REFERENCES

1. de Silva AP, Gunaratne HQN, Gunnlaugsson T, Huxley AJM, McCoy CP, Rademacher JT, et al. Signaling recognition events with fluorescent sensors and switches. *Chem Rev* 1997;**97**:1515–66.
2. Johnson I. Fluorescent probes for living cells. *Histochem J* 1998;**30**:123–40.
3. Haugland RP, Spence MTZ, Johnson ID, Basey A. *The handbook: a guide to fluorescent probes and labeling technologies*. 10th ed. Eugene, OR: Molecular Probes; 2005.
4. Lakowicz JR. *Principles of fluorescence spectroscopy*. 3rd ed. New York, NY: Springer; 2006.
5. Lavis LD, Raines RT. Bright ideas for chemical biology. *ACS Chem Biol* 2008;**3**:142–55.
6. Kiyose K, Kojima H, Nagano T. Functional near-infrared fluorescent probes. *Chem Asian J* 2008;**3**:506–15.
7. Drake CR, Miller DC, Jones EF. Activatable optical probes for the detection of enzymes. *Curr Org Synth* 2011;**8**:498–520.
8. Puliti D, Warther D, Orange C, Specht A, Goeldner M. Small photoactivatable molecules for controlled fluorescence activation in living cells. *Bioorg Med Chem* 2011;**19**:1023–9.
9. Jun ME, Roy B, Ahn KH. "Turn-on" fluorescent sensing with "reactive" probes. *Chem Commun* 2011;**47**:7583–601.
10. Herschel JFW. On a case of superficial colour presented by a homogeneous liquid internally colourless. *Philos Trans R Soc Lond* 1845;**135**:143–5.

11. Stokes GG. On the change of refrangibility of light. *Philos Trans R Soc Lond* 1852;**142**:463–562.
12. Udenfriend S. Development of the spectrophotofluorometer and its commercialization. *Protein Sci* 1995;**4**:542–51.
13. Coons AH, Kaplan MH. Localization of antigen in tissue cells. II. Improvements in a method for the detection of antigen by means of fluorescent antibody. *J Exp Med* 1950;**91**:1–13.
14. Stryer L, Haugland RP. Energy transfer: a spectroscopic ruler. *Proc Natl Acad Sci USA* 1967;**58**:719–26.
15. Sapsford KE, Berti L, Medintz IL. Materials for fluorescence resonance energy transfer analysis: beyond traditional donor–acceptor combinations. *Angew Chem Int Ed* 2006;**45**:4562–88.
16. Mitnaul LJ, Tian J, Burton C, Lam MH, Zhu Y, Olson SH, et al. Fluorogenic substrates for high-throughput measurements of endothelial lipase activity. *J Lipid Res* 2007;**48**:472–82.
17. Gagey N, Neveu P, Benbrahim C, Goetz B, Aujard I, Baudin JB, et al. Two-photon uncaging with fluorescence reporting: evaluation of the *o*-hydroxycinnamic platform. *J Am Chem Soc* 2007;**129**:9986–98.
18. Cho SY, Song YK, Kim JG, Oh SY, Chung CM. Photoconversion of *o*-hydroxycinnamates to coumarins and its application to fluorescence imaging. *Tetrahedron Lett* 2009;**50**:4769–72.
19. Greenspan P, Fowler S. Spectrofluorometric studies of the lipid probe, Nile red. *J Lipid Res* 1985;**26**:781–9.
20. Greenspan P, Mayer EP, Fowler SD. Nile Red: a selective fluorescent stain for intracellular lipid droplets. *J Cell Biol* 1985;**100**:965–73.
21. Minta A, Kao JP, Tsien RY. Fluorescent indicators for cytosolic calcium based on rhodamine and fluorescein chromophores. *J Biol Chem* 1989;**264**:8171–8.
22. Tsien RY. New calcium indicators and buffers with high selectivity against magnesium and protons: design, synthesis, and properties of prototype structures. *Biochemistry* 1980;**19**:2396–404.
23. Adams SR. How calcium indicators work. In: Yuste R, Konnerth A, editors. *Imaging in neuroscience and development: a laboratory manual*. Cold Spring Harbor, New York: CSHL Press; 2005. p. 239–44.
24. Sun W-C, Gee KR, Haugland RP. Synthesis of novel fluorinated coumarins: excellent UV-light excitable fluorescent dyes. *Bioorg Med Chem Lett* 1998;**8**:3107–10.
25. Jones G, Jackson WR, Choi CY, Bergmark WR. Solvent effects on emission yield and lifetime for coumarin laser dyes. Requirements for a rotatory decay mechanism. *J Phys Chem* 1985;**89**:294–300.
26. Gee KR, Sun W-C, Bhalgat MK, Upson RH, Klaubert DH, Latham KA, et al. Fluorogenic substrates based on fluorinated umbelliferones for continuous assays of phosphatases and β-galactosidases. *Anal Biochem* 1999;**273**:41–8.
27. Lavis LD, Chao TY, Raines RT. Latent blue and red fluorophores based on the trimethyl lock. *ChemBioChem* 2006;**7**:1151–4.
28. Jacks T, Kircher H. Fluorometric assay for the hydrolytic activity of lipase using fatty acyl esters of 4-methylumbelliferone. *Anal Biochem* 1967;**21**:279–85.
29. Gilham D, Lehner R. Techniques to measure lipase and esterase activity in vitro. *Methods* 2005;**36**:139–47.
30. Guilbault GG, Hieserman J. Fluorometric substrate for sulfatase and lipase. *Anal Chem* 1969;**41**:2006–9.
31. Yamazaki H, Inoue K, Mimura M, Oda Y, Guengerich FP, Shimada T. 7-Ethoxycoumarin O-deethylation catalyzed by cytochromes P450 1A2 and 2E1 in human liver microsomes. *Biochem Pharmacol* 1996;**51**:313–9.

32. Gao W, Xing B, Tsien RY, Rao J. Novel fluorogenic substrates for imaging β-lactamase gene expression. *J Am Chem Soc* 2003;**125**:11146–7.
33. van Tilbeurgh H, Claeyssens M. Detection and differentiation of cellulase components using low molecular mass fluorogenic substrates. *FEBS Lett* 1985;**187**:283–8.
34. Leroy E, Bensel N, Reymond J-L. A low background high-throughput screening (HTS) fluorescence assay for lipases and esterases using acyloxymethylethers of umbelliferone. *Bioorg Med Chem Lett* 2003;**13**:2105–8.
35. Babiak P, Reymond J-L. A high-throughput, low-volume enzyme assay on solid support. *Anal Chem* 2005;**77**:373–7.
36. Uttamapinant C, White KA, Baruah H, Thompson S, Fernández-Suárez M, Puthenveetil S, et al. A fluorophore ligase for site-specific protein labeling inside living cells. *Proc Natl Acad Sci USA* 2010;**107**:10914–9.
37. Zimmerman M, Ashe B, Yurewicz EC, Patel G. Sensitive assays for trypsin, elastase, and chymotrypsin using new fluorogenic substrates. *Anal Biochem* 1977;**78**:47–51.
38. Salisbury CM, Maly DJ, Ellman JA. Peptide microarrays for the determination of protease substrate specificity. *J Am Chem Soc* 2002;**124**:14868–70.
39. Amsberry KL, Borchardt RT. The lactonization of 2'-hydroxyhydrocinnamic acid-amides: a potential prodrug for amines. *J Org Chem* 1990;**55**:5867–77.
40. Yee DJ, Balsanek V, Sames D. New tools for molecular imaging of redox metabolism: development of a fluorogenic probe for 3α-hydroxysteroid dehydrogenases. *J Am Chem Soc* 2004;**126**:2282–3.
41. Zhao YR, Zheng Q, Dakin K, Xu K, Martinez ML, Li WH. New caged coumarin fluorophores with extraordinary uncaging cross sections suitable for biological imaging applications. *J Am Chem Soc* 2004;**126**:4653–63.
42. Guo YM, Chen S, Shetty P, Zheng G, Lin R, Li W. Imaging dynamic cell-cell junctional coupling in vivo using Trojan-LAMP. *Nat Methods* 2008;**5**:835–41.
43. Sokkalingam P, Lee C-H. Highly sensitive fluorescence "turn-on" indicator for fluoride anion with remarkable selectivity in organic and aqueous media. *J Org Chem* 2011;**76**:3820–8.
44. Lemieux GA, de Graffenried CL, Bertozzi CR. A fluorogenic dye activated by the Staudinger ligation. *J Am Chem Soc* 2003;**125**:4708–9.
45. Loudet A, Burgess K. BODIPY dyes and their derivatives: syntheses and spectroscopic properties. *Chem Rev* 2007;**107**:4891–932.
46. Ulrich G, Ziessel R, Harriman A. The chemistry of fluorescent BODIPY dyes: versatility unsurpassed. *Angew Chem Int Ed* 2008;**47**:1184–201.
47. Boens N, Leen V, Dehaen W. Fluorescent indicators based on BODIPY. *Chem Soc Rev* 2012;**41**:1130–72.
48. Thompson VF, Saldaña S, Cong J, Goll DE. A BODIPY fluorescent microplate assay for measuring activity of calpains and other proteases. *Anal Biochem* 2000;**279**:170–8.
49. Rurack K, Kollmannsberger M, Daub J. A highly efficient sensor molecule emitting in the near infrared (NIR): 3,5-distyryl-8-(p-dimethylaminophenyl) difluoroboradiaza-s-indacene. *New J Chem* 2000;**25**:289–92.
50. Koutaka H, Kosuge J, Fukasaku N, Hirano T, Kikuchi K, Urano Y, et al. A novel fluorescent probe for zinc ion based on boron dipyrromethene (BODIPY) chromophore. *Chem Pharm Bull* 2004;**52**:700–3.
51. Kamiya M, Johnsson K. Localizable and highly sensitive calcium indicator based on a BODIPY fluorophore. *Anal Chem* 2010;**82**:6472–9.
52. Baeyer A. Ueber eine neue Klasse von Farbstoffen. *Ber Dtsch Chem Ges* 1871;**4**:555–8.
53. Lavis LD, Rutkoski TJ, Raines RT. Tuning the p$K_a$ of fluorescein to optimize binding assays. *Anal Chem* 2007;**79**:6775–82.

54. Sun W-C, Gee KR, Klaubert DH, Haugland RP. Synthesis of fluorinated fluoresceins. *J Org Chem* 1997;**62**:6469–75.

55. Leonhardt H, Gordon L, Livingston R. Acid-base equilibriums of fluorescein and 2′,7′-dichlorofluorescein in their ground and fluorescent states. *J Phys Chem* 1971;**75**:245–9.

56. Fleming GR, Knight AWE, Morris JM, Morrison RJS, Robinson GW. Picosecond fluorescence studies of xanthene dyes. *J Am Chem Soc* 1977;**99**:4306–11.

57. Rotman B, Papermaster BW. Membrane properties of living mammalian cells as studied by enzymatic hydrolysis of fluorogenic esters. *Proc Natl Acad Sci USA* 1966;**55**:134–41.

58. Lavis LD, Chao TY, Raines RT. Synthesis and utility of fluorogenic acetoxymethyl ethers. *Chem Sci* 2011;**2**:521–30.

59. Possel H, Noack H, Augustin W, Keilhoff G, Wolf G. 2, 7-Dihydrodichlorofluorescein diacetate as a fluorescent marker for peroxynitrite formation. *FEBS Lett* 1997;**416**:175–8.

60. Rotman B, Zderic JA, Edelstein M. Fluorogenic substrates for β-D-galactosidases and phosphatases derived from fluorescein (3,6-dihydroxyfluoran) and its mono-methylether. *Proc Natl Acad Sci USA* 1963;**50**:1–6.

61. Huang Z, Wang QP, Ly HD, Gorvindarajan A, Scheigetz J, Zamboni R, et al. 3,4-Fluorescein diphosphate: a sensitive fluorogenic and chromogenic substrate for protein tyrosine phosphatases. *J Biomol Screen* 1999;**4**:327–34.

62. Scheigetz J, Gilbert M, Zamboni R. Synthesis of fluorescein phosphates and sulfates. *Org Prep Proc Int* 1997;**29**:561–8.

63. MacGregor GR, Nolan GP, Fiering S, Roederer M, Herzenberg LA. Use of *E. coli lacZ* (β-galactosidase) as a reporter gene. *Methods Mol Biol* 1991;**7**:217–35.

64. Rukavishnikov A, Gee K, Johnson I, Corry S. Fluorogenic cephalosporin substrates for β-lactamase TEM-1. *Anal Biochem* 2011;**419**:9–16.

65. Krafft GA, Sutton WR, Cummings RT. Photoactivable fluorophores. 3. Synthesis and photoactivation of fluorogenic difunctionalized fluoresceins. *J Am Chem Soc* 1988;**110**:301–3.

66. Hong Y, Cho M, Yuan YC, Chen S. Molecular basis for the interaction of four different classes of substrates and inhibitors with human aromatase. *Biochem Pharmacol* 2008;**75**:1161–9.

67. Wysocki LM, Grimm JB, Tkachuk AN, Brown TA, Betzig E, Lavis LD. Facile and general synthesis of photoactivatable xanthene dyes. *Angew Chem Int Ed* 2011;**50**:11206–9.

68. Urano Y, Kamiya M, Kanda K, Ueno T, Hirose K, Nagano T. Evolution of fluorescein as a platform for finely tunable fluorescence probes. *J Am Chem Soc* 2005;**127**:4888–94.

69. Tian L, Yang Y, Wysocki LM, Arnold AC, Hu A, Ravichandran B, et al. A selective esterase–ester pair for targeting small molecules with cellular specificity. *Proc Natl Acad Sci USA* 2012;**109**:4756–61.

70. Mitchison TJ, Sawin KE, Theriot JA, Gee K, Mallavarapu A, Marriott G. Caged fluorescent probes. *Methods Enzymol* 1998;**291**:63–78.

71. Kojima H, Nakatsubo N, Kikuchi K, Kawahara S, Kirino Y, Nagoshi H, et al. Detection and imaging of nitric oxide with novel fluorescent indicators: diaminofluoresceins. *Anal Chem* 1998;**70**:2446–53.

72. Song F, Garner AL, Koide K. A highly sensitive fluorescent sensor for palladium based on the allylic oxidative insertion mechanism. *J Am Chem Soc* 2007;**129**:12354–5.

73. Sullivan E, Tucker EM, Dale IL. Measurement of $[Ca^{2+}]$ using the fluorometric imaging plate reader (FLIPR). *Methods Mol Biol* 1999;**114**:125–33.

74. Stosiek C, Garaschuk O, Holthoff K, Konnerth A. In vivo two-photon calcium imaging of neuronal networks. *Proc Natl Acad Sci USA* 2003;**100**:7319–24.

75. Ceresole, M. Production of new red coloring matter. U.S. Patent 377,349, January 31, 1888.

76. Fay IW. *The chemistry of the coal-tar dyes*. New York, NY: D. Van Nostrand; 1919.
77. Ioffe IS, Otten VF. Rhodamine dyes and related compounds. I. Progenitor of rhodamines, its preparation and properties. *Zh Obshch Khim* 1961;**31**:1511–6.
78. Lavis LD, Chao T-Y, Raines RT. Fluorogenic label for biomolecular imaging. *ACS Chem Biol* 2006;**1**:252–60.
79. Grimm JB, Lavis LD. Synthesis of rhodamines from fluoresceins using Pd-catalyzed C–N cross-coupling. *Org Lett* 2011;**13**:6354–7.
80. Karstens T, Kobs K. Rhodamine B and rhodamine 101 as reference substances for fluorescence quantum yield measurements. *J Phys Chem* 1980;**84**:1871–2.
81. Gee KR, Weinberg ES, Kozlowski DJ. Caged Q-rhodamine dextran: a new photoactivated fluorescent tracer. *Bioorg Med Chem Lett* 2001;**11**:2181–3.
82. Knauer KH, Gleiter R. Photochromism of rhodamine derivatives. *Angew Chem Int Ed* 1977;**16**:113.
83. Belov VN, Bossi ML, Fölling J, Boyarskiy VP, Hell SW. Rhodamine spiroamides for multicolor single molecule switching fluorescent nanoscopy. *Chem Eur J* 2009;**15**:10762–76.
84. Jun ME, Ahn KH. Fluorogenic and chromogenic detection of palladium species through a catalytic conversion of a rhodamine B derivative. *Org Lett* 2010;**12**:2790–3.
85. Lee MH, Kim HJ, Kim JS, Yoon J. A new trend in rhodamine-based chemosensors: application of spirolactam ring-opening to sensing ions. *Chem Soc Rev* 2008;**37**:1465–72.
86. Leytus SP, Melhado LL, Mangel WF. Rhodamine-based compounds as fluorogenic substrates for serine proteinases. *Biochem J* 1983;**209**:299–307.
87. Liu J, Bhalgat M, Zhang C, Diwu Z, Hoyland B, Klaubert DH. Fluorescent molecular probes V: a sensitive caspase-3 substrate for fluorometric assays. *Bioorg Med Chem Lett* 1999;**9**:3231–6.
88. Levine MN, Raines RT. Sensitive fluorogenic substrate for alkaline phosphatase. *Anal Biochem* 2011;**418**:247–52.
89. Huang ST, Lin YL. New latent fluorophore for DT diaphorase. *Org Lett* 2006;**8**:265–8.
90. Yatzeck MM, Lavis LD, Chao TY, Chandran SS, Raines RT. A highly sensitive fluorogenic probe for cytochrome P450 activity in live cells. *Bioorg Med Chem Lett* 2008;**18**:5864–6.
91. Zhang J, Shibata A, Ito M, Shuto S, Ito Y, Mannervik B, et al. Synthesis and characterization of a series of highly fluorogenic substrates for glutathione transferases, a general strategy. *J Am Chem Soc* 2011;**133**:14109–19.
92. Fölling J, Belov V, Kunetsky R, Medda R, Schönle A, Egner A, et al. Photochromic rhodamines provide nanoscopy with optical sectioning. *Angew Chem Int Ed* 2007;**46**:6266–70.
93. Belov VN, Wurm CA, Boyarskiy VP, Jakobs S, Hell SW. Rhodamines NN: a novel class of caged fluorescent dyes. *Angew Chem Int Ed* 2010;**49**:3520–3.
94. Koide Y, Urano Y, Hanaoka K, Terai T, Nagano T. Evolution of Group 14 rhodamines as platforms for near-infrared fluorescence probes utilizing photoinduced electron transfer. *ACS Chem Biol* 2011;**6**:600–8.
95. Egawa T, Hanaoka K, Koide Y, Ujita S, Takahashi N, Ikegaya Y, et al. Development of a far-red to near-infrared fluorescence probe for calcium ion and its application to multicolor neuronal imaging. *J Am Chem Soc* 2011;**133**:14157–9.
96. Dujols V, Ford F, Czarnik AW. A long-wavelength fluorescent chemodosimeter selective for Cu(II) ion in water. *J Am Chem Soc* 1997;**119**:7386–7.
97. Kenmoku S, Urano Y, Kojima H, Nagano T. Development of a highly specific rhodamine-based fluorescence probe for hypochlorous acid and its application to real-time imaging of phagocytosis. *J Am Chem Soc* 2007;**129**:7313–8.
98. Bueno C, Villegas ML, Bertolotti SG, Previtali CM, Neumann MG, Encinas MV. The excited-state interaction of resazurin and resorufin with amines in aqueous solutions. Photophysics and photochemical reaction. *Photochem Photobiol* 2002;**76**:385–90.

99. Theriot JA, Mitchison TJ. Actin microfilament dynamics in locomoting cells. *Nature* 1991;**352**:126–31.
100. Zhou M, Diwu Z, Panchuk-Voloshina N, Haugland RP. A stable nonfluorescent derivative of resorufin for the fluorometric determination of trace hydrogen peroxide: applications in detecting the activity of phagocyte NADPH oxidase and other oxidases. *Anal Biochem* 1997;**253**:162–8.
101. Jose J, Burgess K. Benzophenoxazine-based fluorescent dyes for labeling biomolecules. *Tetrahedron* 2006;**62**:11021–37.
102. Cohen BE, Pralle A, Yao X, Swaminath G, Gandhi CS, Jan YN, et al. A fluorescent probe designed for studying protein conformational change. *Proc Natl Acad Sci USA* 2005;**102**:965–70.
103. Magde D, Brannon JH, Cremers TL, Olmsted J. Absolute luminescence yield of cresyl violet. A standard for the red. *J Phys Chem* 1979;**83**:696–9.
104. Van Noorden CJ, Boonacker E, Bissell ER, Meijer AJ, van Marle J, Smith RE. Ala-Pro-cresyl violet, a synthetic fluorogenic substrate for the analysis of kinetic parameters of dipeptidyl peptidase IV (CD26) in individual living rat hepatocytes. *Anal Biochem* 1997;**252**:71–7.
105. Koller E, Wolfbeis OS. Syntheses and spectral properties of longwave absorbing and fluorescing substrates for the direct and continuous kinetic assay of carboxylesterases, phosphatases, and sulfatases. *Monatsh Chem* 1985;**116**:65–75.
106. Burke MD, Thompson S, Weaver RJ, Wolf CR, Mayer RT. Cytochrome P450 specificities of alkoxyresorufin O-dealkylation in human and rat liver. *Biochem Pharmacol* 1994;**48**:923–36.
107. Kramer DN, Guilbault GG. Resorufin acetate as a substrate for determination of hydrolytic enzymes at low enzyme and substrate concentrations. *Anal Chem* 1964;**36**:1662–3.
108. Kitson TM. Studies on the chymotrypsin-catalysed hydrolysis of resorufin acetate and resorufin bromoacetate. *Biochim Biophys Acta* 1998;**1385**:43–52.
109. Hofmann J, Sernetz M. Immobilized enzyme kinetics analyzed by flow-through microfluorimetry: resorufin-β-galactopyranoside as a new fluorogenic substrate for β-galactosidase. *Anal Chim Acta* 1984;**163**:67–72.
110. Porcal GV, Previtali CM, Bertolotti SG. Photophysics of the phenoxazine dyes resazurin and resorufin in direct and reverse micelles. *Dyes Pigments* 2009;**80**:206–11.
111. O'Brien J, Wilson I, Orton T, Pognan F. Investigation of the Alamar Blue (resazurin) fluorescent dye for the assessment of mammalian cell cytotoxicity. *Eur J Biochem* 2000;**267**:5421–6.
112. Boonacker E, Elferink S, Bardai A, Fleischer B, Van Noorden CJF. Fluorogenic substrate [Ala-Pro]$_2$-cresyl violet but not Ala-Pro-rhodamine 110 is cleaved specifically by DPPIV activity: a study in living Jurkat cells and CD26/DPPIV-transfected Jurkat cells. *J Histochem Cytochem* 2003;**51**:959–68.
113. Lee BW, Johnson GL, Hed SA, Darzynkiewicz Z, Talhouk JW, Mehrotra S. DEVDase detection in intact apoptotic cells using the cell permeant fluorogenic substrate, (Z-DEVD)$_2$-cresyl violet. *Biotechniques* 2003;**35**:1080–5.
114. Ho N, Weissleder R, Tung CH. Development of water-soluble far-red fluorogenic dyes for enzyme sensing. *Tetrahedron* 2006;**62**:578–85.
115. Miller EW, Albers AE, Pralle A, Isacoff EY, Chang CJ. Boronate-based fluorescent probes for imaging cellular hydrogen peroxide. *J Am Chem Soc* 2005;**127**:16652–9.
116. Adams SR, Campbell RE, Gross LA, Martin BR, Walkup GK, Yao Y, et al. New biarsenical ligands and tetracysteine motifs for protein labeling in vitro and in vivo: synthesis and biological applications. *J Am Chem Soc* 2002;**124**:6063–76.
117. Sharonov A, Hochstrasser RM. Wide-field subdiffraction imaging by accumulated binding of diffusing probes. *Proc Natl Acad Sci USA* 2006;**103**:18911–6.

118. Corey PF, Trimmer RW, Biddlecom WG. A new chromogenic β-galactosidase substrate: 7-β-D-galactopyranosyloxy-9,9-dimethyl-9H-acridin-2-one. *Angew Chem Int Ed* 1991;**30**:1646–8.
119. Warther D, Bolze F, Leonard J, Gug S, Specht A, Puliti D, et al. Live-cell one- and two-photon uncaging of a far-red emitting acridinone fluorophore. *J Am Chem Soc* 2010;**132**:2585–90.
120. Leira F, Vieites J, Vieytes M, Botana L. Characterization of 9H-(1,3-dichlor-9,9-dimethylacridin-2-ona-7-yl)-phosphate (DDAO) as substrate of PP-2A in a fluorimetric microplate assay for diarrhetic shellfish toxins (DSP). *Toxicon* 2000;**38**:1833–44.
121. Rush JS, Beatty KE, Bertozzi CR. Bioluminescent probes of sulfatase activity. *ChemBioChem* 2010;**11**:2096–9.
122. Waggoner A. Dye indicators of membrane potential. *Ann Rev Biophys Bioeng* 1979;**8**:47–68.
123. Rye HS, Yue S, Wemmer DE, Quesada MA, Haugland RP, Mathies RA, et al. Stable fluorescent complexes of double-stranded DNA with bis-intercalating asymmetric cyanine dyes: properties and applications. *Nucleic Acids Res* 1992;**20**:2803–12.
124. Mujumdar RB, Ernst LA, Mujumdar SR, Lewis CJ, Waggoner AS. Cyanine dye labeling reagents: sulfoindocyanine succinimidyl esters. *Bioconjugate Chem* 1993;**4**:105–11.
125. Waggoner A. Fluorescent labels for proteomics and genomics. *Curr Opin Chem Biol* 2006;**10**:62–6.
126. Waggoner A, Kenneth S. Covalent labeling of proteins and nucleic acids with fluorophores. *Methods Enzymol* 1995;**246**:362–73.
127. Thomas, N., Michael, N. P., Millar, V., Davies, B. & Briggs, M. S. J. Fluorescent detection method and reagent. U.S. Patent 7,662,973, February 16, 2010.
128. Kiyose K, Aizawa S, Sasaki E, Kojima H, Hanaoka K, Terai T, et al. Molecular design strategies for near-infrared ratiometric fluorescent probes based on the unique spectral properties of aminocyanines. *Chem Eur J* 2009;**15**:9191–200.
129. Maurel D, Banala S, Laroche T, Johnsson K. Photoactivatable and photoconvertible fluorescent probes for protein labeling. *ACS Chem Biol* 2010;**5**:507–16.
130. Song F, Peng X, Lu E, Wang Y, Zhou W, Fan J. Tuning the photoinduced electron transfer in near-infrared heptamethine cyanine dyes. *Tetrahedron Lett* 2005;**46**:4817–20.
131. Karton-Lifshin N, Segal E, Omer L, Portnoy M, Satchi-Fainaro R, Shabat D. A unique paradigm for a turn-ON near-infrared cyanine-based probe: non-invasive intravital optical imaging of hydrogen peroxide. *J Am Chem Soc* 2011;**133**:10960–5.
132. van de Linde S, Krstić I, Prisner T, Doose S, Heilemann M, Sauer M. Photoinduced formation of reversible dye radicals and their impact on super-resolution imaging. *Photochem Photobiol Sci* 2010;**10**:499–506.
133. Dempsey GT, Bates M, Kowtoniuk WE, Liu DR, Tsien RY, Zhuang X. Photoswitching mechanism of cyanine dyes. *J Am Chem Soc* 2009;**131**:18192–3.
134. Dempsey GT, Vaughan JC, Chen KH, Bates M, Zhuang X. Evaluation of fluorophores for optimal performance in localization-based super-resolution imaging. *Nat Methods* 2011;**8**:1027–36.
135. Adie E, Kalinka S, Smith L, Francis M, Marenghi A, Cooper M, et al. A pH-sensitive fluor, CypHer-5, used to monitor agonist-induced G protein-coupled receptor internalization in live cells. *Biotechniques* 2002;**33**:1152–7.
136. Oushiki D, Kojima H, Terai T, Arita M, Hanaoka K, Urano Y, et al. Development and application of a near-infrared fluorescence probe for oxidative stress based on differential reactivity of linked cyanine dyes. *J Am Chem Soc* 2010;**132**:2795–801.
137. Ozmen B, Akkaya EU. Infrared fluorescence sensing of submicromolar calcium: pushing the limits of photoinduced electron transfer. *Tetrahedron Lett* 2000;**41**:9185–8.
138. Stewart WW. Synthesis of 3,6-disulfonated 4-aminonaphthalimides. *J Am Chem Soc* 1981;**103**:7615–20.

139. Wolfbeis OS, Fürlinger E, Kroneis H, Marsoner H. Fluorimetric analysis. *Fresenius J Anal Chem* 1983;**314**:119–24.

140. Naleway JJ, Fox CMJ, Robinhold D, Terpetschnig E, Olson NA, Haugland RP. Synthesis and use of new fluorogenic precipitating substrates. *Tetrahedron Lett* 1994;**35**:8569–72.

141. Whitaker JE, Haugland RP, Ryan D, Hewitt PC, Haugland RP, Prendergast FG. Fluorescent rhodol derivatives: versatile, photostable labels and tracers. *Anal Biochem* 1992;**207**:267–79.

142. Yuan L, Lin W, Yang Y, Chen H. A unique class of near-infrared functional fluorescent dyes with carboxylic-acid-modulated fluorescence ON/OFF switching: rational design, synthesis, optical properties, theoretical calculations, and the applications for fluorescence imaging in living animals. *J Am Chem Soc* 2011;**134**:1200–11.

143. Larison KD, BreMiller R, Wells KS, Clements I, Haugland RP. Use of a new fluorogenic phosphatase substrate in immunohistochemical applications. *J Histochem Cytochem* 1995;**43**:77–83.

144. Paragas VB, Zhang YZ, Haugland RP, Singer VL. The ELF-97 alkaline phosphatase substrate provides a bright, photostable, fluorescent signal amplification method for FISH. *J Histochem Cytochem* 1997;**45**:345–57.

145. Aw J, Shao Q, Yang Y, Jiang T, Ang C, Xing B. Synthesis and characterization of 2-(2'-hydroxy-5'-chlorophenyl)-6-chloro-4(3H)-quinazolinone-based fluorogenic probes for cellular imaging of monoamine oxidases. *Chem Asian J* 2010;**5**:1317–21.

146. Li J, Yao SQ. "Singapore Green": a new fluorescent dye for microarray and bioimaging applications. *Org Lett* 2009;**11**:405–8.

147. Kamiya M, Asanuma D, Kuranaga E, Takeishi A, Sakabe M, Miura M, et al. β-Galactosidase fluorescence probe with improved cellular accumulation based on spirocyclized rhodol scaffold. *J Am Chem Soc* 2011;**133**:12960–3.

148. Jasuja R, Keyoung J, Reid GP, Trentham DR, Khan S. Chemotactic responses of *Escherichia coli* to small jumps of photoreleased L-aspartate. *Biophys J* 1999;**76**:1706–19.

149. Bloodgood BL, Sabatini BL. Neuronal activity regulates diffusion across the neck of dendritic spines. *Science* 2005;**310**:866–9.

150. Trigo FF, Corrie JET, Ogden D. Laser photolysis of caged compounds at 405 nm: photochemical advantages, localisation, phototoxicity and methods for calibration. *J Neurosci Methods* 2009;**180**:9–21.

151. Zhang JF, Lim CS, Bhuniya S, Cho BR, Kim JS. A highly selective colorimetric and ratiometric two-photon fluorescent probe for fluoride ion detection. *Org Lett* 2011;**13**:1190–3.

152. Srikun D, Miller EW, Domaille DW, Chang CJ. An ICT-based approach to ratiometric fluorescence imaging of hydrogen peroxide produced in living cells. *J Am Chem Soc* 2008;**130**:4596–7.

153. Duke RM, Veale EB, Pfeffer FM, Kruger PE, Gunnlaugsson T. Colorimetric and fluorescent anion sensors: an overview of recent developments in the use of 1,8-naphthalimide-based chemosensors. *Chem Soc Rev* 2010;**39**:3936–53.

CHAPTER TWO

# Fluorescent Environment-Sensitive Dyes as Reporters of Biomolecular Interactions

## Andrey S. Klymchenko, Yves Mely

Laboratoire de Biophotonique et Pharmacologie, UMR 7213 CNRS, Université de Strasbourg, Faculté de Pharmacie, Illkirch Cedex, France

## Contents

## Abstract

Monitoring biomolecular interactions is a fundamental issue in biosensing, with numerous applications ranging from biological research to clinical diagnostics. Fluorescent dyes capable of changing their color and brightness in response to changes of their environment properties, the so-called environment-sensitive dyes, have recently emerged as reporters of these interactions. The most well established of these are dyes that undergo excited-state charge transfer showing red shift of their single emission band with increase in the solvent polarity. The other promising class are dyes of the

*Progress in Molecular Biology and Translational Science*, Volume 113
ISSN 1877-1173
http://dx.doi.org/10.1016/B978-0-12-386932-6.00002-8

35

3-hydroxychromone family that undergo excited-state intramolecular proton transfer and show solvent-sensitive dual emission. Examples of existing solvatochromic dyes and their biosensing applications are given, with particular focus on the 3-hydroxychromones. It is shown that solvatochromic dyes are powerful tools for monitoring conformation changes of proteins and their interactions with nucleic acids, proteins, and lipid membranes.

## 1. INTRODUCTION

Monitoring biomolecular interactions is a fundamental issue in biosensing, with numerous applications ranging from basic biological research to clinical diagnostics. Fluorescence techniques are particularly well suited for this purpose.[1,2] The most established one is the Förster resonance energy transfer (FRET)-based approach, where interacting partners are labeled with donor and acceptor molecules.[3,4] The interaction event results in an energy transfer between the proximal donor and acceptor, providing the analytical signal. Though the approach is robust, it requires double labeling, which is complicated and cannot be realized in many screening assays. Therefore, single fluorescence labeling techniques, where only one of the partners is labeled, are of high interest. The most well-established single-labeling approach is based on fluorescence anisotropy, which follows changes in the mobility of the fluorescent label that is grafted to one of the interacting partners.[5] The other approach, which has emerged only recently, is utilization of environment-sensitive dyes.[6,7]

Unlike "classical" dyes, environment-sensitive dyes can change their fluorescence properties, fluorescence intensity or emission color, in response to changes in the physicochemical properties of their molecular environment.[8] While classical dyes are perfect markers of biological molecules, the environment-sensitive dyes are "smart molecules" that can be used as sensors for probing the local biological environment and monitoring biomolecular interactions. Within this approach, the interaction between the molecules changes the properties of the local site of interaction, which in turn affects the fluorescence properties (emission maximum or intensity) of the environment-sensitive labels (Fig. 2.1).[6,7]

The response of environment-sensitive dyes to the environment is driven by excited-state reactions (conformational change, charge, electron and proton transfer, etc.) and noncovalent interactions with the surrounding, such as universal interactions (van der Waals, dipole–dipole, dipole–external electric

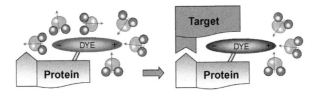

**Figure 2.1** Monitoring biomolecular interactions by a solvatochromic fluorescent dye. (For color version of this figure, the reader is referred to the online version of this chapter.)

field, etc.), and specific H-bonding interactions. Here, we do not consider pH- and ion-sensitive dyes as environment sensitive, as the response of these dyes is associated with changes in their chemical structure: protonation/deprotonation or formation of a complex with an ion.

Several types of environment-sensitive fluorophores are of particular interest for biomolecular applications: molecular rotors and solvatochromic fluorescent dyes. Molecular rotors are an interesting class of environment-sensitive dyes, which change their emission intensity in response to the change of the solvent viscosity.[9] These dyes feature high rotational flexibility of their conjugated system so that they are poorly emissive in nonviscous environments (such as water or organic solvents). In viscous environments, such as biological membranes and biomacromolecules, their rotation mobility is restricted, which dramatically increases their fluorescence quantum yield. Thus, these dyes can turn on their fluorescence in response to interactions that rigidify their environment.[10] Though these dyes were largely applied as probes in biological membranes, their application for detection of biomolecular interactions is still poorly explored and therefore will not be reviewed here. A much better established class comprises solvatochromic fluorescent dyes. They exhibit shifts in their emission maxima and sometimes change in their fluorescence quantum yield as a function of polarity and hydration of their environment.[11,12] In these dyes, the dipole moment increases dramatically upon electronic excitation ($S_0 \rightarrow S_1$ transition) because of an intramolecular charge transfer from the electron-donor group to the electron–acceptor group (Fig. 2.2). Polar solvents relax efficiently the excited molecules to the $S_1^{solv}$ state as a result of polarization of the solvent dipoles around the fluorophore dipole. An increase in the solvent polarity decreases the energy of the $S_1^{solv}$ state, resulting in a red shift in the emission spectra (Fig. 2.2). In addition, protic solvents (which contain hydrogen atoms bound to oxygen (hydroxyl) or to nitrogen (amine, amide, etc.)) interact with the

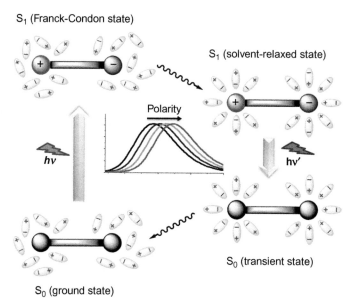

**Figure 2.2** Simplified diagram explaining the phenomenon of solvatochromism. (See Color Insert.)

fluorophore through H-bonding and thus can also decrease the energy of $S_1^{solv}$ state. Therefore, both dipole–dipole and H-bonding interactions in polar solvents can shift the emission of these dyes to the red. Water is a highly dipolar molecule as well as an exceptionally strong H-bond donor. Therefore, its effect on the emission color of the solvatochromic dyes is particularly drastic and can be used for detection of biomolecular interactions as shown in Fig. 2.1. An additional important property of most environment-sensitive dyes is their poor fluorescence intensity in water, because of intermolecular electron or proton transfer. Therefore, incorporation of these dyes into proteins and lipid membranes usually increases strongly their fluorescence intensity as a result of efficient screening of these molecules from bulk water.[6,8,13] An unusual class is the two-band solvatochromic fluorescent dyes based on 3-hydroxychromone (3HC) derivatives. Because of excited-state intramolecular proton transfer (ESIPT), they show two emission bands, which change their relative intensities in response to solvent polarity.[8] This chapter will briefly present the design and applications of single-band environment-sensitive dyes, with the main focus on the dyes based on 3HC. For more details on single-band solvatochromic dyes and their biological applications, the reader should see the excellent review by Imperiali et al.[6]

## 2. SINGLE-BAND SOLVATOCHROMIC DYES: DESIGN AND APPLICATIONS

Representative examples of single-band solvatochromic dyes are shown in Fig. 2.3. All these molecules bear an electron-donor and electron-acceptor group attached to a conjugated (usually aromatic) system. These dyes present a single emission band, which changes its position and sometimes intensity in response to changes in the environment polarity. Their spectroscopic and solvatochromic properties are shown in Table 2.1. As a general trend, it can be noticed that these dyes are characterized by an absorption in the violet or ultraviolet (UV) region as well as by a relatively low extinction coefficient, compared to the classical fluorescent dyes (rhodamine, fluorescein, cyanines, etc.). Two dye families, phenoxazine and nitrobenzoxadiazole (NBD) derivatives, present red-shifted absorption but are characterized by relatively weak fluorescence solvatochromism. Most solvatochromic dyes show relatively high fluorescence quantum yields in apolar solvents, which drop drastically in polar solvents, especially in water. From Table 2.1, it is clear that, though there is a large variety of solvatochromic dyes, there is none that could feature excellent

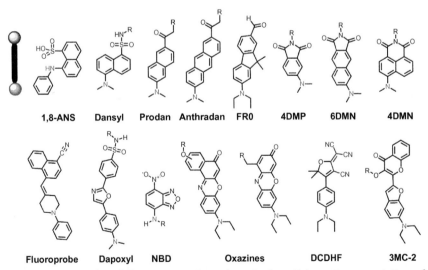

**Figure 2.3** Examples of fluorescent solvatochromic dyes. Schematic presentation of fluorophore structure: electron-donor and -acceptor groups are shown in red and blue, respectively. Group R indicates the common point of the dye conjugation. (See Color Insert.)

**Table 2.1** Properties of common solvatochromic fluorescent dyes[a]

| Dye | $\lambda_{Abs}$ (nm) (MeOH) | $\varepsilon_{max}$ (MeOH) | Polar solvent (MeOH) | | Apolar solvent (Toluene) | | $\Delta\nu_{Fluo}$ (cm$^{-1}$) | Refs. |
| | | | $\lambda_{Fluo}$ (nm) | QY (%) | $\lambda_{Fluo}$ (nm) | QY (%) | | |
|---|---|---|---|---|---|---|---|---|
| Fluoroprobe | 308[b] | 12,000 | 695[c] | <0.1[c] | 476 | 46 | 6620 | 14 |
| FR0 | 396 | 43,000 | 570 | 19 | 434 | 98 | 5500 | 15 |
| Dapoxyl® derivatives | 373 | 28,000 | 584 | 39 | 457 | 86 | 4760 | 16 |
| Prodan | 361 | 18,400 | 498 | 51 | 417 | 55 | 3900 | 17 |
| 3MC-2 | 445 | 29,000 | 597 | 1.4 | 485 | 68 | 3870 | 18 |
| 6DMN | 382 | 8000 | 589 | 1.2 | 491 | 21 | 3390 | 19 |
| Anthradan | 456 | 12,100[d] | 604 | 41 | 507 | 58 | 3170 | 20 |
| 4DMP | 396 | 6500 | 534 | 12[e] | 457[c] | 62[c] | 3160 | 21 |
| Dansyl derivatives | 335 | 4600 | 526 | 49 | 471 | 81 | 2220 | 22 |
| Nile Red | 553 | 45,000 | 632 | 38 | 569 | 90 | 1750 | 23 |
| NBD | 465 | 22,000 | 541 | – | 529 | – | 420 | 24 |

[a]$\lambda_{Abs}$, absorption maximum; $\varepsilon_{max}$, absorption coefficient (M$^{-1}$cm$^{-1}$); $\lambda_{Fluo}$, fluorescence maximum; QY, fluorescence quantum yield; MeOH, methanol; $\delta_{\nu Fluo}$, band shift in response to change of solvent from toluene to methanol.
[b]Data in cyclohexane.
[c]Tetrahydrofuran.
[d]Acetonitrile.
[e]Dioxane.

brightness and red-shifted absorption, together with high fluorescence solvatochromism.

To label biomolecules with solvatochromic dyes, their amino or thiol-reactive derivatives are commonly used. In solvatochromic labels, the fluorophore should be connected to the reactive group through the shortest possible linker, in order to localize it precisely at the labeling site of a biomolecule of interest. However, the most efficient method of site-specific protein labeling corresponds to the direct introduction of the amino acid derivative of the solvatochromic dye into the peptide. This can be achieved by solid-state peptide synthesis or by using cellular protein synthesis machinery, where the genetic code is extended to an unnatural fluorescent amino acid.[25,26] We will briefly present the most common classes of single-band solvatochromic dyes and examples of their applications.

## 2.1. Naphthalene sulfonic acid derivatives

One of the first solvatochromic dyes, which still remains an essential tool for protein and membrane studies, is 1-Anilinonaphthalene-8-Sulfonic Acid (1,8-ANS).[13,27] In addition to the strong solvent-dependent shift in its emission spectrum, it shows a dramatic increase in fluorescence intensity on binding to biomolecules. In bulk water, the fluorescence of this dye is strongly quenched, while, on being bound to proteins or lipid membranes, it is efficiently screened from water, resulting in a strong increase in fluorescence. The other naphthalene sulfonic acid analogues that have found even more applications in biology are Dansyl derivatives (Fig. 2.3). The reactive derivative Dansyl chloride was commonly used to label amino groups in proteins and lipids. Similar to 1,8-ANS, the emission color and emission intensity of the Dansyl moiety are highly sensitive to solvent polarity.[28,29] This fluorophore was one of the first used for protein labeling, but currently it is used only rarely because of its UV absorption and rather weak solvatochromism.

## 2.2. Prodan

Prodan is one of the best classical examples of solvatochromic dyes,[17] and has found numerous biological applications because of its remarkable sensitivity to solvent polarity together with its relatively small size. Reactive derivatives of Prodan such as Acrylodan (6-acryloyl-2-dimethylamino-naphthalen)[22,30] or Badan (6-bromoacetyl-2-dimethylaminonaphthalene)[31] have been attached covalently to proteins via reaction with thiol groups. For instance, Acrylodan attached to the N-terminus of peptide ligands was used to monitor their interactions with the holecystokinin receptor.[30] In another representative study, six cysteine-substituted sites of mouse acetylcholinesterase were labeled individually with Acrylodan and the kinetics of substrate hydrolysis and inhibitor binding were examined. While some sites of labeling located far from the active center did not show any spectral changes of Acrylodan on inhibitor binding, the sites located near the perimeter of the gorge showed blue shifts, reflecting the exclusion of solvent and creation of a hydrophobic environment by the associated ligand.[32] This study is a nice demonstration how a conformation change upon enzyme–inhibitor interaction can be specifically "mapped" over all protein. To localize better the fluorophore on the peptide backbone, an amino acid derivative of Prodan, namely, 6-dimethylaminonaphtoyl alanine (Aladan), was synthesized independently by two research groups.[33,34] The probe allowed monitoring binding of S-peptide with ribonuclease S[34] and estimation of the local dielectric constant of the B1 domain of the staphylococcal protein G at different sites.[33] Later, this amino acid was successfully applied to the study of delta-opioid receptor antagonist binding.[35]

## 2.3. Improved Prodan analogues

The key weakness of Prodan is its absorption in UV (360 nm), which limits its cellular applications. In order to shift the absorbance of Prodan to the red, Lu et al.[20] synthesized its benzo-analogue, 2-propionyl-6-dimethylaminoanthracene, Anthradan. This dye showed the desired red-shifted absorption (around 430 nm), but its brightness was limited because of its low absorption coefficient. Recently, we extended the electronic conjugation of Prodan by substituting its naphthalene core with fluorene.[15] The obtained fluorene derivative, FR0 (Fig. 2.3), showed a

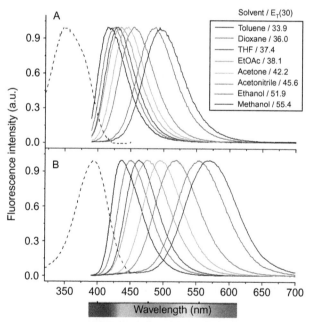

**Figure 2.4** Absorption (dash) and fluorescence (solid) spectra of Prodan (A) and FR0 (B) in organic solvents of different polarities. Absorption spectra were recorded in toluene. $E_T(30)$—empirical polarity index.[11] Data from Ref. 15. (See Color Insert.)

red-shifted absorption (close to 400 nm)(Fig. 2.4), with twice as large an absorption coefficient and a manifold larger two-photon absorption cross section (400 Goeppert-Mayer) compared to Prodan. Moreover, studies in organic solvents have revealed a much stronger dependence of its emission maximum on solvent polarity (Fig. 2.4), which is connected with its twice as large change in the dipole moment (14 D).

## 2.4. 4-Aminophthalimide analogues

The other important family of solvatochromic dyes, which attracted attention only recently, is 4-aminophthalimide analogues. The simplest of them, 4-N,N-dimethylaminophthalimide (4DMP), can be considered as one of the first chromophores for polarity sensing. This relatively small and rigid molecule presents a very strong solvent sensitivity.[21] However, similar to Prodan, it shows absorption in the UV region (380–390 nm) and a very low absorption coefficient. To improve its properties, the 4DMP fluorophore was extended, giving a new environment-sensitive dye 6-N,N-dimethylamino-2,3-naphthalimide (6DMN)[19] (Fig. 2.3). This fluorophore exhibits interesting fluorescence properties with emission in the 500–600 nm range and combined (fluorescence intensity and position of the maximum) response to changes in the environment polarity, though its absorption properties are not considerably improved. In this respect, a recently developed analogue of 4DMP, 4-N,N-dimethylamino-1,8-naphthalimide (4DMN), constitutes a significant improvement, showing a shifted absorption maximum at 440 nm.[36] It should be noted that, similar to 1,8-ANS, 4DMP and all its analogues are nearly nonfluorescent in water but become highly fluorescent in aprotic media. These properties are important for intensiometric detection of molecular interactions.[36] However, their poor fluorescence in water makes them inefficient in the cases where the label shows significant water exposure at all steps of interaction.

To apply the 4DMP fluorophore for protein studies, Fmoc-protected amino acid bearing this fluorophore was synthesized and introduced into an octapeptide using standard solid-phase synthesis. The label was able to sense phosphorylation-dependent binding of the synthesized peptide to the 14-3-3bp protein.[37] Peptides labeled with a 6DMN-based amino acid were used for monitoring protein–protein interactions, as exemplified in studies with the SH2 phosphotyrosine binding domains.[19] The same labeled amino acid was used for sensing peptide binding to proteins of a major histocompatibility complex at the cell surface.[38] Furthermore, an amino acid analogue of 4DMN, featuring improved spectroscopic properties and chemical stability, was introduced into a peptide that is recognized by calmodulin. Remarkably, the interaction event between these two molecules yielded > 900-fold increase in the fluorescence intensity of the dye.[36]

## 2.5. Fluoroprobe

In the search for advanced environment-sensitive dyes, a bichromophoric dye Fluoroprobe (Fig. 2.3) was developed.[14] This dye exhibits a charge transfer through space, which generates an exceptional change in the dipole moment (27 D) and, thus, solvent sensitivity. For the moment, Fluoroprobe remains the most solvatochromic fluorescent dye. However, it has found no applications in biology, because of the extremely strong quenching of its fluorescence in polar media, UV absorption (308 nm), and very low extinction coefficient (Table 2.1).

## 2.6. Dapoxyl® derivatives

It is also worth mentioning a Dapoxyl® dye (Fig. 2.3) showing remarkable fluorescent solvatochromism (up to 200 nm red shift from hexane to water/acetonitrile (4/1) mixture), as well as high fluorescence quantum yield and extinction coefficient.[16] However, its absorption in the UV range remains a disadvantage. Moreover, its applications are limited so far to only a few examples, such as biosensor development[39] and FRET-based assays in peptides.[40]

## 2.7. NBD

An important example of red-shifted environment-sensitive dye is NBD. Its small-sized fluorophore and absorption around 480 nm are very convenient for biological applications. However, its solvatochromism is very small, which limits its applications to the cases where the changes in the environment are drastic. NBD was one of the first environment-sensitive dyes (together with Dansyl) used as a protein label. NBD helped revealing a mechanism for the transduction of ligand-induced protein conformational changes[24] through the distinct maltose-dependent fluorescence response observed with three individual cysteine mutants of *Escherichia coli* maltose-binding protein, covalently labeled with NBD. The results provided insights for designing fluorescent biosensors.

## 2.8. Phenoxazine derivatives

A representative example of phenoxazine-based dyes is Nile Red.[23] It combines red-shifted absorption (around 530 nm) and emission together with high brightness. Nevertheless, its solvatochromism is moderate, being larger than that of NBD, but smaller than that of the UV/blue analogues described above (Table 2.1). In the recent years, a variety of Nile Red derivatives for biomolecule labeling have been reported.[41,42] Thiol-reactive Nile Red derivatives were successfully applied for labeling a galactose/glucose binding protein. The obtained conjugate was validated as a fluorescence biosensor for glucose detection.[43] Phenoxazine derivatives, which are more compact analogues of Nile Red, were also used for designing the thiol-reactive probe, aminophenoxazone maleimide (APM), which was used for reporting protein conformational changes.[44] APM has a short linker between the probe and the protein, ensuring that it can closely follow the motions the water-exposed domain of the β2 adrenergic receptor during its interactions with ligands. In addition, APM was shown to sense the conformational changes underlying voltage sensing in the Shaker potassium channel.[44]

## 2.9. New advanced solvatochromic dyes

This overview on the existing environment-sensitive dyes shows that there is a strong need for new solvatochromic fluorescent dyes presenting both strong solvatochromism and good fluorescence properties. Most of the dyes described above (except Nile Red and NBD analogues) absorb in the UV range (Table 2.1), which is generally not suitable for biological applications. Excitation at longer wavelengths would decrease significantly the photo-damage of the biological samples and decrease their autofluorescence. The photostability of most of environment-sensitive dyes is also limited so that they are not compatible with some modern techniques, such as single-molecule fluorescence detection.[45] Moreover, the low fluorescence quantum yield in aqueous media of most of the described dyes limits their application for the investigation of small peptides, where the label is exposed to water. Finally, sensitivity to solvent polarity of these dyes is frequently not enough to detect subtle changes in the environment of the biomolecule of interest. Therefore, the research in this field is now focused on improving the solvent sensitivity of the dyes, shifting their emission wavelength to the red and increasing fluorescence brightness and photostability. For example, the 2-dicyanomethylene-3-cyano-2,5–dihydrofuran family of fluorophores (see DCDHF in Fig. 2.3), featuring red-shifted absorption and good solvatochromism, appears promising for single-molecule experiments because of their very high photostability.[46,47] The recently introduced 3-methoxychromones (3MC-2) appear also of interest, as in addition to good photostability and fluorescence quantum yields, they present very high solvatochromism.[18] We should also mention polymethine dyes displaying near-IR emission and solvent-dependence of their lifetime,[48] as well as BODIPY derivatives, capable of switching on–off their fluorescence in response to solvent polarity changes.[49] Although these fluorophores have not yet been applied for biomolecular studies, they appear as attractive building blocks for future high-performance polarity-sensitive labels of biomolecules.

## 3. TWO-BAND SOLVATOCHROMIC DYES BASED ON ESIPT

An alternative mechanism of solvent sensitivity can be realized by utilizing the so-called ESIPT. Particularly interesting ESIPT dyes are 3-hydroxychromones (3HCs), presenting dual emission originating from the normal excited state (N*) and the ESIPT tautomer (T*) (Fig. 2.5).[50] The pathway for ESIPT in 3HCs is provided by the intramolecular H-bond through a five-membered cycle, which is much weaker than the six-membered cycle presented by other ESIPT systems. Therefore, it can be easily perturbed by H-bonding interactions, thus modulating the dual emission of 3HCs.

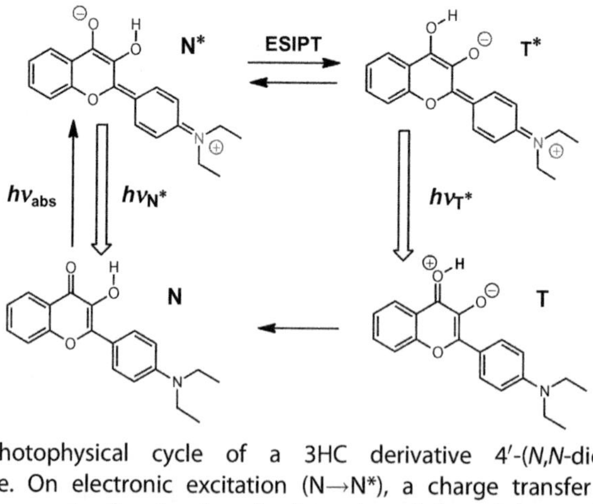

**Figure 2.5** Photophysical cycle of a 3HC derivative 4'-(N,N-diethylamino)-3-hydroxyflavone. On electronic excitation (N→N*), a charge transfer from the 4'-dialkylamino group to the 4-carbonyl takes place followed by an ESIPT process (N*→T*). After T*→T transition, the proton remains at the 4-carbonyl group, producing a zwitterionic T state that rapidly converts into the stable N state. (For color version of this figure, the reader is referred to the online version of this chapter.)

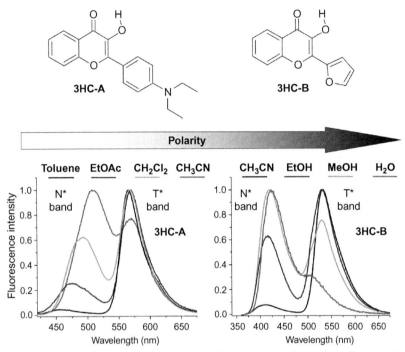

**Figure 2.6** Chemical structure of 3HC dyes and their solvent-dependent dual emission. Reproduced from Ref. 7. (See Color Insert.)

Among the 3HCs developed so far, two derivatives are of particular interest: 4′-(dialkylamino)-3-hydroxyflavone (3HC-A) and 2-(2-furyl)-3-hydroxychromone (3HC-B) (Fig. 2.6). Owing to the 4′-dialkylamino group, the N⋆ excited state of 3HC-A exhibits a large dipole moment, where the electronic charge is transferred from the dialkylamino group to the chromone moiety (Fig. 2.5).[51] In contrast, the ESIPT product T⋆ state exhibits much lower charge separation and, thus, lower dipole moment. Therefore, only the N⋆ state shows a significant shift to the red on increase in solvent polarity (Fig. 2.6). This red shift is accompanied by an increase in the relative intensity of the N⋆ band, because this state becomes energetically more favorable than the T⋆ state.[52] Therefore, the intensity ratio of the N⋆ and T⋆ bands, that is, N⋆/T⋆, is an important indicator of solvent polarity.[51] However, 3HC-A shows dual emission only in the range of low–polar and polar aprotic solvents (Fig. 2.6). In polar protic solvents, including water, the ESIPT is efficiently inhibited so that the T⋆ emission is no more observed.[51] Therefore, this dye could be applied for probing biological environments of relatively low polarity and hydration, such as, for instance, biological membranes (see below).

On the other hand, the 3HC-B dye is much more appropriate for highly polar media. Owing to its much weaker electron-donor 2-aryl group (2-furanyl vs. 4-dialkylaminophenyl in 3HC-A), the dipole moment of its N* state is relatively low. Therefore, this state cannot be stabilized even in highly polar aprotic solvents so that its emission is almost negligible (Fig. 2.6). Moreover, in contrast to 3HC-A, the ESIPT inhibition by protic solvents is not complete so that a clear dual emission depending on solvent polarity is observed.[53] Thus, the 3HC-B dye is suitable for probing polar protic environments characterized by high hydration, which corresponds well to peptides and nucleic acids (see below).

We should note the key differences between single-band solvatochromic fluorescent dyes and 3HC dyes. While the former shift their emission maximum in response to solvent polarity, 3HC dyes may change both the positions of bands as well as their intensity ratio. This ratio is an additional channel of spectroscopic information, which allows more detailed (multiparametric) characterization of the probe environment.[51] Moreover, as a result of ESIPT, 3HC dyes are focused in a narrower polarity range, where they can show higher sensitivity to properties of environment compared to single-band solvatochromic dyes.

## 4. APPLICATIONS OF TWO-COLOR DYES FOR MONITORING BIOMOLECULAR INTERACTIONS

### 4.1. Monitoring conformational changes of proteins

Conformational changes in proteins can result in significant changes in the site exposure to bulk water. This idea was validated using α1-antitrypsin (α1-AT), which, during the complex multistep inhibition of proteinases, undergoes dramatic changes in the tertiary structure (Fig. 2.7). As α1-AT contains only one cysteine (Cys-232), this residue was specifically labeled with a thiol-reactive 3HC-B derivative.[55] The intensity ratio T*/N* of 1.18 for the labeled protein was between that observed for water (0.45) and for ethanol (2.22), suggesting that the label is partially screened from bulk water by the protein environment. The interaction of the labeled α1-AT with pancreatic elastase led to ∼65% change in the T*/N* intensity ratio of the two emission bands, suggesting an increased exposure of the labeled Cys-232 residue to the bulk water on complex formation. Similar experiments with elastase and α1-AT conjugated to a NBD derivative

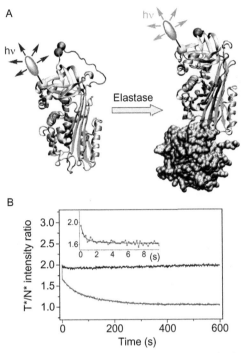

**Figure 2.7** Conformational changes in α1-AT on its reaction with elastase, as monitored by a 3HC probe. The graph A shows the X-ray structures of the free α1-AT and the α1-AT/ elastase complex.[54] The labeled Cys is in yellow. The graph B shows the kinetics of this conformation transition monitored by our 3HC probe. The inset in this figure presents the fast part of this transition.[55] (See Color Insert.)

confirmed these results but led to much smaller modifications in the emission spectrum, indicating superiority of the 3HC label in terms of environment sensitivity. Stopped-flow studies of the reaction between the labeled α1-AT and elastase showed, in addition to the well-described fast step, a new slow step of the inhibition process[55] that is probably associated with a slow structural reorganization aimed at stabilizing the final inhibited complex.

## 4.2. Peptide–nucleic acid interactions

Monitoring interactions of peptides with nucleic acids can be achieved through a single labeling of the peptide partner with a solvatochromic dye. However, in this case, solvent sensitivity of these dyes should be optimized for polar environments and remain fluorescent in the DNA complex. 3HC-B fluorophore meets both requirements according to our previous

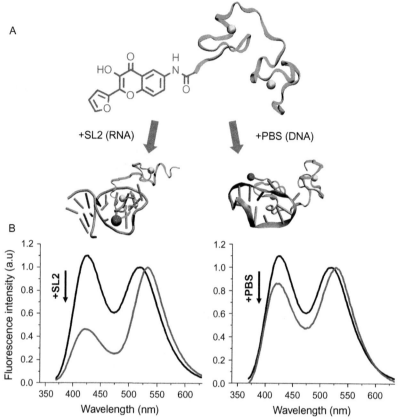

**Figure 2.8** (A) Schematic structure of a conjugate of the 3HC-B label with the (11–55) sequence of the NCp7 protein and its complexes with different oligonucleotides, SL2 and PBS, described by NMR.[56,57] (B) Fluorescence spectra of the fluorescently labeled conjugate (NC-3HC) on addition of SL2 and PBS oligonucleotides.[58] Reproduced from Ref. 7. (For color version of this figure, the reader is referred to the online version of this chapter.)

studies.[53] A carboxylic acid derivative of 3HC-B was attached to the N-terminus of the (11–55) sequence of the nucleocapsid protein (NCp7) from human immunodeficiency virus (HIV-1) using solid-state peptide synthesis (Fig. 2.8). This protein plays an important role in HIV-1 through interactions with the viral RNA and its DNA copies,[59] and notably with cTAR, the complementary DNA sequence of the transactivation response element from the HIV-1 genome.[60] The interaction of the labeled NCp7 fragment with cTAR changed dramatically the emission color of the 3HC probe.[58] Different oligonucleotide (ODN) sequences have been tested and the obtained spectroscopic data were correlated with the known nuclear magnetic resonance (NMR) structure of the peptide–ODN complexes (Fig. 2.8). The results suggested that the 3HC label senses the proximity of the peptide-labeling site (N-terminus) to the ODN bases. This approach allowed us to determine the peptide–ODN binding parameters and distinguish multiple binding sites in ODNs, which is rather difficult using other fluorescence methods. Moreover, this method was found to be more sensitive than steady-state fluorescence anisotropy, in the case of small ODNs. Further synthesis of the analogues of 3HC-B presenting increased brightness and solvatochromism allowed us to significantly improve the sensitivity of this method.[61] Currently, we are developing L-amino acids bearing 3HC fluorophores, which is an important milestone in the probing of peptide–DNA interactions at any desired peptide site.

## 4.3. Protein–protein interactions

Monitoring protein–protein interactions is of key importance in the development of peptide-based biosensors. These interactions result commonly in changes of the environment particularly at their interface and, therefore, can be detected by solvatochromic labels (Fig. 2.1). However, classical solvatochromic dyes (based on NBD, Prodan, 4DMP, etc.) working on the principle of the emission band shift and/or fluorescence quenching are not always so efficient in highly polar media. In this respect, ESIPT-based 3HC labels are attractive alternatives, particularly for applications with small peptides and proteins, where the polarity of the labeling site is high.

To evaluate the possibility of using our 3HC probe for sensing peptide–peptide interactions, the high-affinity model of interaction between the synthetic 18-amino acid peptide pTMVP and a recombinant antibody fragment, Fab57P, was used.[62] The pTMVP peptide, which contains the Fab57P epitope of the tobacco mosaic virus coat protein, was functionalized with a thiol-reactive 3HC-B derivative. The dissociation constant $K_d$ of the interaction between pTMVP and Fab57P was largely preserved upon labeling, as evidenced by the surface plasmon resonance technique. The ratio of the two emission bands (N*/T*) of the pTMVP peptide labeled at its C-terminus was found to change by 40% upon interaction with Fab57P. These changes corresponded to a decrease in the

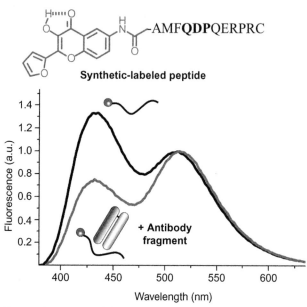

**Synthetic-labeled peptide**

**Figure 2.9** Chemical structure of the peptide biosensor and its two-color fluorescence response to interaction with the antibody fragment.[62] (For color version of this figure, the reader is referred to the online version of this chapter.)

hydration of the peptide labeling site upon its interaction with the target peptide. Following a similar approach, another biosensor was developed (Fig. 2.9), in which peptides labeled at the N-terminus were used for specific detection of an antibody fragment scFv.[63] Studies of different fluorescent peptide conjugates revealed that the response to the interaction event requires optimal distance between the fluorophore and the peptide interaction site. These works show the possibility of transforming a peptide, representing a minimized analyte binding site, into a ratiometric biosensor molecule by functionalization with a fluorophore.

Another important aspect is monitoring protein aggregation, which commonly leads to the dramatic changes in the protein environment at their interface. A current challenge in this field is to monitor the early and intermediate stages of α-synuclein aggregation, a process associated with Parkinson's disease.[64] To this end, an Ala→Cys mutant of α-synuclein was labeled with the thiol-reactive 3HC-B derivative and used as a sensor of the α-synuclein aggregation.[65] Strong changes in the dual emission of the label resulted from protein aggregation, allowing continuous monitoring of this process and detection of the aggregated structures much earlier than other techniques.

## 4.4. Peptide–membrane interactions

Monitoring interactions of proteins with biological membranes is of particular importance for studying membrane proteins and peptide-based toxins and for understanding the peptide transport through cell membranes. These interactions should lead to remarkable changes in the polarity of the peptide environment, as the rather polar peptide interface is substituted with the highly apolar lipid membrane environment. Therefore, the 3HC-A dye is more appropriate for these studies. Its carboxylic acid derivative was attached to the N-terminus of melittin and poly-L-lysine peptides, which interact with lipid membranes in a very different manner. Binding of these peptides to lipid vesicles induced a strong fluorescence increase, which enabled the quantification of the peptide–membrane interactions.[66] Moreover, the dual emission of the label in these peptides correlated well with the depth of its insertion, measured by the parallax quenching method (Fig. 2.10). Thus, in melittin, which shows deep insertion of its N-terminus, the label presented a dual emission corresponding to a low-polar environment, while the environment of the poly-L-lysine N-terminus was rather polar, in line with its binding at the surface of the lipid head groups. Moreover, imaging of labeled peptides bound to giant vesicles gave some clues on the orientation of the label within the membrane, which could help estimating the peptide orientation.

This label was further successfully applied for monitoring the interaction of α-synuclein with model membranes, as this interaction was hypothesized to play a role in the pathological misfolding and aggregation of this protein during Parkinson's disease. Systematic studies of α-synuclein labeled with 3HC-A fluorophore revealed the influence of charge, phase, curvature, defects, and lipid unsaturation on binding of α-synuclein to model membranes[67] and its further conformational changes.[68]

## 5. CONCLUSIONS

Solvatochromic dyes, because of their ability to undergo excited-state reactions (charge and proton transfer), can change their emission color and intensity in response to variation of solvent polarity. Though a number of

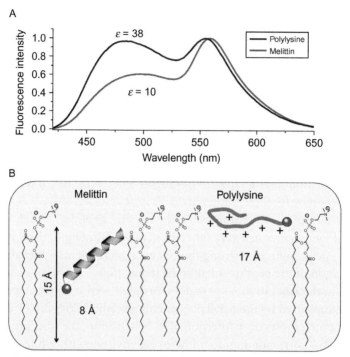

**Figure 2.10** Monitoring the interaction of melittin and polylysine labeled with the 3HC-A fluorophore with lipid membranes. (A) Fluorescence spectra of labeled melittin and polylysine bound to model lipid membranes. (B) Schematic presentation of the insertion of the peptides into the lipid membrane.[66] (For color version of this figure, the reader is referred to the online version of this chapter.)

solvatochromic dyes has already been developed, there is a clear need for new dyes presenting high solvatochromism, brightness, and photostability. 3HCs are particularly interesting examples of solvatochromic dyes because of the strong solvent sensitivity of their dual emission generated by ESIPT Sensitivity of the dyes to the environment polarity and hydration can be applied for probing the interaction between biomolecules. Being covalently attached to peptides, they enable monitoring conformation changes of proteins and their interactions with nucleic acids, proteins, and lipid membranes. Therefore, solvatochromic dyes could become a universal tool for detecting almost any kind of biomolecular interactions. However, the further success of the approach will strongly rely on the development of new improved dyes.

# REFERENCES

1. Borisov SM, Wolfbeis OS. Optical biosensors. *Chem Rev* 2008;**108**:423–61.
2. Demchenko AP. *Introduction to fluorescence sensing*. Springer; 2008.
3. Schäferling M, Nagl S. Förster resonance energy transfer methods for quantification of protein-protein interactions on microarrays. *Methods Mol Biol* 2011;**723**:303–20.
4. Boute N, Jockers R, Issad T. The use of resonance energy transfer in high-throughput screening: BRET versus FRET. *Trends Pharmcol Sci* 2002;**23**:351–4.
5. Owicki JC. Fluorescence polarization and anisotropy in high throughput screening. perspectives and primer. *J Biomol Screen* 2000;**5**:297–306.
6. Loving GS, Sainlos M, Imperiali B. Monitoring protein interactions and dynamics with solvatochromic fluorophores. *Trends Biotechnol* 2010;**28**:73–83.
7. Klymchenko AS. Solvatochromic fluorescent dyes as universal tools for biological research. *Actual Chim* 2012;**359**:20–6.
8. Demchenko AP, Mely Y, Duportail G, Klymchenko AS. Monitoring biophysical properties of lipid membranes by environment-sensitive fluorescent probes. *Biophys J* 2009;**96**:3461.
9. Haidekker MA, Theodorakis EA. Molecular rotors—fluorescent biosensors for viscosity and flow. *Org Biomol Chem* 2007;**5**:1669–78.
10. Hwan MK, Byeong HJ, Hyon JY, Myoung JA, Mun SS, Jin HH, et al. Two-photon fluorescent turn-on probe for lipid rafts in live cell and tissue. *J Am Chem Soc* 2008;**130**:4246–7.
11. Reichardt C. Solvatochromic dyes as solvent polarity indicators. *Chem Rev* 1994;**94**:2319–58.
12. Lippert EL. Laser-spectroscopic studies of reorientation and other relaxation processes in solution. In: Birks JB, editor. *Organic and biomolecular chemistry*, vol. 2. New York: Wiley; 1975. p. 1–31.
13. Slavik J. Anilinonaphthalene sulfonate as a probe of membrane-composition and function. *Biochim Biophys Acta* 1982;**694**:1–25.
14. Mes GF, De Jong B, Van Ramesdonk HJ, Verhoeven JW, Warman JM, De Haas MP, et al. Excited-state dipole moment and solvatochromism of highly fluorescent rod-shaped bichromophoric molecules. *J Am Chem Soc* 1984;**106**:6524.
15. Kucherak OA, Didier P, Mely Y, Klymchenko AS. Fluorene analogues of prodan with superior fluorescence brightness and solvatochromism. *J Phys Chem Lett* 2010;**1**:616.
16. Diwu Z, Lu Y, Zhang C, Klaubert DH, Haugland RP. Fluorescent molecular probes II. The synthesis, spectral properties and use of fluorescent solvatochromic dapoxyla¢ dyes. *Photochem Photobiol* 1997;**66**:424.
17. Weber G, Farris FJ. Synthesis and spectral properties of a hydrophobic fluorescent probe: 6-propionyl-2-(dimethylamino)naphthalene. *Biochemistry* 1979;**18**:3075–8.
18. Kucherak OA, Richert L, Mély Y, Klymchenko AS. Dipolar 3-methoxychromones as bright and highly solvatochromic fluorescent dyes. *Phys Chem Chem Phys* 2012;**14**:2292–300.
19. Vázquez ME, Blanco JB, Imperiali B. Photophysics and biological applications of the environment-sensitive fluorophore 6-N, N-dimethylamino-2,3-naphthalimide. *J Am Chem Soc* 2005;**127**:1300–6.
20. Lu Z, Lord SJ, Wang H, Moerner WE, Twieg RJ. Long-wavelength analogue of PRODAN: synthesis and properties of Anthradan, a fluorophore with a 2,6-donor-acceptor anthracene structure. *J Org Chem* 2006;**71**:9651–7.
21. Soujanya T, Fessenden RW, Samanta A. Role of nonfluorescent twisted intramolecular charge transfer state on the photophysical behavior of aminophthalimide dyes. *J Phys Chem* 1996;**100**:3507.

22. Lehrer SS, Ishii Y. Fluorescence properties of acrylodan-labeled tropomyosin and tropomyosin-actin: evidence for myosin subfragment 1 induced changes in geometry between tropomyosin and actin. *Biochemistry* 1988;**27**:5899–906.
23. Greenspan P, Mayer EP, Fowler SD. Nile red: a selective fluorescent stain for intracellular lipid droplets. *J Cell Biol* 1985;**100**:965.
24. Dattelbaum JD, Logger LL, Benson DE, Sali KM, Thompson RB, Hellinga HW. Analysis of allosteric signal transduction mechanisms in an engineered fluorescent maltose biosensor. *Protein Sci* 2005;**14**:284–91.
25. Chin JW, Cropp TA, Anderson JC, Mukherji M, Zhang Z, Schultz PG. An expanded eukaryotic genetic code. *Science* 2003;**301**:964–7.
26. Hyun SL, Guo J, Lemke EA, Dimla RD, Schultz PG. Genetic incorporation of a small, environmentally sensitive, fluorescent probe into proteins in Saccharomyces cerevisiae. *J Am Chem Soc* 2009;**131**:12921–3.
27. Lakowicz JR. *Principles of fluorescence spectroscopy*. New York: Kluwer Academic; 1999.
28. Goncalves MS. Fluorescent labeling of biomolecules with organic probes. *Chem Rev* 2009;**109**:190–212.
29. Holmes-Farley SR, Whitesides GM. Fluorescence properties of dansyl groups covalently bonded to the surface of oxidatively functionalized low-density polyethylene film. *Langmuir* 1986;**2**:266.
30. Harikumar KG, Pinon DI, Wessels WS, Prendergast FG, Miller LJ. Environment and mobility of a series of fluorescent reporters at the amino terminus of structurally related peptide agonists and antagonists bound to the cholecystokinin receptor. *J Biol Chem* 2002;**277**:18552–60.
31. Hiratsuka T. ATP-induced opposite changes in the local environments around Cys(697) (SH2) and Cys(707) (SH1) of the myosin motor domain revealed by the prodan fluorescence. *J Biol Chem* 1999;**274**:29156–63.
32. Shi J, Boyd AE, Radic Z, Taylor P. Reversibly bound and covalently attached ligands induce conformational changes in the omega loop, Cys69-Cys96, of mouse acetylcholinesterase. *J Biol Chem* 2001;**276**:42196.
33. Cohen BE, McAnaney TB, Park ES, Jan YN, Boxer SG, Jan LY. Probing protein electrostatics with a synthetic fluorescent amino acid. *Science* 2002;**296**:1700–3.
34. Nitz M, Mezo AR, Ali MH, Imperiali B. Enantioselective synthesis and application of the highly fluorescent and environment-sensitive amino acid 6-(2-dimethylaminonaphthoyl) alanine (DANA). *Chem Commun* 2002;**17**:1912–3.
35. Chen H, Chung NN, Lemieux C, Zelent B, Vanderkooi JM, Gryczynski I, et al. [Aladan3]TIPP: a fluorescent delta-opioid antagonist with high delta-receptor binding affinity and delta selectivity. *Biopolymers* 2005;**80**:325–31.
36. Loving G, Imperiali B. A versatile amino acid analogue of the solvatochromic fluorophore 4-N, N-dimethylamino-1,8-naphthalimide: a powerful tool for the study of dynamic protein interactions. *J Am Chem Soc* 2008;**130**:13630–8.
37. Eugenio Vazquez M, Rothman DM, Imperiali B. A new environment-sensitive fluorescent amino acid for Fmoc-based solid phase peptide synthesis. *Org Biomol Chem* 2004;**2**:1965–6.
38. Venkatraman P, Nguyen TT, Sainlos M, Bilsel O, Chitta S, Imperiali B, et al. Fluorogenic probes for monitoring peptide binding to class II MHC proteins in living cells. *Nat Chem Biol* 2007;**3**:222–8.
39. Bailey DM, Hennig A, Uzunova VD, Nau WM. Supramolecular tandem enzyme assays for multiparameter sensor arrays and enantiomeric excess determination of amino acids. *Chem Eur J* 2008;**14**:6069–77.
40. Brea RJ, Eugenio Vázquez M, Mosquera M, Castedo L, Granja JR. Controlling multiple fluorescent signal output in cyclic peptide-based supramolecular systems. *J Am Chem Soc* 2007;**129**:1653–7.

41. Sherman DB, Pitner JB, Ambroise A, Thomas KJ. Synthesis of thiol-reactive, long-wavelength fluorescent phenoxazine derivatives for biosensor applications. *Bioconjugate Chem* 2006;**17**:387–92.

42. Jose J, Burgess K. Benzophenoxazine-based fluorescent dyes for labeling biomolecules. *Tetrahedron* 2006;**62**:11021–37.

43. Thomas KJ, Sherman DB, Amiss TJ, Andaluz SA, Pitner JB. A long-wavelength fluorescent glucose biosensor based on bioconjugates of galactose/glucose binding protein and Nile Red derivatives. *Diabetes Technol Ther* 2006;**8**:261–8.

44. Cohen BE, Pralle A, Yao X, Swaminath G, Gandhi CS, Jan YN, et al. A fluorescent probe designed for studying protein conformational change. *Proc Natl Acad Sci USA* 2005;**102**:965–70.

45. Moerner WE, Fromm DP. Methods of single-molecule fluorescence spectroscopy and microscopy. *Rev Sci Instrum* 2003;**74**:3597–619.

46. Lu Z, Liu N, Lord SJ, Bunge SD, Moerner WE, Twieg RJ. Bright, red single-molecule emitters: synthesis and properties of environmentally sensitive dicyanomethylenedihydrofuran (DCDHF) fluorophores with bisaromatic conjugation. *Chem Mater* 2009;**21**:797–810.

47. Lord SJ, Conley NR, Lee HLD, Nishimura SY, Pomerantz AK, Willets KA, et al. DCDHF fluorophores for single-molecule imaging in cells. *Chemphyschem* 2009;**10**:55–65.

48. Berezin MY, Lee H, Akers W, Achilefu S. Near infrared dyes as lifetime solvatochromic probes for micropolarity measurements of biological systems. *Biophys J* 2007;**93**:2892–9.

49. Sunahara H, Urano Y, Kojima H, Nagano T. Design and synthesis of a library of BODIPY-based environmental polarity sensors utilizing photoinduced electron-transfer-controlled fluorescence ON/OFF switching. *J Am Chem Soc* 2007;**129**:5597–604.

50. Sengupta PK, Kasha M. Excited state proton-transfer spectroscopy of 3-hytdroxyflavone and quercetin. *Chem Phys Lett* 1979;**68**:382–5.

51. Klymchenko AS, Demchenko AP. Multiparametric probing of intermolecular interactions with fluorescent dye exhibiting excited state intramolecular proton transfer. *Phys Chem Chem Phys* 2003;**5**:461–8.

52. Shynkar VV, Mely Y, Duportail G, Piemont E, Klymchenko AS, Demchenko AP. Picosecond time-resolved fluorescence studies are consistent with reversible excited-state intramolecular proton transfer in 4′-(dialkylamino)-3-hydroxyflavones. *J Phys Chem A* 2003;**107**:9522.

53. Das R, Klymchenko AS, Duportail G, Mély Y. Unusually slow proton transfer dynamics of a 3-hydroxychromone dye in protic solvents. *Photochem Photobiol Sci* 2009;**8**:1583–9.

54. Huntington JA, Read RJ, Carrell RW. Structure of a serpin-protease complex shows inhibition by deformation. *Nature* 2000;**407**:923–6.

55. Boudier C, Klymchenko AS, Mely Y, Follenius-Wund A. Local environment perturbations in alpha1-antitrypsin monitored by a ratiometric fluorescent label. *Photochem Photobiol Sci* 2009;**8**:814.

56. Amarasinghe GK, De Guzman RN, Turner RB. Summers, MF NMR structure of stem-loop SL2 of the HIV-1 ψ RNA packaging signal reveals a novel A-U-A base-triple platform. *J. Mol. Biol.* 2000;**299**:145–56.

57. Bourbigot S, Ramalanjaona N, Boudier C, Salgado GFJ, Roques BP, Mély Y, Bouaziz S, Morellet N. How the HIV-1 Nucleocapsid Protein Binds and Destabilises the (-)Primer Binding Site During Reverse Transcription. *J. Mol. Biol.* 2008;**383**: 1112–28.

58. Shvadchak VV, Klymchenko AS, De Rocquigny H, Mely Y. Sensing peptide-oligonucleotide interactions by a two-color fluorescence label: application to the HIV-1 nucleocapsid protein. *Nucleic Acids Res* 2009;**37**:e25.

59. Darlix J-L, Godet J, Ivanyi-Nagy R, Fossé P, Mauffret O, Mély Y. Flexible nature and specific functions of the HIV-1 nucleocapsid protein. *J. Mol. Biol.* 2011;**410**:565–81.
60. Bernacchi S, Stoylov S, Piémont E, Ficheux D, Roques BP, Darlix JL, et al. HIV-1 nucleocapsid protein activates transient melting of least stable parts of the secondary structure of TAR and its complementary sequence. *J Mol Biol* 2002;**317**:385–99.
61. Zamotaiev OM, Postupalenko VY, Shvadchak VV, Pivovarenko VG, Klymchenko AS, Mély Y. Improved hydration-sensitive dual-fluorescence labels for monitoring peptide-nucleic acid interactions. *Bioconjugate Chem* 2011;**22**:101–7.
62. Enander K, Choulier L, Olsson AL, Yushchenko DA, Kanmert D, Klymchenko AS, et al. A peptide-based, ratiometric biosensor construct for direct fluorescence detection of a protein analyte. *Bioconjugate Chem* 2008;**19**:1864.
63. Choulier L, Shvadchak VV, Naidoo A, Klymchenko AS, Mely Y, Altschuh D. A peptide-based fluorescent ratiometric sensor for quantitative detection of proteins. *Anal Biochem* 2010;**401**:188–95.
64. Baba M, Nakajo S, Tu PH, Tomita T, Nakaya K, Lee VMY, et al. Aggregation of α-synuclein in Lewy bodies of sporadic Parkinson's disease and dementia with Lewy bodies. *Am J Pathol* 1998;**152**:879–84.
65. Yushchenko DA, Fauerbach JA, Thirunavukkuarasu S, Jares-Erijman EA, Jovin TM. Fluorescent ratiometric MFC probe sensitive to early stages of α-Synuclein aggregation. *J Am Chem Soc* 2010;**132**:7860–1.
66. Postupalenko VY, Shvadchak VV, Duportail G, Pivovarenko VG, Klymchenko AS, Mely Y. Monitoring membrane binding and insertion of peptides by two-color fluorescent label. *Biochim Biophys Acta* 2010;**1808**:424.
67. Shvadchak VV, Falomir-Lockhart LJ, Yushchenko DA, Jovin TM. Specificity and kinetics of α-synuclein binding to model membranes determined with fluorescent excited state intramolecular proton transfer (ESIPT) probe. *J Biol Chem* 2011;**286**:13023–32.
68. Shvadchak VV, Yushchenko DA, Pievo R, Jovin TM. The mode of α-synuclein binding to membranes depends on lipid composition and lipid to protein ratio. *FEBS Lett* 2011;**585**:3513–9.

# CHAPTER THREE

# Rational Design of Fluorophores for *In Vivo* Applications

**Marcin Ptaszek**
Department of Chemistry and Biochemistry, University of Maryland Baltimore County, Baltimore, Maryland, USA

## Contents

## Abstract

Several classes of small organic molecules exhibit properties that make them suitable for fluorescence *in vivo* imaging. The most promising candidates are cyanines, squaraines, boron dipyrromethenes, porphyrin derivatives, hydroporphyrins, and phthalocyanines.

*Progress in Molecular Biology and Translational Science*, Volume 113
ISSN 1877-1173
http://dx.doi.org/10.1016/B978-0-12-386932-6.00003-X

The recent designing and synthetic efforts have been dedicated to improving their optical properties (shift the absorption and emission maxima toward longer wavelengths and increase the brightness) as well as increasing their stability and water solubility. The most notable advances include development of encapsulated cyanine dyes with increased stability and water solubility, squaraine rotaxanes with increased stability, long-wavelength-absorbing boron dipyrromethenes, long-wavelength-absorbing porphyrin and hydroporphyrin derivatives, and water-soluble phthalocyanines. Recent advances in luminescence and bioluminescence have made self-illuminating fluorophores available for *in vivo* applications. Development of new types of hydroporphyrin energy-transfer dyads gives the promise for further advances in *in vivo* multicolor imaging.

# 1. INTRODUCTION

Fluorescence spectroscopy, which has been proved to be an extremely powerful tool in analytical chemistry and cell biology,[1,2] has reached the stage when it can be used to visualize biological processes and detect biologically relevant species in tissues and in whole living animals. It opens a particularly fascinating opportunity to noninvasively diagnose disease stages at the molecular level, which can revolutionize medicinal diagnosis. The progress in applications of fluorescence spectroscopy for *in vivo* imaging relies on both advances in excitation and detection technologies and development of molecular probes suitable for visualization of molecular processes in tissue or whole body. The fluorescent molecular probes must consist of two components: a reporter (fluorophore), that is, the unit that emits the light upon excitation, and a recognition unit, which can selectively recognize the given molecular process or species of interest and translate the recognition event in a well-defined manner into changes of the fluorescence properties of a reporter.[1,2] Thus, fluorophore as a reporter is a centerpiece of every fluorescence molecular probe. Fluorophores for *in vivo* applications must fulfill a set of requirements as for their optical, chemical, and biological properties. The most critical properties are summarized below.[3–5]

For *in vivo* applications, fluorophores must absorb and emit in the red and near–infrared (near–IR) spectral window, preferably in the range 650–900 nm. In this window, the tissue absorbance and autofluorescence, as well as light scattering, are diminished, while below 650 nm, tissue and cellular components strongly absorb the light and above 900 nm water absorbs (e.g., see discussion in Ref. 5). Fluorophores should possess high brightness, which is a product of the fluorophore excitation coefficient $\varepsilon$

and the fluorescence quantum yield $\Phi_f$; thus preferably fluorophores should have both high absorbance and fluorescence quantum yield. They should possess a Stokes shift (i.e., spacing between excitation and emission wavelengths) large enough to avoid detection of scattered light from the excitation beam. They must retain their bright fluorescence in the biological milieu, so, ideally, fluorophores should be soluble in water and should not aggregate in aqueous solution, or at least one should be able to formulate them in a form in which they stay fluorescent in the biological environment. It is also important that fluorophores should not interact (or should not change their optical characteristic upon interaction) with biomolecules, particularly serum proteins. High chemo- and photostability and lack of (photo) toxicity are other important characteristics of fluorophores. In addition, fluorophores should be synthetically available, easy to functionalize, and able to be attached to biomolecules or targeting/recognition units.

For a chemist, development of fluorophores for *in vivo* imaging is a particular challenge, as integration of all required attributes in one molecular framework is highly demanding, if at all possible. Rational design of optimal fluorophores starts from the selection of a molecular platform with suitable optical properties and then requires gaining a deep insight into their structure–property relationship in order to fine-tune their optical characteristics. The next step is usually optimization of their chemical properties, such as water solubility, bioconjugation, attaching targeting or recognition moiety, and necessary chemical modification to improve their performance *in vivo* (such as proper blood circulation, cell permeability, accumulation in the target tissue, etc., depending on the desired application). The design of efficient methods for synthesis and chemical modification of the target systems is also a key part of fluorophore development. The lack of robust synthetic methods is often the limiting factor in determining the structure–property relationship and optimizing the properties of certain fluorophores. The whole process of fluorophore development requires deep insight into the electronic structure of the given molecules, robust synthetic methods for their preparation and functionalization, and understanding the biological processes that fluorophores undergo *in vivo*, thus requiring expertise from various fields such as physics, chemistry, biochemistry, and biology.

There are several classes of fluorophores that have already been used or can be potentially applied, *in vivo*, including small organic and inorganic molecules, fluorescent proteins, conjugated polymers, and inorganic nanoparticles. There are excellent up-to-date reviews covering fluorophores for biological applications,[6,7] fluorescent molecular probes for biomedical applications,[5] fluorophores for medical *in vivo* imaging,[3,4,8–11] near-IR fluorophores,[12,13] and fluorophores for labeling of biomolecules.[14] This chapter is particularly focused on the recent progress in the small organic fluorophores for *in vivo* imaging and specifically highlights the chemical approaches to achieve the properties of optimal fluorophores listed above. *In vivo* application is defined here as an application in whole tissue or whole body, rather than in a single cell.

Several classes of organic fluorophores are discussed here, those being commonly used for *in vivo* imaging (such as cyanines, squaraines) as well as the less commonly used ones but showing properties making them promising candidates for future use (such as boron dipyrromethenes (BDP) and tetrapyrrolic macrocycles). Finally, two special classes of fluorophores of growing importance for *in vivo* applications are discussed: self-illuminating fluorophores and fluorophores for multicolor *in vivo* imaging.

Small organic molecules comprise one of the major classes of fluorophores for *in vivo* applications. They offer virtually unlimited diversity of structures, have a broad range of methods for modification of their physicochemical properties, and are relatively easily available, inexpensive, and usually nontoxic.[6] On the other hand, small organic fluorophores suffer from serious imperfection of optical properties, such as low quantum yields (especially for near-IR fluorophores), moderate excitation coefficients, and often moderate photostability.[6] For these reasons, intensive research effort has been made to find alternative fluorophores with improved optical properties and stability, which has led to the development of other classes of materials suitable for *in vivo* imaging. These classes of fluorophores already investigated for *in vivo* applications, such as near-IR fluorescent proteins,[15,16] fluorescent polymer nanoparticles,[17] quantum dots,[18–20] carbon dots,[21] dye-doped nanoparticles,[22] and carbon nanotubes,[23,24] as well as commercially available near-IR organic fluorophores, with unrevealed structures (such as Alexa Fluors[25]), are not discussed here. Readers interested in these topics are referred to the reviews and original articles listed above. This review covers the literature published approximately till the end of 2011.

## 2. CYANINE AND RELATED FLUOROPHORES

### 2.1. General characterization

Carbocyanine dyes (cyanines) remain the most prevalent fluorophores used for *in vivo* imaging. The cyanine derivative indocyanine green (ICG, IR-125) is the only near-IR fluorophore approved by the Food and Drug

Administration for clinical use and is considered as a "gold standard" for fluorophores for *in vivo* applications.

Cyanines are formally compounds with two nitrogen atoms linked by an odd number of methene units.[26–28] The nitrogen atoms are parts of the heterocyclic units (such as indole, benzoxazol, or benzothiazol). The structures and optical properties of representative cyanine dyes used for *in vivo* imaging are presented in Chart 3.1. Cyanines are characterized by long wavelength, tunable absorption and emission, very high extinction coefficient (up to 300,000 $M^{-1}$ $cm^{-1}$), good water solubility, and relatively straightforward synthesis. The wavelengths of absorption and emission in cyanines can be tuned and shifted toward longer wavelengths either by changing the number of carbon atoms in the polymethine chain or by expanding the aromatic part of the terminal heterocyclic units. The increase of polymethine chain by two carbon atoms shifts bathochromically the absorption band by $\sim 100$ nm, whereas fusing the benzo ring at the terminal indole moiety shifts the absorption band by about 30 nm (see data in Chart 3.1, as well as Refs. 29 and 30). The long-wavelength absorbing and emitting cyanine fluorophores suitable for *in vivo* applications are (a) pentamethine cyanines with an additional benzene ring fused to the terminal indole moieties (e.g., Cy5.5, Chart 3.1), with absorption and emission at $\sim 675/695$ nm[29]; (b) heptamethine cyanines, with an indole terminal moiety, absorbing/emitting at 750–790 and 780–820 nm, respectively (e.g., Cy7,[30] cybate,[34,35] and NIR-820[31]); and (c) heptamethine cyanines with benzoindole as terminal moieties absorbing/emitting around 780–822 and 810–847 nm (e.g., ICG,[32,33] cypate[34,35], and CyTE-822[36]). The main design and synthetic efforts have been recently dedicated to preparing derivatives suitable for further modifications (e.g., mono- and polyvalent bioconjugatable cyanines) and methods to improve water solubility and chemical and photochemical stability of cyanines.

## 2.2. Bioconjugation

*In vivo* application of any fluorophores often requires attaching a targeting group, or conjugation to biomolecules (such as antibodies), which ensures selective localization of fluorophores in the target cells, tissues, or organs.[14] Therefore, availability of fluorophore derivatives with reactive functional groups that are suitable for bioconjugation is an important issue in designing new fluorophores. Bioconjugation of any molecule is usually achieved by attaching an amino-reactive *N*-hydroxysuccinimide ester (formed by

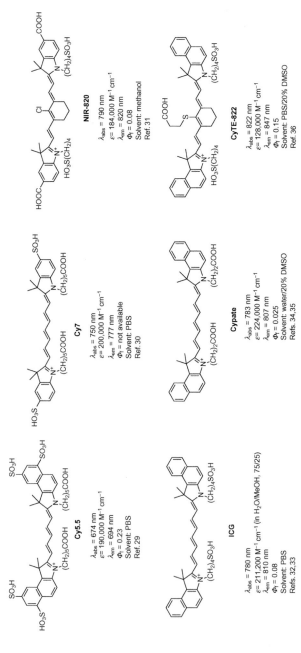

**Cy5.5**

$\lambda_{abs}$ = 674 nm
$\varepsilon$= 190,000 M$^{-1}$ cm$^{-1}$
$\lambda_{em}$ = 694 nm
$\Phi_f$ = 0.23
Solvent: PBS
Ref. 29

**Cy7**

$\lambda_{abs}$ = 750 nm
$\varepsilon$= 200,000 M$^{-1}$ cm$^{-1}$
$\lambda_{em}$ = 777 nm
$\Phi_f$ = not available
Solvent: PBS
Ref. 30

**NIR-820**

$\lambda_{abs}$ = 790 nm
$\varepsilon$ = 184,000 M$^{-1}$ cm$^{-1}$
$\lambda_{em}$ = 820 nm
$\Phi_f$ = 0.08
Solvent: methanol
Ref. 31

**ICG**

$\lambda_{abs}$ = 780 nm
$\varepsilon$= 211,200 M$^{-1}$ cm$^{-1}$ (in H$_2$O/MeOH, 75/25)
$\lambda_{em}$ = 810 nm
$\Phi_f$ = 0.08
Solvent: PBS
Refs. 32,33

**Cypate**

$\lambda_{abs}$ = 783 nm
$\varepsilon$ = 224,000 M$^{-1}$ cm$^{-1}$
$\lambda_{em}$ = 807 nm
$\Phi_f$ = 0.025
Solvent: water/20% DMSO
Refs. 34,35

**CyTE-822**

$\lambda_{abs}$ = 822 nm
$\varepsilon$= 128,000 M$^{-1}$ cm$^{-1}$
$\lambda_{em}$ = 847 nm
$\Phi_f$ = 0.15
Solvent: PBS/20% DMSO
Ref. 36

**Chart 3.1** Structures and optical properties of representative red and near-IR emitting cyanines.

derivatization of carboxylic group), a thiol-reactive iodoacetyl substituent, or a thiol-reactive maleimide group.[38] The available synthetic routes allow the introduction of a variety of substituents at both nitrogen atoms and the aromatic parts of terminal indole moieties (Fig. 3.1). The bioconjugatable polyvalent cyanine derivatives have been prepared by the introduction of suitable reactive groups as substituents either at the nitrogen atoms (R and R[1] groups) or on the aromatic indole moiety (X and X[1], Fig. 3.1). Bioconjugatable Cy5.5[29,30] and Cy7[29] having hexanoic acid substituents (terminated by carboxylic acid group) at both indole nitrogen atoms have been prepared by Waggoner and coworkers. The polyvalent analog of ICG with propionic acid at both benzoindole nitrogen atoms (cypate),[35,37,39,40] as well as the

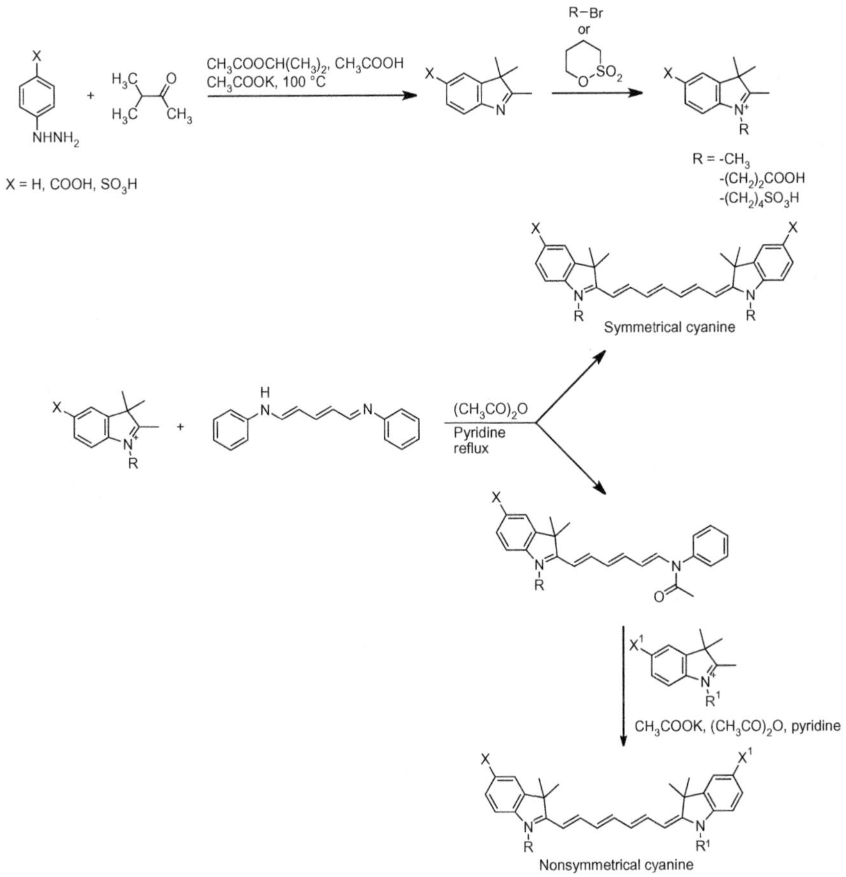

**Figure 3.1** Synthesis of symmetrical and nonsymmetrical cyanines (see text for references).

analogous derivatives of Cy7 (cybate),[34] have been prepared and characterized by Achilefu and coworkers. The carboxylate groups in cypate have been subsequently used for attachment of a variety of functional moieties, such as polycarboxylic acids,[35,41] glucosoamine,[35,42] peptides,[37,39,42,43] and poly (ethylene oxide) dendrimers.[44] Cyanine derivatives containing carboxylate groups at the benzo ring of each terminal indole moiety have been prepared by Licha[45] and Tung (compound NIR–820, Chart 3.1)[31] and used for attaching glucosoamid[45] and transferrin, respectively.[31]

The polyvalent symmetrical derivatives described above are equipped with two identical carboxylic groups. These groups can be selectively functionalized, so that only one is activated and derivatized[31,42,45,46]; however, having two identical carboxylic groups complicates the derivatization and purification procedure because of the formation of the bis–derivatized ester as a side product. Therefore, monovalent derivatives (i.e., derivatives with one bioconjugatable group) have been developed. One strategy for the preparation of the monovalent cyanine derivatives entails the preparation of nonsymmetrical derivatives, where the functional derivatizable group is located only on the one indole moiety (either as a substituent on indole nitrogen or on the benzo ring, Fig. 3.1). A series of nonsymmetrical carboxylate derivatives of Cy5.5,[47] Cy7,[48–50] and benzo-heptamethine cyanines[51,52] have been prepared and conjugated with proteins,[49] peptides,[50,53,54] poly(ethylene) glycols,[55] sugars,[56,57] fluorescence quenchers,[58] and other targeting agents.[59] Tung and coworkers prepared the nonsymmetrical monovalent carboxylate-substituted hybrid cyanine with benzoindole moiety on one side and carboxylate-substituted indole on the other side of the heptamethine chain.[60] This derivative has been subsequently conjugated to the PEGylated graft polymer (to make enzyme-activatable fluorescent probe)[60] or to the folate residue (to prepare folate receptor expressing cancer cells).[61]

Preparation of nonsymmetrical cyanine derivatives causes purification problems because of the inevitable formation of the corresponding symmetrical side product. Therefore, recently a new strategy for the preparation of mono-functional cyanine fluorophores has been pursued (Fig. 3.2). This strategy relies on the synthesis and derivatization of cycloheptamethine cyanines (**Cl–cyclo-Cy**) with chlorine atom on the cyclohexene ring.[62–64] In these new cyanine derivatives, in which part of the polymethine chain is embedded in the cyclohexane ring, the chlorine substituent on the cyclohexane ring can be further derivatized by nucleophilic substitution, using phenolates (to form C—O bond), thiols (to form C—S bond), or amines (to form C—N

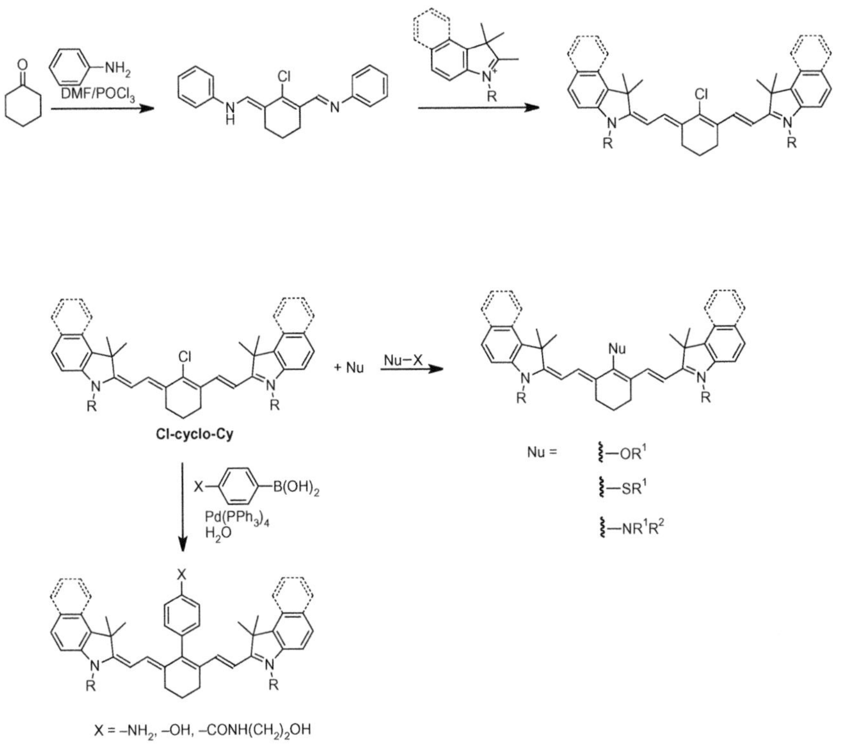

**Figure 3.2** Synthesis and derivatization of cycloheptamethine cyanines (see text for references).

bond).[65] Alternatively, palladium-catalyzed cross-coupling reaction has been employed to form the more robust C—C bond.[66,68] Both approaches have been used for preparing a broad range of derivatives, including those equipped with bioconjugatable groups.[36,47,65–70] Nucleophilic substitution of chlorine atom in cycloheptamethine and benzocycloheptamethine cyanines **Cl-cyclo-Cy** with oxygen, nitrogen, or sulfur nucleophiles has also been widely used for the attachment of groups to improve water solubility,[71–73] prevent aggregation,[74] and alter the cyanine affinity to albumin,[75] as well as to target or recognize motifs such as peptides,[69,75] glucosamine,[76] photosensitizers,[77] and zinc dipicolylamine group.[78,79]

## 2.3. Improvement of water solubility and chemical stability

One of the main problems with cyanine fluorophores is their chemical and photochemical instability, especially under physiological conditions. Cyanines undergo many complex physicochemical transformations in solution, which alter their optical properties. Cyanines in solution are prone to pho-tobleaching, oxidation, solvatochromic effects, and nonspecific interactions with plasma proteins (see Refs. 27, 28, and introduction to Ref. 80 and references cited therein). Cyanine itself has a low fluorescence quantum yield because of the competitive internal conversion and photoisomerization.[33] In addition, in aqueous solutions, cyanines undergo different types of aggregation,[45] and in a biological environment, they nonspecifically interact with biomolecules.[28] All of these effects cause diminishing of the fluorescence quantum yield and altering of the maxima of both absorption and emission bands of cyanines *in vivo*. Hence, much effort has been devoted to improving the chemical and photochemical stability of cyanines; minimizing the nonspecific interactions of cyanines with blood, plasma, or cellular components; diminishing their aggregation in aqueous solution; and increasing their quantum yields of fluorescence. Many of the reported methods collectively improve the optical properties, stability and water solubility, or aggregation behavior, so all of them will be discussed here in the same subchapter. Two general directions have been followed to improve the stability and fluorescence properties of cyanines: chemical modification of the cyanine structure and encapsulation of cyanines inside dendrimers, cyclodextrines, or nanoparticles.

### 2.3.1 Chemical modifications

Improvement of chemical stability and moderate increase in quantum yields of fluorescence have been observed for cyanines with a polymethine chain rigidified by the cyclohexene ring.[31] However, cycloheptamethine deriva-tives substituted at cyclohexene ring with electron-donating amine groups exhibit significantly reduced stability.[81,82] It has been subsequently demonstrated that installing an acetyl substituent on the nitrogen, which reduces electron density on the cyanine scaffold, significantly stabilizes the *N*-acyl-substituted derivatives.[82,83]

An another strategy to improve chemical stability has been pursued by Armitage and coworkers, who observed that polyfluorinated pentamethine cyanine having benzothiazolium units at both termini exhibits increased fluorescence quantum yield, reduced tendency to aggregate in aqueous solution, higher chemical stability, and greater resistance toward photo-bleaching compared to its nonfluorinated analog.[84] The authors pointed out that fluorination might be a general strategy for the improvement of properties of cyanines for imaging applications, but this strategy has not yet been tested for other cyanines or *in vivo*. Similarly, substitution of the mer-ocyanine dye (i.e., cyanine analog having an oxygen atom at one of the terminal heterocycles instead of nitrogen) with an electron-withdrawing cyano group increases its chemo- and photostability.[85]

Contrary to the results described above, where chemo- and photo-stability have been improved by substitution of cyanines with electron-withdrawing groups, it has also been reported that substitution of cycloheptamethine with electron-donating groups at the nitrogen slightly inhibits the photobleaching of the resulting fluorophores.[86]

## 2.3.2 Encapsulation

Encapsulation of cyanines inside larger organic molecules, such as dendri-mers or cyclodextrins, or inside nanoparticles would protect the fluorophore from interacting with external agents that may quench fluorescence or decompose the molecule, such as solvents, oxygen, degrading enzymes, proteins, etc. Moreover, encapsulation would prevent cyanines from aggregation and improve their water solubility and, in many cases, pharma-codynamic properties. As the simplest way of encapsulation is PEGylation, attachment of poly(ethyleneglycol) chains can be considered. Brechbiel demonstrated that conjugation of the amine-reactive ICG analog with PEG of average molecular weight of 3400 g/mol improves the water solubility and prevents aggregation of cyanine, both in the free form as well as when the resulting PEGylated fluorophore is attached to the antibody panitumumab.[55] PEGylation of cyanine does not affect the targeting ability of the cyanine–antibody conjugate; hence PEGylation seems to be a viable strategy to improve the performance of cyanine fluorophores for *in vivo* applications.

Properties of cyanines can be also improved by the attachment of large water-solubilizing dendrons. Fréchet and Achilefu have examined the cypate encapsulated within a covalently attached polyester dendrimer with poly(eth-ylene oxide) branches on its periphery.[44] Such nanoencapsulated fluorophore exhibits higher water solubility, diminished aggregation, and higher quantum yield in aqueous solution than cypate itself. Moreover, it shows improved pharmacokinetics (longer plasma circulation, low accumulation in normal tis-sue) and better stability toward cytochrome P450-catalyzed oxidation com-pared to the clinically approved ICG. Weck studied the properties of cycloheptamethine cyanines substituted at the central carbon atoms with both short-chain substituents and large dendrons.[74] The results indicate that the aggregation behavior of resulting conjugates depends on the terminal groups present on the periphery of the dendrons.

Another strategy to improve the water solubility, stability, and optical properties of cyanines is encapsulation of fluorophore inside the cyclodextrin cavity. Complexes of cyclodextrins with pentamethine[87,88] and heptamethine[89] cyanines have been reported. The complexation generally reduces the fluorescence quantum yields but substantially increases photostability up to nine times compared to the uncomplexed cyanines, both in solution as well as in cells.[87,88] Cyanine–cyclodextrin complexes have not been examined *in vivo* yet.

The encapsulation inside the nanostructures has been intensively studied recently as a method for improving stability as well as photochemical and pharmacodynamic properties of cyanines. For example, silica nanoparticles have received considerable attention as potential fluorophore stabilizers and nanocarriers *in vivo* because of the nontoxicity of silica and its optical transparency. ICG has been incorporated inside mesoporous silica nanoparticles, where dye molecules have been entrapped inside nanopores by electrostatic interaction between negatively charged sulfonic groups and tetraalkylammonium-modified silica.[90] The resulting fluorescent nanoparticles with diameters of 50–100 nm are relatively stable under physiological conditions, without substantial fluorophore leakage, and show maximum fluorescence at 800 µg of ICG per gram of silica (whereas the maximum fluorescence of free ICG in solution is achieved at concentration of 2 µg/mL); thus encapsulation of ICG allows much brighter fluorescence. *In vivo* biodistribution of the resulting nanoparticles has been examined as well.[90] The cyanine fluorophore Dy776 has been also encapsulated inside ultrafine (< 20 nm diameter) organically modified silica nanoparticles, and their *in vivo* biodistribution has been examined.[91] The use of organically modified silica as a nanocarrier offers an additional capability to attach the targeting agent.[91]

ICG has also been encapsulated inside surface-modified calcium phosphate or calcium phosphosilicate composite nanoparticles.[80,92] phosphate buffered saline (PBS) suspension of ICG-doped calcium phosphate (average diameter of 16 nm) nanoparticles showed twofold higher quantum yield of fluorescence per molecule of ICG and five times longer fluorescence half-life than a solution of the free fluorophore. Both calcium phosphate and calcium phosphosilicate nanoparticles with encapsulated ICG have been used for *in vivo* cancer imaging.[80,92]

The stability, water solubility, blood circulation time, and tumor accumulation of ICG (or its analogs) have also been improved by encapsulation in

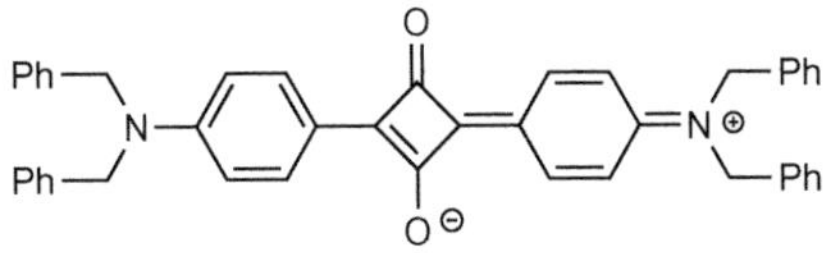

**SQ1**

$\lambda_{abs}$ = 637 nm
$\varepsilon$ = 257,000 M$^{-1}$ cm$^{-1}$
$\lambda_{em}$ = 660 nm
$\Phi_f$ = 0.45
Solvent: THF/H$_2$O
Ref. 109

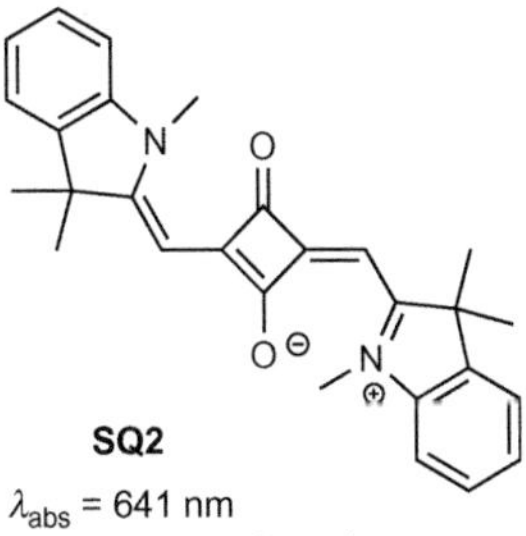

**SQ2**

$\lambda_{abs}$ = 641 nm
$\varepsilon$ = 280,000 M$^{-1}$ cm$^{-1}$
$\lambda_{em}$ = 652 nm
$\Phi_f$ = 0.20
Solvent: DMSO
Ref. 108

**Chart 3.2** Structures and optical properties of representative squaraine dyes.

nanoparticles made from organic polymers: poly(lactic-*co*-glycolic acid)[93,94] and poly-(DL-lactic-*co*-glycolic acid)–poly(vinyl alcohol) systems,[95] lipid nanoparticles,[96] self-assembled polyethylene glycol–phospholipid nanoparticles,[97] self-assembled polyallylamine hydrochloride–dihydrogen phosphate nanoaggregates,[98,99] poly-(γ-glutamic acid),[100] low-density lipoprotein (LDL) nanoparticles,[101] liposomes,[102] frozen ionic liquid nanoparticles,[103] perfluorocarbon nanoparticles,[104] and peptosomes.[105] Many of these systems have been examined *in vivo*. Encapsulation of cyanine fluorophores in nanoparticles seems to be an excellent method for improving their chemical and optical properties, and given the significant progress in the formulation of a variety of self-assembled nanostructures, this direction is likely to be intensively investigated in the future.

# 3. SQUARAINES

Squaraines (squarilium dyes) are polymethine fluorophores containing a hydroxyoxocyclobutene core with electron-donating substituents on both sides.[106,107] Squaraines possess a strong, narrow absorption band in the red and near-IR spectral region, and in many cases, they are intensely fluorescent. The structures and optical properties of the most representative squaraines are presented in Chart 3.2.[108,109] The absorption maxima, extinction coefficients, and fluorescence quantum yields of squaraines depend on the nature of aromatic groups flanking the oxabutene core. Synthesis of symmetrical squaraines is relatively straightforward and entails one-step condensation of commercially available squaric acid with electron-rich aromatic or heteroaromatic compounds with simultaneous removal of the water formed during the reaction (Fig. 3.3).[106,107] Two-step condensation affords nonsymmetrical squaraines containing two different (hetero)aromatic groups.[106,107]

Despite the favorable optical properties of squaraines, their application for *in vivo* imaging suffers from two major limitations. While squaraines exhibit often very high quantum yield of fluorescence in nonpolar solvents, they extensively aggregate in polar organic solvents or in water and the aggregation dramatically changes both absorption and emission properties: broadens their absorption spectra and reduces significantly their fluorescence quantum yields.[110] The electron-deficient hydroxyoxocyclobutene core of squaraines is prone to reacting with nucleophiles, including those present in biological media (especially thiols). The nucleophilic addition breaks the electronic conjugation and the resulting product loses near-IR absorption and fluorescence.[111]

The breakthrough that has overcome these problems and opened the door for the broad application of squaraines for *in vivo* imaging was the discovery that the stability of squaraines can be significantly improved upon formation of rotaxanes (for a review of the early development of fluorescent

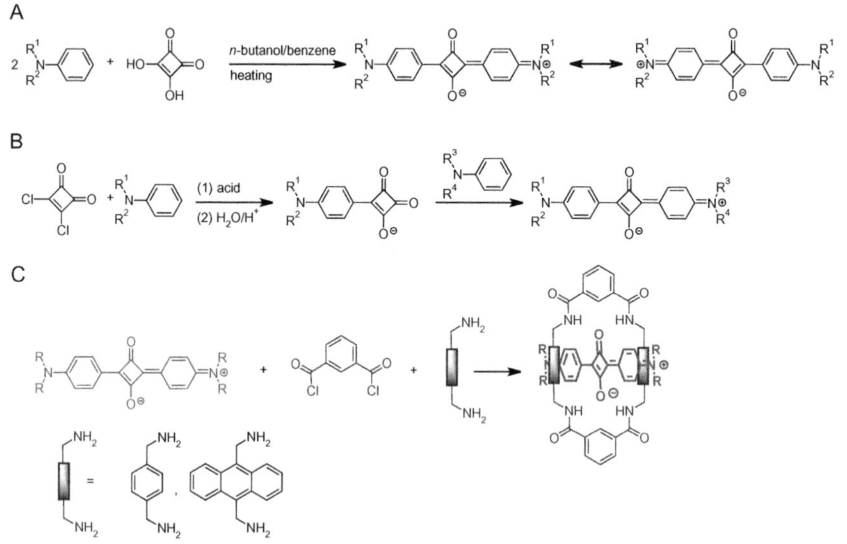

**Figure 3.3** General schemes for synthesis: (A) symmetrical squaraine, (B) nonsymmetrical squaraines, and (C) squaraine rotaxanes.[109] (For color version of this figure, the reader is referred to the online version of this chapter.)

squaraine rotaxanes, see Ref. 112). Rotaxanes are the supramolecular structures in which a rode-like molecule is threaded through the macrocyclic component, and the bulky groups located on both ends of the axle prevent de-threading.[112] Squaraine rotaxanes have been assembled by the "clipping method," that is, macrocyclic amide was synthesized from diamine and isophthalic chloride *in situ* in the presence of squaraines containing bulky terminal substituents (Fig. 3.3C).[109,113–115] The macrocycle formed *in situ* "clips" the axle (squaraine) and the process is facilitated by the hydrogen bonding between amide N–H in the intermediate acyclic precursor of macrocycle and oxygen atoms from the oxabutadiene core of squaraine (template effect). In an alternative synthetic strategy (capping method), macrocyclic lactam is synthesized first and forms a reversible host–guest complex with squaraine (pseudo-rotaxane). In subsequent chemical reactions, bulky groups (stoppers) are installed at both ends of the threaded squaraine and lock rotaxane irreversibly.[116]

The resulting squaraine rotaxanes possess essentially the same optical properties as free squaraines (though they sometimes exhibit lower fluorescence quantum yields than the corresponding free squaraines), but they are substantially more inert toward nucleophiles in solution and less prone to aggregation in polar solvents.[109] For example, both near-IR absorption and emission of the free squaraine **SQI** are reduced two times within 5 min in the presence of cysteine (due to the nucleophilic attack of thiol), whereas the corresponding rotaxanes react substantially more slowly.[109] Similarly, whereas **SQI** exhibits a very broad absorption in a dimethylsulfoxide/water mixture (due to the aggregation), the corresponding rotaxanes show only minor broadening under similar conditions.[109] Both effects, namely, reduced reactivity toward nucleophiles and diminished tendencies for aggregation in polar media, are attributed to steric protection of the macrocycle on the squaraine core.[109] The macrocyclic teralactam surrounds the squaraine, makes it less accessible by nucleophiles, and prevents it from chromophore–chromophore interaction, leading to aggregation.

Squaraine rotaxanes have been converted into water-soluble derivatives by attaching multiple sulfonic, carboxylic, guanidine, and quarternary ammonium groups as terminal "capping" groups.[117] In a similar fashion, multivalent bioconjugatable derivatives have been prepared.[115] Squaraine rotaxanes have been utilized subsequently for intracellular[118] and *in vivo* imaging.[117,119]

An increase in the application of squaraine can be expected in the future. There is a wealth of squaraine derivatives described in the literature with absorption above 700 nm, and their relatively straightforward synthesis, recent advances in chemistry of squaraine rotaxanes, and recently discovered chemiluminescence properties of squaraine rotaxanes (see Section 6.2), all together make this class of fluorophores promising candidates for a broad variety of applications.

## 4. BORON DIPYRROMETHENE AND RELATED FLUOROPHORES

### 4.1. General characterization

Another class of organic fluorophores potentially useful for *in vivo* imaging are boron dipyrroethenes.[120–122] BDP (4,4-difluoro-4-bora-3a4a-diaza-*s*-indacene, BODIPY) are a class of neutral organic fluorophores containing a conjugated system of two pyrrole rings linked by methine bridge and complexed by the difluoroboron moiety. Boron dipyromethenes exhibit strong absorption and emission in the visible and near-IR spectral window, and their absorption and emission are relatively insensitive to the solvent polarity and pH. BDPs show also remarkable photo- and chemostability compared to cyanines. The rigid BDP molecular framework makes BDP derivatives less prone to nonradiative decay of the excited state; thus their fluorescence quantum yield is usually high. Synthesis of BDP typically entails an acid-catalyzed condensation between pyrrole and aldehyde, oxidation of resulting dipyrromethane, and finally, complexation of the resulting dipyrrin with boron trifluoride.[120,122]

BDPs with their strong absorption in visible region, high fluorescence quantum yields, relatively high chemical and photochemical stability, and the very rich chemistry that allows their versatile structural modification and enables further fine-tuning of their chemical and optical properties represent an excellent platform to develop fluorescence probes for *in vitro* and intracellular imaging (for the most recent review of the application of BDP as a fluorescent probe, see Ref. 121). However, there are fewer examples of *in vivo* applications of BDP derivatives.[123–125]

One of the major reasons is that simple BDPs exhibit rather short-wavelength absorption and emission bands ($\sim$500–600 nm), which are unsuitable for *in vivo* applications. Therefore, several strategies have been developed to shift the absorption and emission toward longer wavelengths. Representative examples of red- and near-IR-emitting BDP derivatives that illustrate the design strategies to shift the absorption/emission bands toward longer wavelengths are given in Chart 3.3. Substitution on the pyrrole subunits with conjugated substituents, such as styryl or arylethynyl, with electron-donating groups shifts absorption/emission above 800 nm (e.g., compounds **BDP-I**, **BDP-II**, and **BDP-III**).[126,132–134] BDP derivatives with a fused benzene or naphthalene ring at the pyrrole subunits exhibit absorption/emission bands around 700 nm (e.g., compounds **BDP-IV** and **BDP-V**).[127,135,136] Replacement of *meso* carbon atom with nitrogen leads to the formation of aza-BDP derivatives whose absorption and emission bands are shifted toward longer wavelengths by about 150 nm compared to the corresponding BDP analogs.[128] Aza-BDP derivatives substituted at the α-pyrrolic position with aromatic substituents possessing an electron-donating group exhibit absorption/emission at ∼688/715 nm (**aza-BDP-VI**).[128,137,138] An additional bathochromic shift of about 50 nm and an increase in extinction coefficient have been achieved by rigidifying the aryl substituents, either by embedding the aryl substituent into a cyclohexane ring (**aza-BDP-VII**)[129] or by the formation of a boron–oxygen bond (**aza-BDP-VIII**).[130] Fusing of the benzo ring on the pyrrole moieties allows shifting of the absorption/emission bands above 800 nm.[139,140]

BDP derivatives with fused furan rings and *p*-methoxyphenyl substituents absorb and emit in the red or near-IR spectral window, depending on the substituent present at the meso position (**BDP-IX** and **BDP-X**).[131] The latter derivatives exhibit exceptionally high extinction coefficients and high quantum yield of fluorescence, both of which make them probably the brightest near-IR organic fluorophores.[131]

An inspection of the literature data, exemplified in Chart 3.3, indicates that the optical properties of BDP can be broadly tuned and optimized by careful molecular design, whereby derivatives with long absorption wavelength and high extinction coefficient and quantum yield of fluorescence can be obtained.

Another problem associated with use of BDPs for *in vivo* imaging is their inherent hydrophobic character and the lack of water solubility of their simple derivatives. Water solubility, however, can be imparted by the attachment of hydrophilic groups to the BDP core. Several water-soluble, highly fluorescent BDP derivatives have been prepared containing hydrophilic groups, such as sulfonates,[141,142] carboxylates,[141,143,144] phosphonates,[145] quarternary ammonium salts,[141] di(hydroxyethyl) amine,[146] oligoethylene glycol chains,[134,147–150] sulfonated peptides,[151]

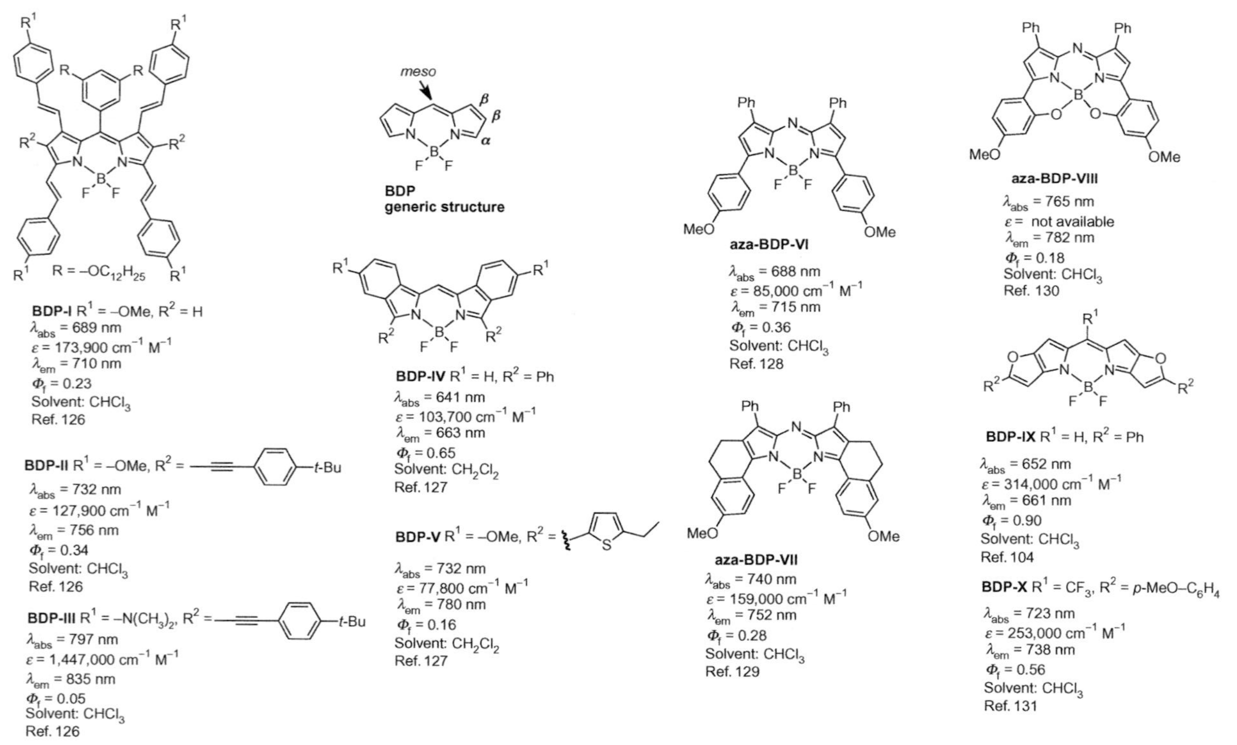

**aza-BDP-VIII**
$\lambda_{abs}$ = 765 nm
$\varepsilon$ = not available
$\lambda_{em}$ = 782 nm
$\Phi_f$ = 0.18
Solvent: CHCl$_3$
Ref. 130

**BDP-IX** R$^1$ = H, R$^2$ = Ph
$\lambda_{abs}$ = 652 nm
$\varepsilon$ = 314,000 cm$^{-1}$ M$^{-1}$
$\lambda_{em}$ = 661 nm
$\Phi_f$ = 0.90
Solvent: CHCl$_3$
Ref. 104

**BDP-X** R$^1$ = CF$_3$, R$^2$ = $p$-MeO-C$_6$H$_4$
$\lambda_{abs}$ = 723 nm
$\varepsilon$ = 253,000 cm$^{-1}$ M$^{-1}$
$\lambda_{em}$ = 738 nm
$\Phi_f$ = 0.56
Solvent: CHCl$_3$
Ref. 131

**aza-BDP-VI**
$\lambda_{abs}$ = 688 nm
$\varepsilon$ = 85,000 cm$^{-1}$ M$^{-1}$
$\lambda_{em}$ = 715 nm
$\Phi_f$ = 0.36
Solvent: CHCl$_3$
Ref. 128

**aza-BDP-VII**
$\lambda_{abs}$ = 740 nm
$\varepsilon$ = 159,000 cm$^{-1}$ M$^{-1}$
$\lambda_{em}$ = 752 nm
$\Phi_f$ = 0.28
Solvent: CHCl$_3$
Ref. 129

**BDP generic structure**

**BDP-IV** R$^1$ = H, R$^2$ = Ph
$\lambda_{abs}$ = 641 nm
$\varepsilon$ = 103,700 cm$^{-1}$ M$^{-1}$
$\lambda_{em}$ = 663 nm
$\Phi_f$ = 0.65
Solvent: CH$_2$Cl$_2$
Ref. 127

**BDP-V** R$^1$ = –OMe, R$^2$ =
$\lambda_{abs}$ = 732 nm
$\varepsilon$ = 77,800 cm$^{-1}$ M$^{-1}$
$\lambda_{em}$ = 780 nm
$\Phi_f$ = 0.16
Solvent: CH$_2$Cl$_2$
Ref. 127

R = –OC$_{12}$H$_{25}$

**BDP-I** R$^1$ = –OMe, R$^2$ = H
$\lambda_{abs}$ = 689 nm
$\varepsilon$ = 173,900 cm$^{-1}$ M$^{-1}$
$\lambda_{em}$ = 710 nm
$\Phi_f$ = 0.23
Solvent: CHCl$_3$
Ref. 126

**BDP-II** R$^1$ = –OMe, R$^2$ =
$\lambda_{abs}$ = 732 nm
$\varepsilon$ = 127,900 cm$^{-1}$ M$^{-1}$
$\lambda_{em}$ = 756 nm
$\Phi_f$ = 0.34
Solvent: CHCl$_3$
Ref. 126

**BDP-III** R$^1$ = –N(CH$_3$)$_2$, R$^2$ =
$\lambda_{abs}$ = 797 nm
$\varepsilon$ = 1,447,000 cm$^{-1}$ M$^{-1}$
$\lambda_{em}$ = 835 nm
$\Phi_f$ = 0.05
Solvent: CHCl$_3$
Ref. 126

**Chart 3.3** Structures and optical properties of representative red and near-IR BDP fluorophores.

nitrilotriacetic acid residue,[152] nucleotides,[153] and sugars.[154] Most of them show similar optical properties in aqueous solutions as their water-insoluble counterparts in organic solvents. Despite the fact that most of the reported water-soluble BDPs are the ones absorbing at rather short wavelength (<650 nm, with a few examples of water-soluble BDPs or aza-BDPs absorbing in red and near-IR),[134,141] the water-solubilizing groups reported so far, in principle, can be used for long-wavelength absorbing derivatives. Monovalent bioconjugatable BDP and aza-BDP derivatives have been also prepared in straightforward manner.[155]

## 4.2. Energy-transfer dyads for increasing the Stokes shift of BDPs

The inherent optical property of BDPs that potentially may hamper their application *in vivo* is their small Stokes shift (20–30 nm). This can be a potential problem, given the relatively narrow absorption bands in BDPs and aza-BDPs. Hence, to achieve efficient excitation, BDPs need to be excited very close to their absorption maxima. While modifications of BDP structures reported so far do not allow substantial increase of the Stokes shift, a potential solution can be the assembling of two different BDP derivatives in energy-transfer dyads. In an energy-transfer dyad, two chromophores are covalently connected by a nonconjugated bridge so that each chromophore retains the optical properties that it had as a monomer.[156–158] The excitation of the chromophore absorbing at the shorter wavelength (donor) causes energy transfer to the chromophore with the longer wavelength of absorption (acceptor) and consequently emission of the acceptor. If the quantum efficiency of energy transfer is high and there are no other competitive processes (such as electron transfer), an energy-transfer dyad behaves as a single chromophore with the excitation wavelength of the donor and the emission wavelength of the acceptor (see Fig. 3.4).

The critical aspect in construction of energy-transfer dyads is the linker connecting the donor and acceptor, which determines the mechanism of energy transfer and thereby the choice of the donor and the acceptor. In dyads in which the donor and the acceptor are connected by a fully nonconjugated linker (such as an alkyl or a peptide chain), the dominant mechanism of energy transfer is through-space Förster resonance energy transfer (commonly referred to as a FRET). Efficient FRET requires a large spectral overlap, that is, overlap between the emission band of the donor and the absorption band of the acceptor, and thus limits the choice of both pairs of chromophores that fulfill this requirement. On the other hand, a conjugated linker that provides, to some extent,

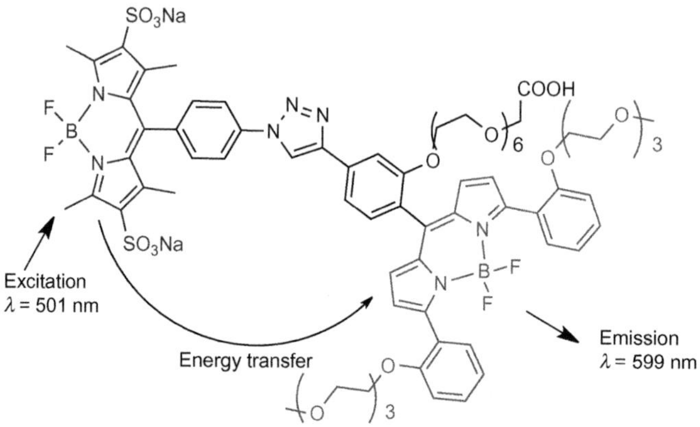

**Figure 3.4** BDP energy-transfer dyad.[158] (For color version of this figure, the reader is referred to the online version of this chapter.)

electronic communication between the donor and the acceptor enables through-bond energy transfer.[156,157] The linker-mediated through-bond energy transfer does not require spectral overlap between the donor and acceptor bands and hence offers a greater flexibility in choosing both molecules. Several energy-transfer arrays containing two different BDP, or aza–BDP subunits, have been reported, and many of them exhibit efficient energy transfer and large difference between the absorption maximum of the donor and the emission maximum of the acceptor (pseudo-Stokes shift).[158–162] Most of the BDP dyads reported so far absorb and emit at shorter wavelength than is required for *in vivo* imaging; however, in principle, energy-transfer dyads can be also constructed from red and near-IR absorbing/emitting derivatives.

Taken together, the excellent optical properties and the rich chemistry that allows synthesis of diverse derivatives and fine-tuning of their chemical and optical properties make BDP and aza–BDP likely candidates for broad *in vivo* applications.

## 4.3. Related fluorophores

Besides BDP, there are classes of related boron complexes with excellent optical properties suitable for *in vivo* applications. The boron complexes of pyrrolopyrrole cyanine dyes (Chart 3.4) exhibit absorption/emission wavelengths in the near-IR (up to 864 nm), a high extinction coefficient, narrow absorption and emission bands, and high fluorescence quantum yield.[163–165] Pyrrolopyrrole cyanines are more chemo- and photostable than classical cyanine dyes. This class of compounds exhibits also relatively long

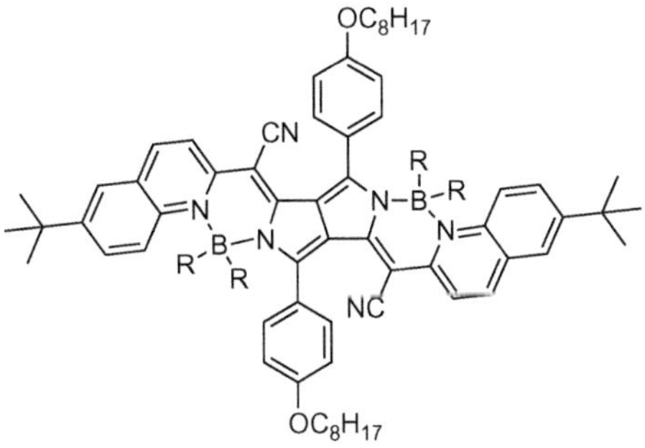

OC₈H₁₇

PP-I

R = F

$\lambda_{abs}$ = 754 nm
$\varepsilon$ = 205,000 M$^{-1}$ cm$^{-1}$
$\lambda_{em}$ = 773 nm
$\Phi_f$ = 0.59
Solvent: chloroform
Refs. 164,165

PP-II

R = Ph

$\lambda_{abs}$ = 819 nm
$\varepsilon$ = 250,000 M$^{-1}$ cm$^{-1}$
$\lambda_{em}$ = 831 nm
$\Phi_f$ = 0.53
Solvent: chloroform
Refs. 164,165

**Chart 3.4** Structures and optical properties of pyrrolopyrrole cyanines utilized for *in vivo* imaging.

fluorescence lifetimes (2.5–3.8 ns, compared to 1.11 ns for ICG) and they have been examined as fluorescence lifetime probes for *in vivo* imaging.[103,166]

# 5. PORPHYRINS, PHTHALOCYANINES, AND RELATED MACROCYCLES

## 5.1. General characterization

Porphyrins are a class of macrocyclic aromatic compounds composed of four pyrrole rings connected by methine bridges (Chart 3.5). Porphyrins are ubiquitous in nature, as a heme cofactor of hemoglobin, cytochromes, and other redox active enzymes, and, as more saturated analogs, in the photosynthetic apparatus in plants and bacteria. Tetrapyrrolic macrocycles have been widely examined for their unique optical and redox properties. In the biomedical field, tetrapyrrolic macrocycles have been mainly investigated as photosensitizers in photodynamic therapy.[167–169] Applications of porphyrins and their

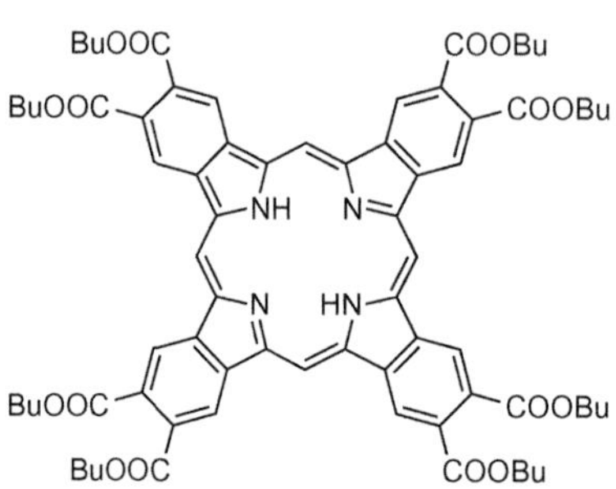

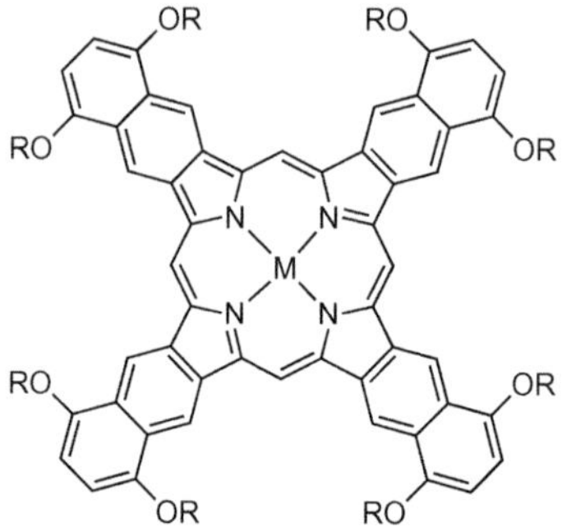

P-I

$\lambda_{abs}$ = 673 nm
$\varepsilon$ = 46,000 cm$^{-1}$ M$^{-1}$
$\lambda_{em}$ = 676,748 nm
$\Phi_f$ = 0.27
Solvent: DMF
Ref. 171

R = –(CH$_2$)$_2$–COOEt

P-II

M = H,H
$\lambda_{abs}$ = 673 nm
$\varepsilon$ = 200,000 cm$^{-1}$ M$^{-1}$
$\lambda_{em}$ = 733, 810 nm
$\Phi_f$ = 0.45
Solvent: pyridine
Ref. 171

P-III

M = Pd
$\lambda_{abs}$ = 696 nm
$\varepsilon$ = 316,000 cm$^{-1}$ M$^{-1}$
$\lambda_{em}$ = 923 nm
$\Phi_{phosphorescence}$ = 0.04
Solvent: pyridine
Ref. 171

**Chart 3.5** Structures and optical properties of representative red and near-IR benzoporphyrins.

analogs for *in vivo* imaging have been less explored, though their optical properties make them also suitable for those applications.

Porphyrins have a unique electronic structure that results in a complex absorption spectrum. Simple porphyrins (such as tetraphenylporphyrin) exhibit a very strong (with $\varepsilon \sim 500{,}000$ M$^{-1}$ cm$^{-1}$) absorption band around 400 nm, a series of much weaker bands in the visible region (500–650 nm), and a very weak absorption band in red spectral window ($\sim 650$ nm).[170] Porphyrins possess also moderate fluorescence quantum yields ($\sim 0.1$).[170] Although simple porphyrins are not suitable for *in vivo* fluorescence imaging because of their weak absorption in red/near-IR spectral window and their rather moderate fluorescence quantum yields, several of their more elaborate derivatives exhibit strong absorption and intense fluorescence in the red and near-IR regions. Derivatives with optical properties most promising for *in vivo* applications are benzoporphyrins, strongly conjugated porphyrin arrays, and hydroporphyrins.

## 5.2. Benzoporphyrins

Extension of the aromatic systems in porphyrins by fusing the benzo (or naphtho) ring at the pyrrolic positions causes a bathochromic shift and substantial intensification of the long-wavelength absorption band (with the absorption wavelength shifted above 700 nm and extinction coefficient up to 200,000 cm$^{-1}$ M$^{-1}$), and significant increase in fluorescence quantum yield (up to 0.45 for naphthoporphyrin; see Chart 3.5).[171] Anthraporphyrins (porphyrins with fused anthracene on the pyrrole units) absorb and emit above 800 nm.[172] Benzo- and naphthoporphyrins upon complexation with palladium (II) and platinum (II) show also an intense and long-lived near-IR phosphorescence.[171] Phosphorescent benzoporphyrins, because of the long lifetime of their phosphorescence, are prone to dynamic quenching by oxygen; thus, palladium and, therefore, platinum benzoporphyrins find application in sensing oxygen *in vivo*.[173,174]

## 5.3. Strongly conjugated porphyrin arrays

An alternative strategy for increasing near-IR absorption and fluorescence of porphyrins is an extension of the porphyrin conjugation by assembling several porphyrin subunits into strongly coupled arrays, that is, arrays where the porphyrin subunits are connected by a linker that provides strong electronic coupling (e.g., acetylene linker,[175,178] or butadiyne linker[176,177]) between the subunits (Chart 3.6). Such arrays composed of two, three, or five zinc complexes of porphyrin connected by an acetylene linker exhibit a progressive bathochromic shift of long-wavelength absorption and emission bands (with emission up to 883 nm), increased extinction coefficient, and good quantum yield of fluorescence.[178,179] The trimeric arrays have been examined for imaging of B16 melanoma cells[180] and labeling of the dendritic cells,[181] and their suitability for *in vivo* imaging in living animals has also been demonstrated.[182]

## 5.4. Chlorins and bacteriochlorins

The partial saturation of pyrrole units in porphyrins leads to formation of new types of tetrapyrrolic macrocycles with distinctive spectral properties: chlorins (with one partially saturated pyrrole ring), bacteriochlorins (two partially saturated pyrrole rings on opposite sites), and isobacteriochlorins (with two partially saturated pyrrole rings on the same site of macrocycle; see Chart 3.7). In contrast to porphyrins, chlorins exhibit a strong absorption in the red region (600–700 nm),[183] whereas bacteriochlorins absorb strongly in the near-IR (700–800 nm).[184] Both classes of macrocycles show also higher fluorescence quantum yields than the corresponding porphyrins (average of $\sim$0.25 for chlorins[185] and $\sim$0.15 for bacteriochlorins[184]). Isobacteriochlorins exhibit strong absorption and emission (with quantum yield

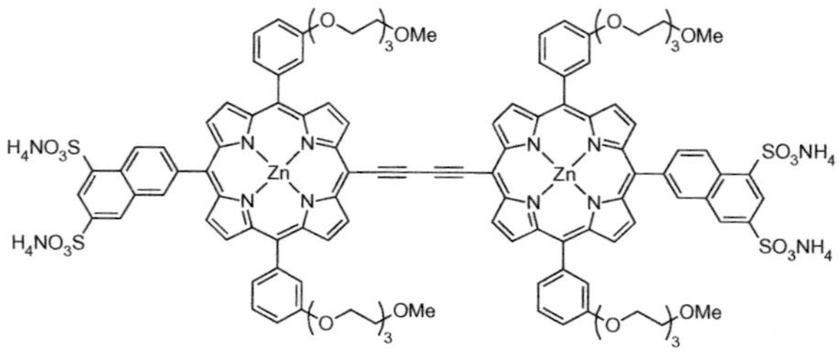

**PP-II**

R$^1$ = 

R$^2$ = 

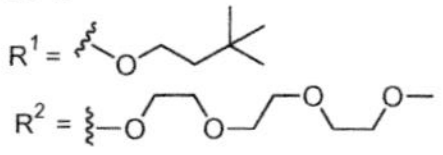

$\lambda_{abs}$ = 770 nm
$\varepsilon$ = 116,000 cm$^{-1}$ M$^{-1}$
$\lambda_{em}$ = 806 nm
$\Phi_f$ = 0.22
Solvent: THF
Ref.178

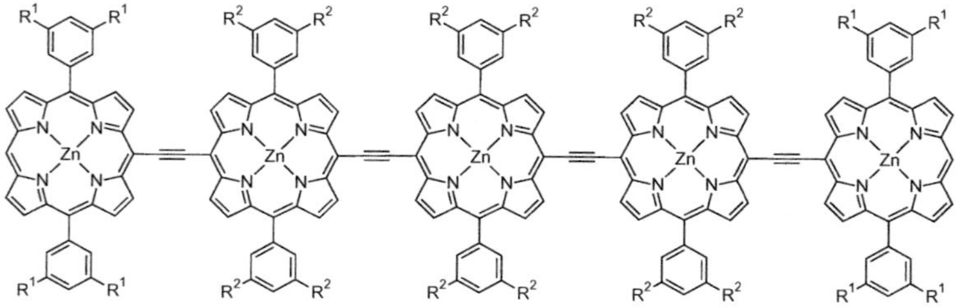

**PA-I**

$\lambda_{abs}$ = 706 nm (DMF/1%pyridine)
$\varepsilon$ = 63,000 cm$^{-1}$ M$^{-1}$ (DMF/15 pyridine)
$\lambda_{em}$ = 725 nm (H$_2$O)
$\Phi_f$ = 0.03 (D$_2$O/0.5%DMSO)
Refs. 176,177

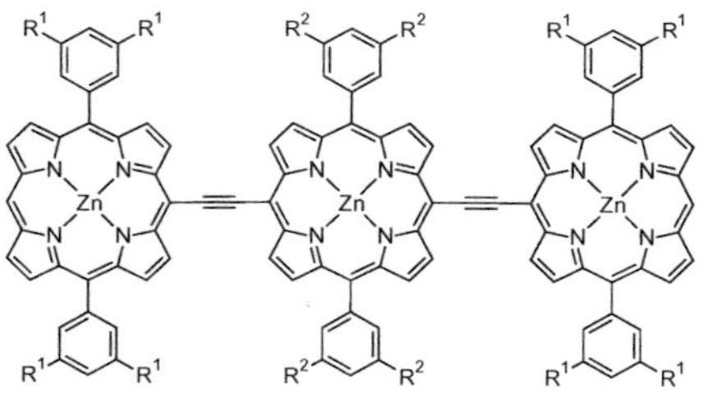

**PP-III**

$\lambda_{abs}$ = 842 nm
$\varepsilon$ = 230,000 cm$^{-1}$ M$^{-1}$
$\lambda_{em}$ = 883 nm
$\Phi_f$ = 0.14
Solvent: THF
Ref. 178

**Chart 3.6** Structures and optical properties of representative strongly conjugated porphyrin arrays.

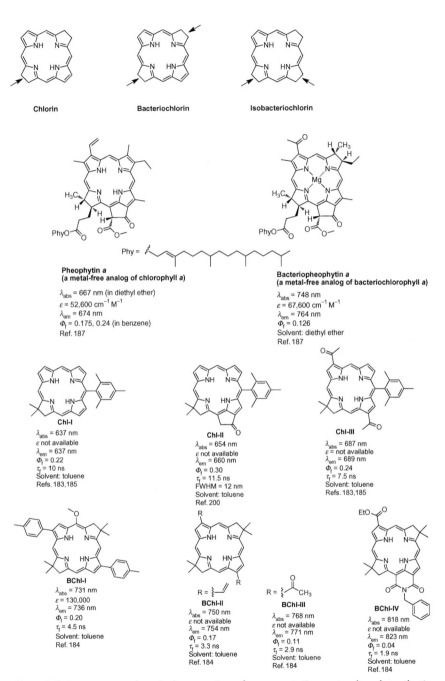

**Chart 3.7** Structures and optical properties of representative natural and synthetic hydroporphyrin fluorophores.

of fluorescence up to 0.70) at somewhat shorter wavelengths ($\sim 600$ nm).[186] Both chlorins and bacteriochlorins are present in the natural photosynthetic apparatus in plants (chlorophyll *a* and *b*, chlorin derivatives) and in bacteria (bacteriochlorophyll *a*, bacteriochlorin derivative).[187] Hence, nature is a convenient and rich source of chlorin and bacteriochlorin derivatives. The chemically modified chlorophyll and bacteriochlorophyll derivatives have been used for *in vivo* imaging of cancer[188,191] or, as conjugates with the fluorescence quencher, have been used *in vivo* to monitor phospholipase activity.[189,190]

Naturally occurring chlorophylls and bacteriochlorophylls are attractive because of their availability, but their use imposes certain problems and limitations. First, naturally occurring derivatives have a full complement of substituents on the macrocycle periphery, and their chemical modification to tune their optical and chemical properties, although possible, is limited. Moreover, naturally occurring bacteriochlorophylls are rather unstable, and outside their natural environment, they undergo oxidation to the more conjugated derivatives with substantially different optical properties. Therefore, a chief effort has been dedicated to developing stable synthetic chlorin and bacteriochlorin analogs that would retain the optical properties of naturally occurring compounds and would be amenable for synthetic modifications and fine-tuning of their physicochemical properties.

The routes developed for fully synthetic hydroporphyrins entail either derivatization of porphyrins (which are usually much easier to prepare)[192] or *de novo* synthesis of the hydroporphyrin macrocycle.[193–200] The latter approach, though more synthetically challenging, is more versatile and enables full control of the position and numbers of substituents on the periphery of the macrocycle (thus allowing also the extensive tuning of the chemical and optical properties of hydroporphyrins). *De novo* routes enable also preparation of more stable hydroporphyrins resistant to oxidation to more conjugated porphyrins. Installation of geminal alkyl groups on the partially saturated pyrroline rings in hydroporphyrins prevents oxidation of chlorins and bacteriochlorins to more conjugated congeners.[193–195]

Synthetic hydroporphyrins display a range of unique photochemical properties, which make them very attractive platforms to develop fluorophores for *in vivo* applications. Both synthetic chlorins and bacteriochlorins exhibit narrow and tunable absorption and emission bands. Their absorption and emission maxima can be broadly tuned by simple substitution on the periphery of the macrocycle, spanning the range of about 635–715 nm for chlorins[183,185,199,200] and 715–823 nm for bacteriochlorins.[184] The wavelengths of absorption and emission can be tuned virtually with nanometer precision by relatively straightforward chemical modification so that one common precursor can be used for the synthesis of a range of derivatives with different emission bands.[193,198] Moreover, chlorins and bacteriochlorins exhibit exceptionally narrow emission bands with full width at half maxima (FWHM) of $\sim$15 nm for chlorins and $\sim$20 nm for bacteriochlorins, which are probably the narrowest emissions among organic compounds.[183–185,198] Most of the chlorin and bacteriochlorin derivatives exhibit also sufficiently high quantum yields of fluorescence and high fluorescence lifetimes: 8–10 ns for chlorins[183] (which is of a magnitude higher than those for typical organic fluorophores) and 4–6 ns for bacteriochlorins.[184] Fluorescence lifetime can be also tuned to some extent by substitution and metalation.[183] Their narrow and tunable emission bands and long and tunable fluorescence lifetimes make them a superior choice for spectral and lifetime *in vivo* multiplexing (see Section 6.1).

## 5.5. Hydroporphyrin arrays for increased Stokes shift and multicolor *in vivo* imaging

The inherent spectroscopic property of hydroporphyrins, both chlorins and bacteriochlorins, is the small Stokes shift, which typically falls in the range 0–10 nm, regardless of the solvent, substitution pattern, and metalation state of the macrocycle.[183,184] To overcome this limitation, which is critical for *in vivo* imaging, hydroporphyrin energy-transfer dyads have been proposed. Holten and coworkers have demonstrated that, in dyads comprising chlorin and bacteriochlorin, efficient energy transfer from chlorin to bacteriochlorin moieties occurs, and dyads behave as a single chromophore with excitation wavelengths of chlorins (650 or 675 nm) and the emission wavelength of bacteriochlorin (760 nm) so that the effective Stokes shift can be increased up to 110 nm (Chart 3.8).[201,202]

## 5.6. Water solubility and aggregation

The planar, aromatic, and highly hydrophobic structure of porphyrins and hydroporphyrins causes difficulties in aqueous solubility. Additionally, porphyrins tend to aggregate in aqueous solution.[203] The aggregation behavior of porphyrins, which is mainly driven by $\pi-\pi$ stacking interactions and hydrophobic forces, is quite complex and depends on the substitution pattern. Introduction of the hydrophilic groups, such as carboxylates, sulfonic acids, quarternary pyridinium, or ammonium groups, only partially solves the problem because such porphyrins still tend to aggregate.[203] The aggregation problem can be overcome by the introduction of a "swallow-tail"

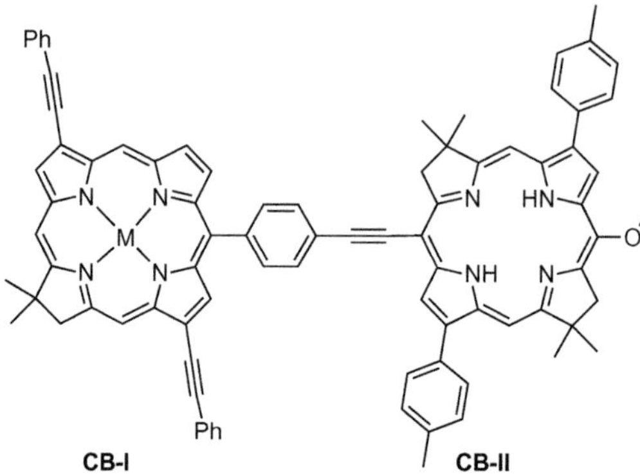

**CB-I**

M = Zn

$\lambda_{abs}$ = 650 nm
$\varepsilon$ = 60,000 M$^{-1}$ cm$^{-1}$
$\lambda_{em}$ = 760 nm
$\Phi_f$ = 0.19
FWHM = 21 nm
Solvent: toluene
Ref. 202

**CB-II**

M = H,H

$\lambda_{abs}$ = 675 nm
$\varepsilon$ = 80,000 M$^{-1}$ cm$^{-1}$
$\lambda_{em}$ = 760 nm
$\Phi_f$ = 0.19
FWHM = 18 nm
Solvent: toluene
Ref. 202

**Chart 3.8** Structures and optical properties of chlorin–bacteriochlorin energy-transfer dyads with tunable apparent Stokes shift.

solubilizing motif, that is, hydrophilic groups that are projected above and below a macrocyclic porphyrinic plane.[204] Such a motif provides water solubility and prevents aggregation by sterically hindering the π-system. Thus, chlorins equipped with swallow-tail substituents with phosphonic acids have been reported, and they show good water solubility, no evidence of aggregation, and good fluorescence quantum yields in water.[204] A similar approach has been developed by Vinogradov and coworkers, who prepared benzoporphyrins with dendritic substituents. Dendrons attached to the benzoporphyrin core possess peripheral hydrophilic groups (such as polyethylene glycol) and provide excellent water solubility, and crowded dendritic substituents protect porphyrin from self-aggregation and interactions with plasma proteins which may alter the optical properties of the fluorophore.[173,174,205,206]

An alternative strategy that has been developed to overcome the lack of water solubility of porphyrinic compounds entails the encapsulation of hydrophobic tetrapyrrolic macrocycles inside nanostructured capsules. Therien and coworkers have embedded strongly coupled porphyrin arrays into the polymersome membranes by cooperative self-assembly of diblock amphiphilic polymers and hydrophobic porphyrins.[179,182,207] As a result, vehicles with diameters of 50 nm–50 μm are formed in which porphyrins are uniformly distributed in the vehicle membrane. As porphyrins stay in the hydrophobic environment of the vehicle membrane, they retain optical properties comparable to those in organic solvents. The resulting highly emissive polymersomes can be freely dispersed in water and have been used for *in vivo* imaging,[182] labeling, and *in vivo* tracking of dendritic cells.[181]

Porphyrin arrays have been also incorporated into the hydrophobic core of the LDL apo forms, forming nanoparticles highly emissive in water. The resulting nanoparticles show tumor specificity, as many tumor cells over-express the LDL receptor.[180] Porphyrin-doped LDLs have been used for imaging B16[180] melanoma cells. Similarly, bacteriopheophytin *a* bisoleate has been incorporated into the hydrophobic core of high-density lipoprotein nanoparticles (size 12 nm) and used for *in vivo* imaging of tumor.[188]

## 5.7. Photocytotoxicity of tetrapyrrolic macrocycle

The use of tetrapyrrolic macrocycles as fluorophores *in vivo* raises concern about their phototoxicity. Porphyrins, hydroporphyrins, and phthalocyanines upon excitation populate the corresponding triplet excited state, which reacts with ambient oxygen to produce highly cytotoxic singlet oxygen and other highly reactive oxygen species.[167–169] This property is very useful in photodynamic therapy, where tetrapyrrolic macrocycles are widely used as photosensitizers, but can be potentially detrimental when one considers using them as fluorophores in living organisms. The photocytotoxicity depends on the many factors, such as the intrinsic photochemical properties of the photosensitizer (quantum yield and lifetime of the triplet state), localization of the photosensitizer inside the tissue, and the intensity of illumination. In general, even good photosensitizers require a higher dosage of light to induce phototoxicity than is typically used in fluorescence imaging experiments. Moreover, it is expected that tetrapyrrolic macrocycles with optical properties optimized for fluorescence imaging (i.e., with high quantum yield of fluorescence) would have a lower quantum yield of the triplet state, as these two processes compete with each other. For example, it has been reported that, in highly conjugated porphyrin arrays, the increase of fluorescence quantum yield is due to the accompanying decrease of triplet state formation, and thus highly conjugated porphyrin arrays are poor photosensitizers.[178] Similarly, there are also suggestions that, for hydroporphyrins, the fluorescence quantum yield increases at the expense of triplet state formation.[185]

Vinogradov and coworkers studied the phototoxicity of phosphorescent dendritic benzoporphyrin used as *in vivo* oxygen probes.[208] In this case, there is a particular concern about phototoxicity of the probe, as the side product of phosphorescent oxygen sensing is a singlet oxygen. They found negligible phototoxicity of their probes, which they attributed to the inability of dendritic benzoporphyrins to penetrate cellular membranes and produce singlet oxygen inside the cellular organelles. Their studies pointed out the importance of probe localization on their toxicity and demonstrated that even a potentially highly photocytotoxic probe can be safely used *in vivo*.

The elegant and general solution to overcome the problem of singlet oxygen generation by tetrapyrrolic fluorophores has been proposed by Moore and coworkers.[209] They prepared the covalently linked carotene–porphyrin dyads and found that carotenoids effectively quenched the triplet state of porphyrins (and other tetrapyrrolic compounds) by energy transfer, thereby making them incapable of producing singlet oxygen (the same principle has been also used for the design of a quenched, protease-activatable, chlorophyll-based photosensitizer for anticancer therapy[210]). This general approach seems to be a viable strategy when phototoxicity is a problem.

## 5.8. Phthalocyanines

Phthalocyanines are benzoporphyrin congeners having nitrogen rather than carbon bridging atoms. Phthalocyanines exhibit strong and sharp absorption in red spectral window and, in contrast to porphyrins, show high fluorescence quantum yields. Expansion of the aromatic system by fusing additional benzo ring gives naphthocyanines with absorption and emission above 700 nm.[211-219] Phthalocyanines exhibit also excellent chemo- and photostability compared to the other near-IR fluorophores. Phthalocyanines have been used as fluorophores in polymerase chain reaction (PCR)[211] and molecular beacons,[212] and broadly examined as photosensitizers in photodynamic therapy,[167] whereas their use as a fluorophore for *in vivo* imaging has been rather neglected.

One of the reasons for this neglect is the fact that phthalocyanines, because of their planar, hydrophobic structure, are difficult to solubilize in water and tend to aggregate in aqueous solution. Their solubility in organic solvents is usually low as well, which makes their purification and handling rather difficult. The few available methods for synthesis and derivatization of phthalocyanines do not offer much flexibility in preparing functional derivatives, especially

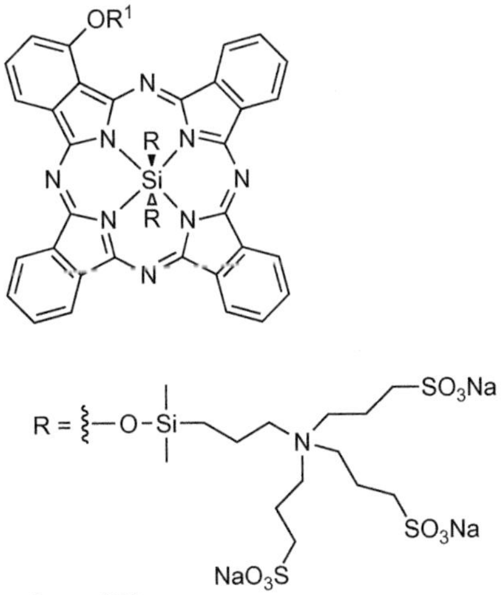

$\lambda_{abs}$ = 689 nm
$\varepsilon$ = 165,000 $M^{-1}$ $cm^{-1}$
$\lambda_{em}$ = 700 nm
$\Phi_f$ = 0.14
Solvent: PBS
Ref. 213

**Chart 3.9** Structure and optical properties of representative phthalocyanine fluorophore.

monofunctional, bioconjugatable derivatives. Finally, similar to porphyrins and hydroporphyrins, phthalocyanines tend to be photocytotoxic.[167] Recent progress in the chemistry of phthalocyanines has partially solved the abovementioned issues. Nonaggregating, water-soluble phthalocyanines bearing neutral, cationic, or anionic hydrophilic groups have been recently reported by Vicente and coworkers[214,215] and Ng and coworkers.[216] The water solubility and suppressed aggregation have been achieved by installing bulky groups, containing poly(ethylene) glycols,[216] quarternary pyridinium (cationic),[215] or carboxylate (anionic)[214] groups on the periphery of the macrocycle. Another notable approach to suppress aggregation and impose water solubility relies on the synthesis of silicon complexes of phthalocyanines and attaching hydrophilic groups as axial ligands to the central silicon atom (see Chart 3.9 for an example).[213,217–219] As axial ligands are situated centrally above and below the phthalocyanine plane, they prevent aggregation by steric repulsion. PEG,[217,219] sulfonic group-terminated alkyl amines,[213] or polyamines[218] have been used as axial hydrophilic groups.

Recently, notable progress has been made on the synthesis of monofunctional phthalocyanine derivatives. Hammer and coworkers reported the solid-phase synthesis of nonsymmetrically substituted phthalocya-nines.[220,221] This type of derivatives was previously prepared by statistical condensation of different phthalonitriles,[222] which led to the formation of complex mixture of products, diminished the yield of desired derivative, and required extensive purification. Solid-phase synthesis significantly simplifies purification and improves the yield of nonsymmetrical derivatives and enables an efficient synthesis of monofunctional, bioconjugatable derivatives.

# 6. SPECIAL TYPES OF FLUOROPHORES

## 6.1. Fluorophores for multicolor imaging

Multicolor imaging (spectral multiplexing) allows targeting simultaneously multiple different markers, processes, physicochemical parameters, cells, or organs. Multicolor imaging requires access to a set of fluorophores in which each fluorophore has a distinctive spectral feature so that fluorescence from each of them can be independently detected in the presence of other fluorophores. Ideally, each fluorophore in such a set should be excited with the same wavelength, and each should exhibit a narrow emission band centered at a different wavelength, without overlap with emission bands from the other fluorophores. Alternatively, each fluorophore should exhibit a narrow absorption band (so that each can be selectively excited at the different wavelengths) and emit at the same wavelength.

Multicolor fluorescence detection has been successfully used, for example, in nucleic acid sequencing (where energy-transfer dyads with a common donor and different acceptors have been used)[223,224] or in flow cytometry.[225] Application of multicolor imaging *in vivo* has, however, certain limitations. The simultaneous use of multiple organic fluorophores *in vivo* is limited because of their broad emission bands (typical FWHM for most organic fluorophores is $>30$ nm) so that only a limited number of fluorophores can be placed in the spectral window suitable for *in vivo* applications without strong overlap of their emission bands.[6] Additionally, in the set of organic fluorophores with different emission wavelengths, each fluorophore usually requires a different excitation wavelength, which makes whole imaging process time consuming and technically complex. Quantum dots, which have a tunable, narrow emission band (FWHM $\sim 30$ nm) and broad absorption bands that allow excitations of the whole set of different quantum dots with the common wavelength are good candidates for spectral multiplexing.[6] The use of the quantum dots, however, raises concerns about their toxicity, as they often contain toxic metals, such as cadmium and selenium.[6] Recently, fluorescent proteins with large Stokes shifts, named Keima, have been developed, which, in combination with other fluorescent proteins, allow multicolor intracellular imaging with single excitation wavelengths (simultaneous imaging using six different fluorescent proteins has been demonstrated).[226] The wavelengths of excitation and emission for Keima ($\lambda_{abs} = 440$ nm, $\lambda_{em} = 620$ nm for the longest emitting variant) make them less suitable for *in vivo* applications.

So far, *in vivo* multicolor imaging has been pursued using a cocktail of organic fluorophores (such as mixture of various rhodamines, coumarines, cyanine dyes, and/or Alexa Fluors),[227–230] quantum dots (also together with organic fluorophores),[231–235] or upconverting nanocrystals.[236,237] Upconverting nanocrystals[238] are a unique class of fluorophores that are composed of, for example, sodium yttrium fluoride and are doped with rare-earth metal cations and they can be excited in the near-IR region (e.g., 980 nm) and emit at a shorter wavelength. The wavelength of emission depends on their composition. Upconverting nanocrystals have been utilized *in vivo* for multicolor imaging alone,[236] or as an energy-transfer donor in conjugation with near-IR organic fluorophores (rhodamines).[237]

Promising fluorophores for *in vivo* multicolor applications are hydro-porphyrin energy-transfer dyads (see Section 5.5 and Chart 3.8). The intrinsic properties of hydroporphyrins, that is, narrow and tunable absorption and emission bands, together with their tunable apparent Stokes shift, which can be achieved by assembling hydroporphyrins in energy-transfer dyads, make them well suited for multicolor imaging. Holten and coworkers studied pairs of model chlorin–bacteriochlorin dyads where each dyad had the same bacteriochlorin acceptor (thus the same emission wavelength) and different chlorin components (thus different excitation wavelengths).[202,239] Because of the narrow absorption bands in chlorins, the 25-nm separation between the absorption maxima of both chlorins in the studied pair is sufficient for selective excitation one dyad in the presence of the other. This selectivity has been demonstrated also *in vivo*[239] and gives the promise for the development of a new class of fluorophores with either a common emission wavelength and different, well-resolved absorption bands, or a common excitation wavelength and different, well-resolved emission wavelengths.

An elegant strategy for fluorophores for multicolor imaging has been developed by Kool and coworkers.[240,241] They assembled deoxyriboside monomers containing small fluorescent organic molecules (aromatic hydrocarbons or small fluorescent heterocycles) in a DNA-like phosphodiester oligomer (oligodeoxyfluorosides). They found that several different interactions between the assembled fluorophores occur (such as excimer and exciplex formation, H-stacking and energy transfer), which result in different fluorescence characteristics for different combinations of the fluorophores. This enabled preparing the sets of short oligomers (containing 1–4 different fluorophores) each with a different fluorescence color, excitable at the common wavelength. The advantages of this approach are the ease of synthesis (as an automated DNA synthesizer can be utilized), water solubility (provided by phosphodiester backbone), and the ability to attach a single bioconjugatable group.[241] The antibody–oligodeoxyfluoroside conjugates have been used for multicolor imaging of living cells[241] and living zebrafish embryo.[240]

## 6.2. Self-illuminating fluorophores

All the fluorophores described so far require photoexcitation prior to fluorescence, that is, the fluorophore must absorb a photon to be excited, which in turn requires illumination of the fluorophore by an external source of light. Such illumination causes excitation of the exogenous tissue or cell fluorophores and causes some background signal which reduces the signal-to-background ratio. Consequently, it diminishes both the sensitivity and the limit of detection, even when the excitation is done in the optimal spectral window.[242]

One of the solutions to overcome this problem is to use fluorophores that do not require photoexcitation but can reach an excited state and subsequently emit fluorescence upon chemical reaction (chemiluminescence) or upon enzymatic reaction (bioluminescence). Chemi- or bioluminescence eliminates the need for an external light source for excitation and therefore eliminates almost completely autofluorescence from the tissue.

The prominent bioluminescent reaction is the luciferase-catalyzed oxidation of luciferin with the concomitant emission of light. Luciferin is a generic name given to the class of small organic molecules that can emit light upon oxidation catalyzed by various types of luciferases, among which firefly luciferin and coelenterazine are the most widely used in biotechnology and, recently, in *in vivo* imaging.[242,243] The luciferin/luciferase pairs emit at relatively short wavelengths (for most luciferins between 480 and 560 nm); therefore, efforts have been made to obtain mutant luciferins or use alternative substrates to shift the resulting bioluminescence toward longer wavelengths.[244,245] In an alternative strategy, the luciferase/luciferin pair is used as an energy-transfer donor in bioluminescence resonance energy transfer (BRET) and transfers the excitation energy to the acceptor emitting at longer wavelength. In one of the first BRET systems developed, luciferase was fused with green fluorescent proteins and this system was used to monitor intracellular molecular events.[242,243] The luciferase–GFP (green fluorescent protein) pair has been also used *in vivo*,[246] but the relatively short wavelength of emission of GFP prompted scientists to seek a BRET acceptor with a longer wavelength of emission.

Recently, a BRET system containing red fluorescent proteins emitting at 635 nm as acceptors has been designed and used for *in vivo* imaging of protein–protein interactions.[247] Alternatively, luciferase has been conjugated with organic near-IR-emitting fluorophores: Alexa Fluors (AF680 and AF750)[245] or cyanine[248] and both exhibit efficient BRET and emission in the red or the near-IR region.

Rao and coworkers have developed a self-illuminating BRET system suitable for *in vivo* application, which consists of the luciferase/coelenterazine pair as donor and quantum dots as acceptors.[249,250] In their system, quantum dots are conjugated to the eight copies of mutated *Renilla reniformis* luciferase (called there Luc8) and, in the presence of coelenterazine, show quantum dot emission, due to the BRET, together with a much weaker emission at 480 from coelenterazine. This system has been examined *in vitro* and *in vivo* in mouse. Luc8 has been subsequently conjugated to quantum dots emitting at different wavelengths, 605, 655, 705, and 800 nm, respectively, and used for *in vivo* spectral multiplexing.[249] A BRET system utilizing the luciferase–quantum dot pair has been also used for *in vivo* cancer detection.[251]

Although chemiluminescence, that is, luminescence occurring upon a chemical, nonenzymatic reaction, is a well-known phenomenon,[252] its application for *in vivo* imaging has been neglected until recently.[253,254] The main reasons are that most of the chemiluminescent reactions typically emit short-wavelength light, utilize unstable, highly reactive compounds (such as peroxides), or require reagents that are toxic or harmful (e.g., hydrogen peroxide).[252] Chemiluminescent molecular probes suitable for *in vivo* applications, where chemiluminescence is activated by temperature, have been developed by Smith and coworkers.[254] They found that squaraine rotaxanes (see Section 3) having an anthracene core in their macrocyclic tetralactam components react with singlet oxygen in cycloaddition reaction to form an adduct, the so-called endoperoxide. This peroxide, upon warming up to the body temperature, undergoes a cycloreversion reaction to produce singlet oxygen. The singlet oxygen decays to the ground-state triplet oxygen, exciting the encapsulated squaraine, which in turn fluoresces in the near-IR region. The exact mechanism of energy transfer from singlet oxygen to squaraine is not known. The system is very convenient, as rotaxane endoperoxide forms quantitatively upon irradiation of squaraine rotaxane solution in the presence of air; the resulting peroxides can be indefinitely stored at $-20\,°C$ and near-IR fluorescence appears upon warming the sample to room temperature.

Scherman and coworkers developed organic nanoparticles with long-lasting (persistent) luminescence.[255] Nanoparticles based on magnesium silicate doped with luminescent cations ($Eu^{2+}$, $Dy^{3+}$, and $Mn^{2+}$) emit upon irradiation red or near-IR luminescence for several hours after irradiation, which can be easily detected from the animal body several hours after injection.

## 7. CONCLUSION

Despite vigorous research efforts, the optimal fluorophore that fulfills all requirements for *in vivo* applications has not been created yet. Given the diverse applications of fluorescence in *in vivo* imaging, it is rather unlikely that a single universal molecular platform can be ever found. Searching for the new fluorescent materials, modification and improvement of the existing fluorophores, and conjugation of different fluorophores to combine their properties—these are the three areas that warrant further progress in the field of fluorophores. The latter approach seems to be particularly powerful for creating new systems with unique properties, as illustrated by some of the recent advances highlighted in this review, such as energy-transfer dyads with tunable Stokes shift, self-illuminating near-IR BRET systems, or near-IR luminescent squaraine rotaxanes. On the other hand, advances in related fields, namely, materials science, nanoscience, and biotechnology, will likely provide the new materials with currently unattainable properties and tools for improvements and expansion of the properties of currently existing fluorophores.

## ACKNOWLEDGMENT

The author wishes to thank the University of Maryland, Baltimore County, for supporting this work (start-up funds and SRAIS award).

# REFERENCES

1. Demchenko AP. *Introduction to fluorescence sensing*. Springer; 2009.
2. Lakowicz JR. *Principles of fluorescence spectroscopy*. 3rd ed. New York: Springer; 2006.
3. Ballou B, Ernst LA, Waggoner AS. Fluorescence imaging of tumors *in vivo*. *Curr Med Chem* 2005;**12**:795–805.
4. Hilderbrandt SA, Weissleder R. Near-infrared fluorescence: application to *in vivo* imaging. *Curr Opin Chem Biol* 2010;**14**:71–9.
5. Kobayashi H, Ogawa M, Alford R, Chouke PL, Urano Y. New strategies for fluorescent probe design in medicinal diagnostic imaging. *Chem Rev* 2010;**110**:2620–40.
6. Resch-Genger U, Grabolle M, Cavaliere-Jaricot S, Nitschke R, Nann T. Quantum dots versus organic dyes as fluorescent labels. *Nat Methods* 2008;**5**:763–75.
7. Lavis LD, Raines RT. Bright ideas for chemical biology. *ACS Chem Biol* 2008;**3**:142–55.
8. Kovar JL, Simpson MA, Schutz-Geschwender A, Olive DM. A systematic approach to the development of fluorescent contrast agents for optical imaging of mouse cancer models. *Anal Biochem* 2007;**367**:1–12.
9. Nolting DD, Gore JC, Pham W. Near-infrared dyes: probe development and applications in optical molecular imaging. *Curr Org Synth* 2011;**8**:521–34.
10. Luo S, Zhang E, Su Y, Cheng T, Shi C. A review of NIR dyes in cancer targeting and imaging. *Biomaterials* 2011;**32**:7127–38.
11. Yong K-T, Roy I, Swihart MT, Prasad PN. Multifunctional nanoparticles as biocompatible targeted probes for human cancer diagnosis and therapy. *J Mat Chem* 2009;**19**:4655–72.
12. Escobedo JO, Rusin O, Lim S, Strongin RM. NIR dyes for bioimaging applications. *Curr Opin Chem Biol* 2010;**14**:64–70.
13. Pansare VJ, Hejazi S, William WJ, Prud'homme RK. Review of long-wavelength optical and NIR imaging materials: contrast agents, fluorophores, and multifunctional nano carriers. *Chem Mat* 2012;**24**:812–27.
14. Gonçalves MST. Fluorescent labeling of biomolecules with organic probes. *Chem Rev* 2009;**109**:190–212.
15. Shcherbo D, Shemiakina II, Ryabova AV, Luker KE, Schmidt BT, Souslova EA, et al. Near-infrared fluorescent proteins. *Nat Methods* 2010;**7**:827–9.
16. Filonov GS, Piatkevich KD, Ting L-M, Zhang J, Kim K, Verkhusha VV. Bright and stable near-infrared fluorescent protein for *in vivo* imaging. *Nat Biotechnol* 2011;**29**:757–61.
17. Kim S, Lim C-K, Na J, Lee Y-D, Kim K, Choi K, et al. Conjugated polymer nanoparticles for biomedical *in vivo* imaging. *Chem Commun* 2010;**46**:1617–9.
18. Aswathy RG, Yoshida Y, Maekawa T, Kumar DS. Near-infrared quantum dots for deep tissue imaging. *Anal Bioanal Chem* 2010;**397**:1417–35.
19. Gao J, Chen X, Cheng Z. Near-infrared quantum dots as optical probes for tumor imaging. *Curr Top Med Chem* 2010;**10**:1147–57.
20. Smith AM, Duan H, Mohs AM, Nie S. Bioconjugated quantum dots for *in vivo* molecular and cellular imaging. *Adv Drug Del Rev* 2008;**60**:1226–40.
21. Yang S-T, Cao L, Luo PG, Lu F, Wang X, Wang H, et al. Carbon dots for optical imaging *in vivo*. *J Am Chem Soc* 2009;**131**:11308–9.
22. Bae SW, Tan W, Hong J-I. Fluorescent dye-doped silica nanoparticles: new tools for bioapplications. *Chem Commun* 2012;**48**:2270–82.
23. Welsher K, Liu Z, Sherlock SP, Robinson JT, Chen Z, Daranciang D, et al. A route to brightly fluorescent carbon nanotubes for near-infrared imaging in mice. *Nat Nanotechnol* 2009;**4**:774–80.
24. Welsher K, Sherlock SP, Dai H. Deep-tissue anatomical imaging of mice using carbon nanotube fluorophores in the second near-infrared window. *Proc Natl Acad Sci USA* 2011;**108**:8943–8.

25. Berlier JE, Rothe A, Buller G, Bradford J, Gray DR, Filanoski BJ, et al. Quantitative comparison of long-wavelength Alexa Fluor dyes to Cy dyes: fluorescence of the dyes and their bioconjugates. *J Histochem Cytochem* 2003;**51**:1699–712.
26. Strekowski L, editor. *Heterocyclic polymethine dyes. Top Heterocycl. Chem.* vol. 14. Springer, Berlin, Heidelberg, New York, 2008.
27. Mishra A, Behera RK, Behera PK, Mishra BK, Behera GB. Cyanine during 1990s: a review. *Chem Rev* 2000;**100**:1973–2011.
28. Levitus M, Ranjit S. Cyanine dyes in biophysical research: the photophysics of polymethine fluorescent dyes in biomolecular environments. *Quart Rev Biophys* 2011;**44**:123–51.
29. Mujumdar SR, Mujumdar RB, Grant CM, Waggoner AS. Cyanine-labelling reagents: sulfobenzindocyanine succinimidyl esters. *Bioconjugate Chem* 1996;**7**:356–62.
30. Mujumdar RB, Ernst LA, Mujumdar SR, Lewis CJ, Waggoner AS. Cyanine dye labeling reagents: sulfoindocyanine succinimidyl esters. *Bioconjugate Chem* 1993;**4**:105–11.
31. Pham W, Lai W-F, Weissleder R, Tung C-H. High efficiency synthesis of a bioconjugatable near-infrared fluorochrome. *Bioconjugate Chem* 2003;**14**:1048–51.
32. Pauli J, Brehm R, Spieles M, Kaiser WA, Hilger I, Resch-Genger U. Novel fluorophores as building blocks for optical probes for in vivo near infrared fluorescence (NIRF) imaging. *J Fluoresc* 2010;**20**:681–93.
33. Soper SA, Mattingly QL. Steady-state and picosecond laser fluorescence studies of non-radiative pathways in tricarbocyanine dyes. Implications to the design of near-IR fluorochromes with high fluorescence efficiencies. *J Am Chem Soc* 1994;**116**:3744–52.
34. Zhang Z, Fan J, Cheney PP, Berezin MY, Edwards WB, Akers WJ, et al. Activatable molecular system using homologous near-infrared fluorescent probes for monitoring enzyme activities *in vitro, in cellulo* and *in vivo*. *Mol Pharmaceutics* 2009;**6**:416–27.
35. Ye Y, Bloch S, Kao J, Achilefu S. Multivalent carbocyanine molecular probes: synthesis and applications. *Bioconjugate Chem* 2005;**16**:51–61.
36. Hilderbrand SA, Kelly KA, Weissleder R, Tung C-H. Monofunctional near-infrared fluorochromes for imaging applications. *Bioconjugate Chem* 2005;**16**:1275–81.
37. Achilefu S, Jimenez HN, Dorshow RB, Bugaj JE, Webb EG, Wilhelm RR, et al. Synthesis, in vitro receptor binding and in vitro evaluation of fluorescein and carbocyanine peptide-based optical contrast agents. *J Med Chem* 2002;**45**:2003–15.
38. Hermanson GT. *Bioconjugate techniques*. 2nd ed. Amsterdam, Boston, Heidelberg, London, New York, Oxford, Paris, San Diego, San Francisco, Singapore, Sydney, Tokio: Academic Press; 2008.
39. Ye Y, Li WP, Anderson CJ, Kao J, Nikiforovich GV, Achilefu S. Synthesis and characterization of a macrocyclic near-infrared optical scaffold. *J Am Chem Soc* 2003;**125**:7766–7.
40. Zhang Z, Berezin MY, Kao JLF, d'Avignon A, Bai M, Achilefu S. Near-infrared dichromic fluorescent carbocyanine molecules. *Angew Chem Int Ed* 2008;**47**:3584–7.
41. Ye Y, Bloch S, Achilefu S. Polyvalent carbocyanine molecular beacon for molecular recognitions. *J Am Chem Soc* 2004;**126**:7740–1.
42. Ye Y, Bloch S, Xu B, Achilefu S. Novel near-infrared fluorescent integrin-targeted DFO analogoue. *Bioconjugate Chem* 2008;**19**:225–34.
43. Berezin MY, Guo K, Akers W, Livingston J, Solomon M, Lee H, et al. Rational approach to select small peptide molecular probes labeled with fluorescent cyanine dyes for *in vivo* optical imaging. *Biochemistry* 2011;**50**:2691–700.
44. Almutairi A, Akers WJ, Berezin MY, Achilefu A, Fréchet JMJ. Monitoring the biodegradation of dendritic near-infrared nanoprobes by in vivo fluorescence imaging. *Mol Pharmaceutics* 2008;**5**:1103–10.

45. Licha K, Riefke B, Ntziachristos V, Becker A, Chance B, Semmler W. Hydrophilic cyanine dyes as contrast agents for near-infrared tumor imaging: synthesis, photophysical properties and spectroscopic *in vivo* characterization. *Photochem Photobiol* 2000;**72**:392–8.

46. Zhang Z, Liang K, Bloch S, Berezin M, Achilefu S. Monomolecular multimodal florescence-radioisotope imaging agents. *Bioconjugate Chem* 2005;**16**:1232–9.

47. Bouteiller C, Clavé G, Bernardin A, Chipon B, Massonneau M, Renard P-Y, et al. Novel water-soluble near-infrared cyanine dyes: synthesis, spectral properties, and use in the preparation of internally quenched fluorescent probes. *Bioconjgate Chem* 2007;**18**:1303–17.

48. Pham W, Medarova Z, Moore A. Synthesis and applications of a water-soluble near-infrared dye for cancer detection using optical imaging. *Bioconjugate Chem* 2005;**16**:735–40.

49. Becker A, Riefke B, Ebert B, Sukowski U, Rinneberger H, Semmler W, et al. Macromolecular contrast agents for optical imaging of tumors: comparison of indotricarbocyanine-labelled human serum albumin and tranferrin. *Photochem Photobiol* 2000;**72**:234–41.

50. Licha K, Hessenius C, Becker A, Henklein P, Bauer M, Wisniewski S, et al. Synthesis, characterization, and biological properties of cyanine-labeled somatostatin analogues as receptor-targeted fluorescent probes. *Bioconjugate Chem* 2001;**12**:44–50.

51. Ito S, Muguruma N, Kakehashi Y, Hayashi S, Okamura S, Shibata H, et al. Development of fluorescence-emitting antibody labeling substance by near-infrared ray detection. *Bioorg Med Chem Lett* 1995;**5**:2689–94.

52. Hirata T, Kogiso H, Morimoto K, Miyamoto S, Taue H, Sano S, et al. Synthesis and reactivities of 3-indocyanine-green-acyl-1,3-thiazolidine-2-thione (ICG-ATT) as a new near-infrared fluorescent-labeling reagent. *Bioorg Med Chem* 1998;**6**:2179–84.

53. Becker A, Hessenius C, Licha K, Ebert B, Sukowski U, Semmler W, et al. *Nat Biotechnol* 2001;**19**:327–31.

54. Cheng Z, Wu Y, Xiong Z, Gambhir SS, Chen X. Near-infrard fluorescent RGD peptides for optical imaging of integrin $a_v b_3$ expression in living mice. *Bioconjugate Chem* 2005;**16**:1433–41.

55. Villaraza AJL, Milenic DE, Brechbiel MW. Improved speciation characteristics of PEGylated indocyanine green-labeled panitumumab: revisiting the solution and spectroscopic properties of a near-infrared emitting anti-HER1 antibody for optical imaging of cancer. *Bioconjugate Chem* 2010;**21**:2305–12.

56. Levi J, Cheng Z, Gheysens O, Patel M, Chan CT, Wang Y, et al. Fluorescent fructose derivatives for imaging breast cancer cells. *Bioconjugate Chem* 2007;**18**:628–34.

57. Ran C, Pantazopoulos P, Medarova Z, Moore A. Synthesis and testing of beta-cell-specific streptozotocin-derived near-infrared imaging probes. *Angew Chem Int Ed* 2007;**46**:8998–9001.

58. Lee S, Park K, Lee S-Y, Ryu JH, Park JW, Ahn HJ, et al. Dark quenched matrix metalloproteinase fluorogenic probe for imaging osteoarthritis development *in vivo*. *Bioconjugate Chem* 2008;**19**:1743–7.

59. Josan JS, Morse DL, Xu L, Trisal M, Baggett B, Davis P, et al. Solid-phase synthetic strategy and bioevaluation of a labeled g-opioid receptor ligand Dmt-Tic-Lys for *in vivo* imaging. *Org Lett* 2009;**11**:2479–82.

60. Lin Y, Weissleder R, Tung C-H. Novel near-infrared cyanine fluorochromes: synthesis, properties, and bioconjugation. *Bioconjugate Chem* 2002;**13**:605–10.

61. Tung C-H, Lin Y, Moon WK, Weissleder R. A receptor-targeted near-infrared fluorescence probe for in vivo tumor imaging. *Chembiochem* 2002;**8**:784–6.

62. Narayanan N, Patonay G. A new method for the synthesis of heptamethine cyanine dyes: synthesis of new near-infrared fluorescent labels. *J Org Chem* 1995;**60**:2391–5.

63. Salon J, Wolinska E, Raszkiewicz A, Patonay G, Strekowski L. Synthesis and Benz[e] indolium heptamethine cyanines containing C-substituent at the central portion of the heptamethine moiety. *J Heterocycl Chem* 2005;**42**:959–61.

64. Strekowski L, Mason C, Lee H, Say M, Patonay G. Water-soluble pH-sensitive 2,6-bis (substituted ethylidene)-cyclohexanone/hydroxy cyanine dyes that absorb in the visible/near-infrared regions. *J Heterocycl Chem* 2004;**41**:227–32.

65. Strekowski L, Lipowska M, Patonay G. Substitution reactions of a nucleofugal group in heptamethine cyanine dyes. Synthesis of an isothiocyanato derivative for labeling of proteins with a near-infrared chromophore. *J Org Chem* 1992;**57**:4578–80.

66. Lee H, Mason C, Achilefu S. Heptamethine cyanine dyes with a robust C-C bond at the central position of the chromophore. *J Org Chem* 2006;**71**:7862–5.

67. Flanagan Jr. J, Khan SH, Menchen S, Soper SA, Hammer RP. Functionalized tricarbocyanine dyes as near-infrared fluorescent probes for biomolecules. *Bioconjugate Chem* 1997;**8**:751–6.

68. Lee H, Mason C, Achilefu S. Synthesis and spectral properties of near-infrared aminophenyl-, hydroxyphenyl-, and phenyl-substituted hepthamethine cyanines. *J Org Chem* 2008;**73**:723–5.

69. Wang W, Ke S, Kwon S, Yallampalli S, Cameron AG, Adams KE, et al. A new optical and nuclear dual-labeled imaging agent targeting interleukin 11 receptor alpha-chain. *Bioconjugate Chem* 2007;**18**:397–402.

70. Zabeer A, Wheat TE, Frangioni JV. IRDye78 conjugates for near-infrared fluorescence imaging. *Mol Imaging* 2002;**1**:354–64.

71. Choi HS, Nasr K, Alayabyev S, Feith D, Lee JH, Kim SH, et al. Synthesis and in vivo fate of zwitterionic near-infrared fluorophores. *Angew Chem Int Ed* 2011;**50**:6258–63.

72. Zhang Z, Achilefu S. Synthesis and evaluation of polyhydroxylated near-infrared carbocyanine molecular probes. *Org Lett* 2004;**6**:2067–70.

73. Licha K, Welker P, Weinhart M, Wegner N, Kern S, Reichert S, et al. Fluorescence imaging with multifunctional polyglycerol sulfates: novel polymeric near-IR probes targeting inflammation. *Bioconjugate Chem* 2011;**22**:2453–60.

74. Ornelas C, Lodescar R, Durandin A, Canary JW, Pennell R, Liebes LF, et al. Combining aminocyanine dyes with polyamide dendrons: a promising strategy for imaging in the near-infrared region. *Chem Eur J* 2011;**17**:3619–29.

75. Lee H, Akers W, Bhushan K, Bloch S, Sudlow G, Tang R, et al. Near-infrared pH-activatable fluorescent probes for imaging primary and metastatic breast tumor. *Bioconjugate Chem* 2011;**22**:777–84.

76. Li C, Greenwood TR, Glunde K. Glucosamine-bound near-infrared fluorescent probes with lysosomal specificity for breast tumor imaging. *Neoplasia* 2008;**10**:389–98.

77. Williams MPA, Ethirajan M, Ohkubo K, Chen P, Pera P, Morgan J, et al. Synthesis and photophysical, electrochemical, tumor-imaging, and phototherapeutic properties of purpurinimide-N-substituted cyanine dyes joined with variable lengths of linkers. *Bioconjugate Chem* 2011;**22**:2283–95.

78. Smith BA, Akers WJ, Leevy WM, Lampkins AJ, Xiao S, Wolter W, et al. Optical imaging of mammary and prostate tumors in living animals using a synthetic near infrared zinc(II)-dipicoylamine probe for anionic cell surface. *J Am Chem Soc* 2010;**132**:67–9.

79. Smith BA, Gammon ST, Xiao S, Wang W, Chapman S, McDermott R, et al. In vivo optical imaging of acute cell death using a near-infrared fluorescent zinc-dipicolylamine probe. *Mol Pharmaceutics* 2011;**8**:583–90.

80. Altinoglu EI, Russin TJ, Kaiser JM, Barth BM, Ekklund PC, Kester M, et al. Near-infrared emitting fluorophore-doped calcium phosphate nanoparticles for *in vivo* imaging of human breast cancer. *ACS Nano* 2008;**2**:2075–84.

81. Peng X, Song F, Lu E, Wang Y, Zhou W, Fan J, et al. Heptamethine cyanine dyes with a large Stokes shift and strong fluorescence: a paradigm for excited-state intramolecular charge transfer. *J Am Chem Soc* 2005;**127**:4170–1.

82. Samanta A, Vendrell M, Das R, Chang Y-T. Development of photostable near-infrared cyanine dyes. *Chem Commun* 2010;**46**:7406–8.

83. Samanta A, Vendrell M, Yun S-W, Guan Z, Xu Q-H, Chang Y-T. A photostable near-infrared protein-labeling dye for in vivo imaging. *Chem Asian J* 2011;**6**:1353–7.

84. Renikuntla BR, Rose HC, Eldo J, Waggoner AS, Armitage BA. Improved photostability and fluorescence properties through polyfluorination of a cyanine dye. *Org Lett* 2004;**6**:909–12.

85. Toutchkine A, Nguyen D-V, Hahn KM. Merocyanine dyes with improved photostability. *Org Lett* 2007;**9**:2775–7.

86. Chen X, Peng X, Cui A, Wang B, Wang L, Zhang R. Photostabilities of novel heptamethine 3*H*-indoleine cyanine dyes with different *N*-substituents. *J Photochem Photobiol* 2006;**181**:79–85.

87. Matsuzawa Y, Tamura SI, Matsuzawa N, Ata M. Light stability of a β-cyclodextrin inclusion complexes of a cyanine dye. *J Chem Soc Faraday Trans* 1994;**90**:3517–20.

88. Guether R, Reddington MV. Photostable cyanine dye β-cyclodextrin conjugates. *Tetrahedron Lett* 1997;**38**:6167–70.

89. Yau CMS, Pascu SI, Odom SA, Warren JE, Klotz EJF, Frampton MJ, et al. Stabilisation of a heptamethine cyanine dye by rotaxane encapsulation. *Chem Commun* 2008;2897–9.

90. Lee C-H, Cheng S-H, Wang Y-J, Chen Y-C, Chen N-T, Souris J, et al. Near-infrared mesoporous silica nanoparticles for optical imaging: characterization and in vivo biodistribution. *Adv Funct Mater* 2009;**19**:215–22.

91. Kumar R, Roy I, Ohulchansky TY, Vathy LA, Bergey EJ, Sajjad M, et al. *In vivo* biodistribution and clearance studies using multimodal organically modified silica nanoparticles. *ACS Nano* 2010;**2**:699–708.

92. Barth BM, Sharma R, Altinoglu EI, Morgan TT, Shanmugavelandy SS, Kaiser JM, et al. Bioconjugation of calcium phosphosilicate composite nanoparticles for selective targeting of human breast and pancreatic cancers *in vivo*. *ACS Nano* 2010;**3**:1279–87.

93. Xu RX, Huang J, Xu JS, Sun D, Hinkle GH, Martin EW, et al. Fabrication of indocyanine green encapsulated biodegradable microbbbles for structural and functional imaging of cancer. *J Biomed Opt* 2009;**14**:034020-1–6.

94. Tosi G, Bondioli L, Ruozi B, Badiali L, Severini GM, Biffi S, et al. NIR-labeled nanoparticles engineered for brain targeting: in vivo optical imaging application and fluorescent microscopy evidences. *J Neural Transm* 2011;**118**:145–53.

95. Saxena V, Sadoqi M, Shao J. Enhanced photo-stability, thermal-stability and aqueous-stability of indocyanine green in polymeric nanoparticulate systems. *J Photochem Photobiol B* 2004;**74**:29–38.

96. Texier I, Goutayer M, Da Silva A, Guyon L, Djaker N, Josserand V, et al. Cyanine-loaded nanoparticles for improved in vivo fluorescence imaging. *J Biomed Opt* 2009;**14**:054005-1–054005-11.

97. Zheng X, Xing D, Zhou F, Wu B, Chen WR. Indocyanine green-containing nanostructure as near infrared dual-function targeting probes for optical imaging and photothermal therapy. *Mol Pharmaceutics* 2011;**8**:447–56.

98. Yu J, Javier D, Yassen MA, Nitin N, Richards-Kortum R, Anvari B, et al. Self-assembly synthesis, tumor cell targeting, and photothermal capabilities of antibody-coated indocyanine green nanocapsules. *J Am Chem Soc* 2010;**132**:272–9.

99. Yu J, Yaseen MA, Anvari B, Wong MS. Synthesis of near-infrared-absorbing nanoparticle-assembled capsules. *Chem Mater* 2007;**19**:1277–84.

100. Noh YW, Park HS, Sung M-H, Lim YT. Enhancement of the photostability and retention time of indocyanine green in sentinel lymph node mapping by anionic polyelectrolytes. *Biomaterials* 2011;**32**:6551–7.
101. Chen J, Corbin IR, Li H, Cao W, Glickson JD, Zheng G. Ligand conjugated low-density lipoprotein nanoparticles for enhanced optical cancer imaging in vivo. *J Am Chem Soc* 2007;**129**:5798–9.
102. Proulx ST, Lucini P, Dierzsi S, Rindkerknecht M, Mumprecht V, Leroux J-C, et al. Quantitative imaging of lymphatic function with liposomal indocyanine green. *Cancer Res* 2010;**70**:7053–62.
103. Bwambok DK, El-Zahab B, Challa SK, Li M, Chandler L, Baker GA, et al. Near-infrared fluorescent nanoGUMBOS for biomedical imaging. *ACS Nano* 2009;**3**:3854–60.
104. Akers WJ, Kim C, Berezin M, Guo K, Fuhrhop R, Lanza GM, et al. Noninvasive photoacoustic and fluorescence sentinel lymph node identification using dye-loaded perfluorocarbon nanoparticles. *ACS Nano* 2011;**5**:173–82.
105. Tanisaka H, Kizaka-Kondoh S, Makino A, Tanaka S, Hiraoka M, Kimura S. Near-infrared fluorescent labeled peptosome for application to cancer imaging. *Bioconjugate Chem* 2008;**19**:109–17.
106. Beverina L, Salice P. Squaraine compounds: tailored design and synthesis towards a variety of material science applications. *Eur J Org Chem* 2010;**2010**:1207–25.
107. Yagi S, Nakazumi H. Squarylium dyes and related compounds. *Top Heterocycl Chem* 2008;**14**:133–81.
108. Terenziani F, Painelli A, Katan C, Charlot M, Blanchard-Desce M. Charge instability in quadrupolar chromophores: symmetry breaking and solvatochromism. *J Am Chem Soc* 2006;**128**:15742–55.
109. Arunkumar E, Forbes CC, Noll BC, Smith BD. Squaraine-derived rotaxanes: sterically protected fluorescent near-IR dyes. *J Am Chem Soc* 2005;**127**:3288–9.
110. Chen H, Farahat MS, Law K-Y, Whitten DG. Aggregation of surfactant squaraine dyes in aqueous solution and microheterogeneous media: correlation of aggregation behavior with molecular structure. *J Am Chem Soc* 1996;**118**:2584–94.
111. Ros-Lis JV, Garcia B, Jiménez D, Martinez-Mañez R, Sancenón F, Soto J, et al. Squaraines as fluoro-chromogenic probes for thiol-containing compounds and their application to the detection of biorelevant thiols. *J Am Chem Soc* 2004;**126**:4064–5.
112. Gassensmith JJ, Baumes JM, Smith BD. Discovery and early development of squaraine rotaxanes. *Chem Commun* 2009;**14**:6329–38.
113. Arunkumar E, Fu N, Smith BD. Squaraine-derived rotaxanes: highly stable, fluorescent near-IR dyes. *Chem Eur J* 2006;**12**:4684–90.
114. Gassensmith JJ, Arunkumar E, Barr L, Baumes JM, DiVittorio KM, Johnson JR, et al. Self-assembly of fluorescent inclusion complexes in competitive media including the interior of living cells. *J Am Chem Soc* 2007;**129**:15054–9.
115. Xiao S, Fu N, Peckham K, Smith BD. Efficient synthesis of fluorescent squaraine rotaxane dendrimers. *Org Lett* 2010;**12**:140–3.
116. Gassensmith JJ, Barr L, Baumes JM, Paek A, Nguyen A, Smith BD. Synthesis and photophysical investigation of squaraine rotaxanes by "clicked capping" *Org Lett* 2008;**10**:3343–6.
117. Cole EL, Arunkumar E, Xiao S, Smith BA, Smith BD. Water-soluble, deep-red fluorescent squaraine rotaxanes. *Org Biomol Chem* 2012;**10**:5769–73.
118. Johnson JR, Fu N, Arunkumar E, Leevy M, Gammon ST, Piwnica-Wrms D, et al. Squaraine rotaxanes: superior substitutes for Cy-5 in molecular probes for near-infrared fluorescence cell imaging. *Angew Chem Int Ed* 2007;**46**:5528–31.
119. White AG, Fu N, Leevy WM, Lee J-J, Blasco MA, Smith BD. Optical imaging of bacterial infection in living mice using deep-red fluorescent squaraine rotaxane probes. *Bioconjugate Chem* 2010;**21**:1297–304.

120. Loudet A, Burgess K. BODIPY dyes and their derivatives: synthesis and spectroscopic properties. *Chem Rev* 2007;**107**:4891–932.
121. Boens N, Leen V, Dehaen W. Fluorescent indicator based on BODIPY. *Chem Soc Rev* 2012;**41**:1130–72.
122. Wood TA, Thompson A. Advances in the chemistry of dipyrrins and their complexes. *Chem Rev* 2007;**107**:1831–61.
123. Hama Y, Urano Y, Koyama Y, Choyke PL, Kobayashi H. Targeted optical imaging of cancer cells using lectin-binding BODIPY conjugated avidin. *Biochem Biophys Res Commun* 2006;**348**:807–13.
124. Urano Y, Asanuma D, Hama Y, Koyama Y, Barrett T, Kamiya M, et al. Selective molecular imaging of viable cancer cells with pH-activatable fluorescence probes. *Nat Med* 2009;**15**:104–9.
125. Kowada T, Kikuta J, Kubo A, Ishii M, Maeda H, Mizzukami S, et al. In vivo fluorescence imaging of bone-resorbing osteoclasts. *J Am Chem Soc* 2011;**133**:17772–6.
126. Buyukcakir O, Bozdemir A, Kolemen S, Erbas S, Akkaya EU. Tetrastyryl-Bodipy dyes: convenient synthesis and characterization of elusive near-IR fluorophores. *Org Lett* 2009;**11**:4644–7.
127. Ulrich G, Goeb S, De Nicola A, Retailleau P, Ziessel R. Chemistry at boron: synthesis and properties of red to near-IR fluorescent dyes based on boron-substituted diisoindolomethene frameworks. *J Org Chem* 2001;**76**:4489–505.
128. Killoran J, Allen L, Gallagher JF, Gallagher WM, O'Shea DF. Synthesis of BF$_2$ chelates of tetraarylazadipyrromethenes and evidence for their photodynamic therapeutic behavior. *Chem Commun* 2002;1862–3.
129. Zhao W, Carreira EM. Conformationally restricted aza-BODIPY: highly fluorescent, stable near-infrared absorbing dyes. *Chem Eur J* 2006;**12**:7254–63.
130. Loudet A, Bandichhor R, Burgess K, Palma A, McDonnell SO, Hall MJ, et al. B, O-Chelated azadipyrromethenes as near-IR probes. *Org Lett* 2008;**10**:4771–4.
131. Umezawa K, Matsui A, Nakamura Y, Citterio D, Suzuki K. Bright, color-tunable fluorescent dyes in the Vis/NIR region: establishment of new "tailor-made" multicolor fluorophores based on borondipyrromethene. *Chem Eur J* 2009;**15**:1096–106.
132. Jiao L, Yu C, Uppal T, Liu M, Li Y, Zhou Y, et al. Long wavelength red fluorescent dyes from 3,5-diiodo-BODIPYs. *Org Biomol Chem* 2010;**8**:2517–9.
133. He H, Ng DKP. A ratiometric near-infrared pH-responsive fluorescent dye based on distyryl BODIPY. *Org Biomol Chem* 2011;**9**:2610–3.
134. He H, Lo P-C, Yeung S-L, Fong W-P, Ng DKP. Preparation of unsymmetrical distyryl BODIPY derivatives and effects of the styryl substituents on their *in vitro* photodynamic properties. *Chem Commun* 2011;**47**:4748–50.
135. Filatov MA, Lebedev AY, Mukhin SN, Vinogradov SA, Cheprakov AV. π-Extended dipyrrins capable of highly fluorogenic complexation with metal ions. *J Am Chem Soc* 2010;**132**:9552–4.
136. Okujama T, Tomimori Y, Nakamura J, Yamada H, Uno H, Ono N. Synthesis of π-expanded BODIPYs and their fluorescent properties in the visible-near-infrared region. *Tetrahedron* 2010;**66**:6895–900.
137. Palma A, Tasior M, Frimansson DO, Vu TT, Méallet-Renault R, O'Shea DF. New on-bead near-infrared fluorophores and fluorescent sensor constructs. *Org Lett* 2009;**11**:3638–41.
138. Bellier Q, Dalier F, Jeanneau E, Maury O, Andraud C. Thiophene-substituted azabodipy as a strategic synthons for the design of near-infrared dyes. *New J Chem* 2012;**36**:768–73.
139. Lu H, Shimizu S, Mack J, Shen Z, Kobayashi N. Synthesis and spectroscopic properties of fused-ring-expanded aza-boradiazaindacenes. *Chem Asian J* 2011;**6**:1026–37.

140. Donyagina VF, Shimizu S, Kobayashi N, Lukyanets EA. Synthesis of N, N-difluoroboryl complexes of 3,3'-diarylazadiisoindolylmethenes. *Tetrahedron Lett* 2008;**49**:6152–4.

141. Tasior M, Murtagh J, Frimannsson DO, McDonnell SO, O'Shea DF. Water-solubilised $BF_2$-chelated tetraarylazadipyrromethenes. *Org Biomol Chem* 2010;**8**:522–5.

142. Li L, Han J, Nguyen B, Burgess K. Syntheses and spectral properties of functionalized, water-soluble BODIPY derivatives. *J Org Chem* 2008;**73**:1963–70.

143. Dodani SC, He Q, Chang CJ. A turn-on fluorescent sensor for detecting nickel in living cells. *J Am Chem Soc* 2009;**131**:18020–1.

144. Dilek Ö, Bane SL. Synthesis, spectroscopic properties and protein labeling of water soluble 3,5-disubstituted boron dipyrromethenes. *Bioorg Med Chem Lett* 2009;**19**:6911–3.

145. Bura T, Ziessel R. Water-soluble phosphonate-substituted BODIPY derivatives with tunable emission channels. *Org Lett* 2011;**13**:3072–5.

146. Jiao L, Li J, Zhang S, Wei C, Hao E, Vicente GH. A selective fluorescent sensor for imaging $Cu^{2+}$ in living cells. *New J Chem* 2009;**33**:1888–93.

147. Zhu S, Zhang J, Vegesna G, Luo F-T, Green SA, Liu H. Highly water-soluble neutral BODIPY dyes with controllable fluorescence quantum yields. *Org Lett* 2011;**13**:438–41.

148. Atilgan S, Ekmekci Z, Dogan AL, Guc D, Akkaya EU. Water soluble distyryl-boradiazaindacenes as efficient photosensitizers for photodynamic therapy. *Chem Commun* 2006;4398–400.

149. Atilgan S, Ozdemir T, Akkaya EU. A sensitive and selective ratiometric near IR fluorescent probe for zinc ions based on the distyryl-Bodipy fluorophore. *Org Lett* 2008;**10**:4065–7.

150. He H, Lo P-C, Yeung S-L, Fong W-P, Ng DKP. Synthesis and in vitro photodynamic activities of pegylated distyryl boron dipyrromethene derivatives. *J Med Chem* 2011;**54**:3097–102.

151. Niu SL, Ulrich G, Ziessel R, Kiss A, Renard P-Y, Romieu A. Water-soluble BODIPY derivatives. *Org Lett* 2009;**11**:2049–52.

152. Brellier M, Duportail G, Baati R. Convenient synthesis of water-soluble nitrilotriacetic acid (NTA) BODIPY dyes. *Tetrahedron Lett* 2010;**51**:1269–72.

153. Giessler K, Griesser H, Göhringer D, Sabirov T, Richert C. Synthesis of 3'-BODIPY-labeled active esters of nucleotides and a chemical primer extension assay on beds. *Eur J Org Chem* 2010;**2010**:3611–20.

154. Murtagh J, Frimannsson DO, O'Shea DF. Azaide conjugatable and pH responsive near-infrared fluorescent imaging probes. *Org Lett* 2009;**11**:5386–5389s.

155. Tasior M, O'Shea DF. $BF_2$-chelated tetraarylazadipyrromethenes as NIR fluorochromes. *Bioconjugate Chem* 2010;**21**:1130–3.

156. Jiao G-S, Thoresen LH, Burgess K. Fluorescent, through-bond energy transfer cassettes for labeling multiple biological molecules in one experiment. *J Am Chem Soc* 2003;**125**:14668–9.

157. Jiao G-S, Thoresen LH, Kim TG, Haaland WC, Gao F, Topp MR, et al. Syntheses, photophysical properties and application of through-bond energy-transfer cassettes for biotechnology. *Chem Eur J* 2006;**12**:7816–26.

158. Wu L, Loudet A, Barhoumi R, Burghardt RC, Burgess K. Fluorescent cassettes for monitoring three-component interactions *in vitro* and in living cells. *J Am Chem Soc* 2009;**131**:9156–7.

159. Coskun A, Akkaya EU. Ion sensing coupled to resonance energy transfer: a highly selective and sensitive ratiometric fluorescent chemosensor for Ag(I) by a modular approach. *J Am Chem Soc* 2005;**127**:10464–5.

160. Atilgan S, Ozdemir T, Akkaya EU. Selective Hg(II) sensing with improved Stokes shift by coupling the internal charge transfer process to excitation energy transfer. *Org Lett* 2010;**12**:4792–5.

161. Yuan M, Yin X, Zheng H, Ouyang C, Zuo Z, Liu H, et al. Light harvesting and efficient energy transfer in dendritic systems: new strategy for functionalized, near-infrared $BF_2$-azadipyrromethenes. *Chem Asian J* 2009;**4**:707–13.

162. Han J, Gonzales O, Aguilar-aguilar A, Peña-Cabreca E, Burgess K. 3- and 5-functionalized BODIPYs *via* Liebeskind-Srogl reaction. *Org Biomol Chem* 2009;**7**:34–6.

163. Fischer GM, Daltrozzo E, Zumbusch A. Selective NIR chromophores: bis(pyrrolopyrrole) cyanines. *Angew Chem Int Ed* 2011;**50**:1406–9.

164. Fischer GM, Isomäki-Krondahl M, Göttker-Schnetmann I, Daltrozzo E, Zumbusch A. Pyrrolopyrrole cyanine dyes: a new class of near-infrared dyes and fluorophores. *Chem Eur J* 2009;**15**:4857–64.

165. Fischer GM, Ehlers AP, Zumbusch A, Daltrozzo E. Near-onfrared dyes and fluorophores based on diketopyrrolopyrroles. *Angew Chem Int Ed* 2007;**46**:3750–3.

166. Berezin MY, Akers WJ, Guo K, Fischer GM, Daltrozzo E, Zumbusch A, et al. Long fluorescence lifetime molecular probes based on near infrared pyrrolopyrrole cyanine fluorophors for in vivo imaging. *Biophys J* 2009;**97**:L22–4.

167. Detty MR, Gibson SL, Wagner SJ. Current clinical and preclinical photosensitizers for use in photodynamic therapy. *J Med Chem* 2004;**47**:3897–915.

168. Ethirajan M, Chen Y, Joshi P, Pandey RK. The role of porphyrin chemistry in tumor imaging and photodynamic therapy. *Chem Soc Rev* 2011;**40**:340–62.

169. Grin MA, Mironov AF, Shtil AA. Bacteriochlorophyll *a* and its derivatives: chemistry and perspectives for cancer therapy. *Anticancer Agents Med Chem* 2008;**8**:683–97.

170. Gouterman M. Optical spectra and electronic structure of porphyrins and related rings. In: Dolphin D, editor. *The porphyrins* Vol. III. New York, San Francisco, London: Academic Press; 1978. p. 1–165.

171. Finikova OS, Cheprakov AV, Vinogradov SA. Synthesis and luminescence of soluble *meso*-unsubstituted tetrabenzo- and tetranaphtoh[2,3]porphyrins. *J Org Chem* 2005;**70**:95862–72.

172. Yamada H, Kuzuhara D, Takahashi T, Shimizu Y, Uota K, Okujima T, et al. Synthesis and characterization of tetraanthroporphyrins. *Org Lett* 2008;**10**:2947–50.

173. Rietveld IB, Kim E, Vinogradov SA. Dendrimers with tetrabenzoporphyrin cores: near infrared phosphors for in vivo oxygen imaging. *Tetrahedron* 2003;**59**:3821–31.

174. Esipova TV, Karagodov A, Miller J, Wilson DF, Busch TM, Vinogradov SA. Two new "protected" oxyphors for biological oximetry: properties and applications in tumor imaging. *Anal Chem* 2011;**83**:8756–65.

175. Lin VS-Y, DiMagno SG, Therien MJ. Highly conjugated, acetylenyl bridged porphyrins: new models for light-harvesting antenna systems. *Science* 1994;**264**:1105–11.

176. Balaz M, Collins HA, Dahlstedt E, Anderson HL. Synthesis of hydrophilic conjugated porphyrin dimers for one-photon and two-photon photodynamic therapy at NIR wavelengths. *Org Biomol Chem* 2009;**7**:874–88.

177. Kuimova MK, Collins HA, Balaz M, Dahlstedt E, Levitt JA, Sergent N, et al. Photophysical properties and intracellular imaging of water-soluble porphyrin dimers, for two-photon excited photodynamic therapy. *Org Biomol Chem* 2009;**7**:889–96.

178. Duncan TV, Susumu K, Sinks LE, Therien MJ. Exceptional near-infrared fluorescence quantum yields and excited-state absorptivity of highly conjugated porphyrin arrays. *J Am Chem Soc* 2006;**128**:9000–1.

179. Ghoroghchian PP, Frail PR, Susumu K, Park T-H, Wu SP, Uyeda HT, et al. Broad spectral domain fluorescence wavelength modulation of visible and near-infrared emissive polymersome. *J Am Chem Soc* 2005;**127**:15388–90.

180. Wu SP, Lee I, Ghoroghchian P, Frail PR, Zheng G, Glickson JD, et al. New-infrared optical imaging of B16 melanoma cells via low-density lipoprotein-mediated uptake and delivery of high emission dipole strength tris[(porphinato)zinc(II)] fluorophores. *Bioconjugate Chem* 2005;**16**:542–50.

181. Christian NA, Milone MC, Ranka SS, Li G, Frail PR, Davis KP, et al. Tat-functionalized near-infrared emissive polymersomes for dendritic cell labeling. *Bioconjugate Chem* 2007;**18**:31–40.

182. Ghoroghchian PP, Frail PR, Susumu K, Blessington D, Brannan AK, Bates FS, et al. Near-infrared-emissive polymersomes: self-assembled soft matter for *in vivo* optical imaging. *Proc Natl Acad Sci USA* 2005;**102**:2922–7.

183. Kee HL, Kirmaier C, Tang Q, Diers JR, Muthiah C, Taniguchi M, et al. Effects of substituents on synthetic analogs of chlorophylls. Part 1: synthesis, vibrational properties and excited-state decay characteristics. *Photochem Photobiol* 2007;**83**:1110–24.

184. Yang E, Kirmaier C, Krayer M, Taniguchi M, Kim H-J, Diers JR, et al. Photophysical properties and electronic structure of stable, tunable synthetic bacteriochlorins: extending the feature of native photosynthetic pigments. *J Phys Chem B* 2011;**115**:10801–16.

185. Kee HL, Kirmaier C, Tang Q, Diers JR, Muthiah C, Taniguchi M, et al. Effects of substituents on synthetic analogs of chlorophylls. Part 2: redox properties, optical spectra and electronic structure. *Photochem Photobiol* 2007;**83**:1125–43.

186. Singh S, Aggarwal A, Thompson S, Tomé JPC, Zhu X, Samaroo D, et al. Synthesis and photophysical properties of thioglycosylated chlorins, isobacteriochlorins, and bacteriochlorins for bioimaging and diagnostics. *Bioconjugate Chem* 2010;**21**:2136–46.

187. Kobayashi M, Akiyama M, Kano H, Kise H. Grimm B, Porra RR, Rüdiger W, Scheer H, editors. Spectroscopy and structure determination. *Chlorophylls and bacteriochlorophylls biochemistry, biophysics, function and applications*. Dordrecht: Springer; 2006. p. 79–94.

188. Cao W, Ng KK, Corbin I, Zhang Z, Ding L, Chen J, et al. Synthesis and evaluation of stable bacteriochlorophyll-analog and its incorporation into high-density lipoprotein nanoparticles for tumor imaging. *Bioconjugate Chem* 2009;**20**:2023–31.

189. Popov AV, Mawn TM, Kim S, Zheng G, Delikatny EJ. Design and synthesis of phospholipase C and A$_2$-activatable near-infrared fluorescent smart probes. *Bioconjugate Chem* 2010;**21**:1724–7.

190. Mawn TM, Popov AV, Beardsley, NJ, Stefflova, K, Milkevitch, M, Zheng, G, Delikatny, EJ. *In vivo* detection of phospholipase C by enzyme-activated near-infrared probes. *Bioconjugate Chem* 2011;**22**:2434–43.

191. Liu TW, Akens MK, Chen J, Wise-Milestone L, Wilson BC, Zheng G. Imaging of specific activation of photodynamic molecular beacons in breast cancer vertebral metastases. *Bioconjugate Chem* 2011;**22**:1021–30.

192. Silva AMG, Cavaleiro JAS. Porphyrins in Diels-Alder and 1,3-dipolar cycloaddition reaction. *Prog Heterocycl Chem* 2008;**19**:44–69.

193. Laha JK, Muthiah C, Taniguchi M, McDowell B, Ptaszek M, Lindsey JS. Synthetic chlorins bearing auxochromes at the 3- and 13-positions. *J Org Chem* 2006;**71**:4092–102.

194. Kim H-J, Lindsey JS. De novo synthesis of stable tetrahydroporphyrinic macrocycles: bacteriochlorins and a tetradehydrocorrin. *J Org Chem* 2005;**70**:5475–86.

195. Krayer M, Ptaszek M, Kim H-J, Meneely KR, Fan D, Secor K, et al. Expanded scope of synthetic bacteriochlorins via improved acid catalysis conditions and diverse dihydrodipyrrin-acetals. *J Org Chem* 2010;**75**:1016–39.

196. O'Neal WG, Jacobi PA. Toward a general synthesis of chlorins. *J Am Chem Soc* 2008;**130**:1102–8.

197. Minehan TG, Kishi Y. Extension of the Eschenmoser sulfide contraction/iminoester cyclization method to the synthesis of tolyporphine chromophore. *Tetrahedron Lett* 1997;**38**:6811–4.

198. Taniguchi M, Cramer DL, Bhise AD, Kee HL, Bocian DF, Holten D, et al. Accessing the near-infrared spectral region with stable, synthetic, wavelength-tunable bacteriochlorins. *New J Chem* 2008;**32**:947–58.

199. Ptaszek M, Lahaye D, Krayer M, Muthiah C, Lindsey JS. *De novo* synthesis of long-wavelength absorbing chlorin-13,15-dicarboximides. *J Org Chem* 2010;**75**:1659–73.

200. Mass O, Taniguchi M, Ptaszek M, Springer JW, Faries KM, Diers JR, et al. Structural characterization that make chlorophyll green: interplay of hydrocarbon skeleton and substituents. *New J Chem* 2010;**34**:1–13.

201. Muthiah C, Kee HL, Diers JR, Fan D, Ptaszek M, Bocian DF, et al. Synthesis and excited-state photodynamics of a chlorin-bacteriochlorin dyad: through-space versus through-bond energy transfer in tetrapyrrole arrays. *Photochem Photobiol* 2008;**84**:786–801.

202. Kee HL, Nothdurft R, Muthiah C, Diers JR, Fan D, Ptaszek M, et al. Examination of chlorin–bacteriochlorin energy-transfer dyads as prototypes for near-infrared molecular imaging probes. *Photochem Photobiol* 2008;**84**:1061–72.

203. Hambright P. Chemistry of water soluble porphyrins. In: Kadish KM, Smith KM, Guilard R, editors. *The porphyrin handbook*, vol. 3. San Diego: Academic Press; 2000. p. 129–210.

204. Borbas KE, Chandrashaker V, Muthiah C, Kee HL, Holten D, Lindsey JS. Design, synthesis, and photophysical characterization of water-soluble chlorins. *J Org Chem* 2008;**73**:3145–58.

205. Briñas RP, Troxler T, Hochstrasser RM, Vinogradov SA. Phosphorescent oxygen sensor with dendritic protection and two-photon absorbing antenna. *J Am Chem Soc* 2005;**127**:11851–62.

206. Finikova O, Galkin A, Rozhkov V, Cordero M, Hägerhäll C, Vinogradov S. Porphyrin and tetrabenzoporphyrin dendrimers: tunable membrane-impermeable fluorescent pH nanosensors. *J Am Chem Soc* 2003;**125**:4882–93.

207. Duncan TV, Ghoroghchian PP, Rubtsov IV, Hammer DA, Therien MJ. Ultrafast excited-state dynamics of nanoscale near-infrared emissive polymersomes. *J Am Chem Soc* 2008;**130**:9733–84.

208. Ceroni P, Lebedev AY, Marchi E, Yuan M, Esipova TV, Bergamini G, et al. Evaluation of phototoxicity of dendritic porphyrin-based phosphorescent oxygen probes: an in vitro study. *Photochem Photobiol* 2011;**10**:1056–65.

209. Tatman D, Liddell PA, Moore TA, Gust D, Moore AL. Carotenohematoporphyrins as tumor-imaging dyes. Synthesis and in vitro photophysical characterization. *Photochem Photobiol* 1998;**68**:459–66.

210. Chen J, Stefflova K, Niedre M, Wilson BC, Chance B, Glickson JD, et al. Protease-triggered photosensitizing beacon based on singlet oxygen quenching and activation. *J Am Chem Soc* 2004;**126**:11450–1.

211. Nesterova IV, Verdree VT, Pakhomov S, Strickler KL, Allen MW, Hammer RP, et al. Metallo-phthalocyanine near-IR fluorophores: oligonucleotide conjugates and their applications in PCR assays. *Bioconjugate Chem* 2007;**18**:2159–68.

212. Nesterova IV, Erdem S, Pakhomov S, Hammer RP, Soper SA. Phthalocyanine dimerization-based molecular beacon using near-IR fluorescence. *J Am Chem Soc* 2009;**131**:2432–3.

213. Peng X, Draney DR, Volcheck WM, Bashford GR, Lamb DT, Grone DL, et al. Phthalocyanine dye as an extremely photostable and highly fluorescent near-infrared labeling reagent. *SPIE-Int Soc Opt Eng* 2006;**6097E**:1–12.

214. Li H, Jensen TJ, Fronczek FR, Vicente MGH. Synthesis and properties of a series of cationic water-soluble phthalocyanines. *J Med Chem* 2008;**51**:502–11.

215. Liu W, Jensen TJ, Fronczek FR, Hammer RP, Smith KM, Vicente MGH. Synthesis and cellular studies of nonaggregated water-soluble phthalocyanines. *J Med Chem* 2005;**48**:1003–41.

216. Bai M, Lo P-C, Ye J, Wu C, Fong W-P, Ng DKP. Facile synthesis of pegylated zinc(II) pthalocyanines via transestrification and their *in vitro* photodynamic activities. *Org Biomol Chem* 2011;**9**:7028–32.

217. Lo P-C, Wang S, Zeug A, Meyer M, Röder B, Ng DKP. Preparation and photophysical properties of halogenated silicon(IV) phthalocyanines substituted axially with poly(ethylene glycol) chains. *Tetrahedron Lett* 2007;**44**:1967–70.

218. Jiang X-J, Yeung S-L, Lo P-C, Fong W-P, Ng DKP. Phthalocyanine-polyamine conjugates as highly efficient photosensitizers for photodynamic therapy. *J Med Chem* 2011;**54**:320–30.

219. Brasseur N, Nguyen T-L, Langlois R, Ouellet R, Marengo S, Houde D, et al. Synthesis and photodynamic activities of silicon 2,3-naphthalocyanine derivatives. *J Med Chem* 1994;**37**:415–20.

220. Erdem SS, Nesterova IV, Soper SA, Hammer RP. Mono-amine functionalized phthalocyanines: microwave-assisted solid-phase synthesis and bioconjugation strategies. *J Org Chem* 2009;**74**:9280–6.

221. Erdem SS, Nesterova IV, Soper SA. Solid-phase synthesis of asymmetrically substituted "AB$_3$-type" phthalocyanines. *J Org Chem* 2008;**73**:5003–7.

222. Hammer RP, Owens CV, Hwang S-H, Sayes CM, Soper SA. Asymmetrical water-soluble phthalocyanine dyes for covalent labeling of oligonucleotides. *Bioconjugate Chem* 2002;**13**:1244–52.

223. Ju J, Ruan C, Fuller CW, Glazer AN, Mathies RA. Fluorescence energy transfer dye-labeled primers for DNA sequencing and snalysis. *Proc Natl Acad Sci USA* 1995;**92**:4347–51.

224. Ju J, Kheterpal I, Scherer JR, Ruan C, Fuller CW, Glazer AN, et al. Design and synthesis of fluorescence energy transfer dye-labeled primers and their application for DNA sequencing and analysis. *Anal Biochem* 1995;**231**:131–40.

225. Baumgarth N, Roederer M. A practical approach to multicolor flow cytometry for immunophenotyping. *J Immun Meth* 2000;**243**:77–97.

226. Kogure T, Karasawa S, Araki T, Saito K, Kinjo M, Miyawaki A. A fluorescent variant of a protein from the stony coral *Montipora* facilitates dual-color single-laser fluorescence cross-correlation spectroscopy. *Nat Biotechnol* 2006;**24**:577–81.

227. Kosaka N, Ogawa M, Longmire MR, Choyke PL, Kobayashi H. Multi-targeted multicolor *in vivo* optical imaging in a model of disseminated peritoneal ovarian cancer. *J Biomed Optics* 2009;**14**:014023-1–8.

228. Koyama Y, Barrett T, Hama Y, Ravizzini PL, Choyke PL, Kobayashi H. *In vivo* molecular imaging to diagnose and subtype tumors through receptor-targeted optically labeled monoclonal antibodies. *Neoplasia* 2007;**9**:1021–9.

229. Barrett T, Koyama Y, Hama Y, Ravizzini G, Shin IS, Jang B-S, et al. *In vivo* diagnosis of epidermal growth factor receptor expression using molecular imaging with a cocktail of optically labeled monoclonal antibodies. *Clin Cancer Res* 2007;**13**:6639–48.

230. Laughlin ST, Bertozzi CR. Imaging the glycome. *Proc Natl Acad Sci USA* 2009;**106**:12–7.

231. Kobayashi H, Ogawa M, Kosaka N, Choyke PL, Urano Y. Multicolor imaging of lymphatic function with two nanomaterials: quantum dot-labeled cancer cells and dendrimer-based optical agents. *Nanomedicine* 2009;**4**:411–9.

232. Hama Y, Koyama Y, Choyke PL, Kobayashi H. Two-color *in vivo* dynamic contrast-enhanced pharmacokinetic imaging. *J Biomed Opt* 2007;**12**:034016-1–7.

233. Hama Y, Koyama Y, Urano Y, Choyke PL, Kobayashi H. Simultaneous two-color spectral fluorescence lymphangiography with near infrared quantum dots to map two lymphatic flows from the breast and the upper extremity. *Breast Cancer Res Treat* 2007;**103**:23–8.

234. Kobayashi H, Hama Y, Koyama Y, Barrett T, Regino CAS, Urano Y, et al. Simultaneous multicolor imaging of five different lymphatic basins using quantum dots. *Nano Lett* 2007;**7**:1711–6.

235. Kobayashi H, Koyama Y, Barrett T, Hama Y, Regino CAS, Shin IS, et al. Multimodal nanoprobes for radionuclide and five-color near-infrared optical lymphatic imaging. *ACS Nano* 2007;**1**:258–64.

236. Kobayashi H, Kosaka N, Ogawa M, Morgan NY, Smith PD, Murray CB, et al. *In vivo* multiple color lymphatic imaging using upconverting nanocrystals. *J Mat Chem* 2009;**19**:6481–4.

237. Cheng L, Yang K, Shao M, Lee S-T, Liu Z. Multicolor in vivo imaging of upconversion nanoparticles with emission tuned by luminescence resonance energy transfer. *J Phys Chem C* 2011;**115**:2686–92.

238. Haase M, Schäfer H. Upconverting nanoparticles. *Angew Chem Int Ed* 2011;**50**:5808–29.

239. Ptaszek M, Kee HL, Muthiah C, Nothdurft R, Akers W, Achilefu C, et al. Near-infrared molecular imaging probes based on chlorin-bacteriochlorin dyads. *SPIE-Int Soc Opt Eng* 2010;**7576E**:1–9.

240. Teo YN, Wilson JN, Kool ET. Polyfluorophores on a DNA backbone: a multicolor set of labels excited at one wavelength. *J Am Chem Soc* 2009;**2009**(131):3923–33.

241. Guo J, Wang S, Dai N, Teo YN, Kool ET. Multispectral labeling of antibodies with polyfluorophores on a DNA backbone and application in cellular imaging. *Proc Natl Acad Sci USA* 2011;**108**:3493–8.

242. Badr CE, Tannous BA. Bioluminescence imaging: progress and applications. *Trends Biotech* 2011;**29**:624–33.

243. Prescher JA, Contag CH. Guided by light: visualizing biomolecular processes in living animals with bioluminescence. *Curr Opin Chem Biol* 2010;**14**:80–9.

244. Loening AM, Wu AM, Gambhir SS. Red-shifted *Renilla feniformis* luciferase variants for imaging in living subjects. *Nat Methods* 2007;**4**:641–3.

245. Branchini BR, Ablamsky DM, Rosenberg JC. Chemically modified luciferase is an efficient source of near-infrared light. *Bioconjugate Chem* 2010;**21**:2023–30.

246. De A, Loening AM, Gambhir SS. An improved bioluminescence resonance energy transfer strategy for imaging intracellular events in single cells and living subjects. *Cancer Res* 2007;**67**:7175–83.

247. Dragulescu-Andrasi A, Chan CT, De A, Massoud TF, Gambhir SS. Bioluminescence resonance energy transfer (BRET) imaging of protein-protein interactions within deep tissue of living objects. *Proc Natl Acad Sci USA* 2011;**108**:12060–5.

248. Wu C, Mino K, Akimoto H, Kawabata M, Nakamura K, Ozaki M, et al. In vivo far-red luminescence imaging of a biomarker based on BRET from *Cypridina* bioluminescence to an organic dye. *Proc Natl Acad Sci USA* 2009;**106**:15599–603.

249. So M-K, Xu C, Loening AM, Gambhir SS, Rao J. Self-illuminating quantum dot conjugates for *in vivo* imaging. *Nat Biotechnol* 2006;**24**:339–43.

250. So M-K, Loening AM, Gambhir SS, Rao J. Creating self-illuminating quantum dot conjugates. *Nat Protoc* 2006;**1**:1160–4.

251. Kosaka N, Mitsunaga M, Bhattacharyya S, Miller SC, Choyke PL, Kobayashi H. Self-illuminating in vivo lymphatic imaging using a bioluminescence resonance energy transfer quantum dot nano-particle. *Contrast Media Mol Imaging* 2011;**6**:55–9.

252. Roda, A. Ed. *Chemiluminescence and bioluminescence:* past, present and future. RSC Publishing, Cambridge, 2010.

253. Dongwon L, Khaja S, Velasquez-Castano JC, Dasari M, Sun C, Petros J, Taylor WR, Murthy N. *In vivo* imaging of hydrogen peroxide with chemiluminescent nanopraticles. *Nat Mat* 2007;**7**:765–9.
254. Baumes JM, Gassensmith JJ, Giblin J, Lee J-J, White AG, Culligan WJ, et al. Storable, thermally activated, near-infrared chemiluminescent dyes and dye-stained microparticles for optical imaging. *Nat Chem* 2010;**2**:1025–30.
255. le Masne de Chermont Q, Chanéac C, Seguin J, Pellé F, Maitrjean S, et al. Nanoprobes with near-infrared persistent luminescence for *in vivo* imaging. *Proc Natl Acad Sci USA* 2007;**104**:9266–71.

CHAPTER FOUR

# Fluorescence Technologies for Monitoring Interactions Between Biological Molecules *In Vitro*

## Sebastien Deshayes, Gilles Divita

Centre de Recherches de Biochimie Macromoléculaire, Department of Chemical Biology and Nanotechnology for Therapeutics, CRBM-CNRS, UMR-5237, UM1-UM2, University of Montpellier, 1919 Route de Mende, Montpellier, France

## Contents

## Abstract

Over the last two centuries, the discovery and understanding of the principle of fluorescence have provided new means of characterizing physical/biological/chemical processes in a noninvasive manner. Fluorescence spectroscopy has become one of the most powerful and widely applied methods in the life sciences, from fundamental research to clinical applications. *In vitro*, fluorescence approaches offer the potential to sense in real-time extra and intracellular molecular interactions and enzymatic reactions, which constitutes a major advantage over other approaches to the study of biomolecular interactions. This technology has been used for the characterization of protein/

*Progress in Molecular Biology and Translational Science*, Volume 113
ISSN 1877-1173
http://dx.doi.org/10.1016/B978-0-12-386932-6.00004-1

protein, protein/nucleic acid, protein/substrate, and biomembrane/biomolecule interactions, which play crucial roles in the regulation of cellular pathways. This chapter
reviews the different fluorescence strategies that have been developed for sensing
molecular interactions *in vitro* at both steady- and pre-steady-state levels.

## ABBREVIATIONS

**ANS** 1-anilinonaphthalene-8-sulfonate
**AP5A** $\alpha,\gamma$-di[adenosine-5′] pentaphosphate
**CPP** cell-penetrating peptide
**FRET** fluorescence resonance energy transfer or Förster resonance energy transfer
**GFP** green fluorescent protein
**GUV** giant unilamellar vesicles
**LUV** large unilamellar vesicles
**Mant** methylanthranylate
**NBD** nitrobenzofurazan
**RT** reverse transcriptase
**SUV** small unilamellar vesicles

# 1. INTRODUCTION TO FLUORESCENCE

Over more than a century, fluorescence methodologies have under-
gone tremendous development, from physical and biological to clinical
applications. Indeed, since the first observations of fluorescence reported
by Sir John Frederick William Herschel in 1845 from a quinine solution
under sunlight,[1] the phenomenon of fluorescence has been continuously investigated. In 1852, Sir George G. Stokes proposed the term "fluorescence"
in honor of the blue-white fluorescent mineral fluorite/fluorspar.[2] Nowadays, fluorescence spectroscopy has become one of the most widely applied
technologies in the study of biological processes.

## 1.1. Fluorescence principle

The simplest definition of fluorescence could be formulated as the emission
of light by a substance that has absorbed light or electromagnetic radiation.[3]
Today, fluorescence might be defined as a luminescent process in which sensitive molecules emit light from electronically excited states induced by a
physical (absorption of light), mechanical (friction), or chemical mechanism.
On the basis of Herschel's and Stokes' observations, Jablonski was the first to
give a clear and scientific explanation of this phenomenon.[4] The Jablonski
diagram indicates the relationship between the ground state ($S_0$), the excited

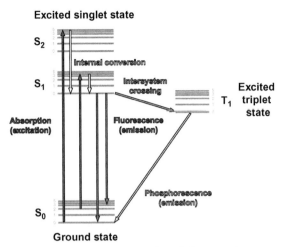

**Figure 4.1** Jablonski energy diagram and Stokes shift. (For color version of this figure, the reader is referred to the online version of this chapter.)

singlet states ($S_1$, $S_2$), and the excited triplet state ($T_1$), and the resulting fluorescence and phosphorescence emission (Fig. 4.1).

Most aromatic molecules with delocalized electrons are theoretically able to undergo luminescence and fluorescence phenomena. They can be associated to different sources in biological molecules, from natural intrinsic fluorescent probes (tryptophan or natural fluorescent protein; GFP, RFP, etc.) to small synthetic chemical dyes (Cyanine, Alexa, Atto, etc.). Fluorescence technology constitutes an ideal noninvasive approach to monitor and characterize in detail specific interactions between biological molecules. A large number of interactions can be investigated at both steady-state and kinetic levels using either intrinsic or extrinsic fluorescence probes. Depending on the type of interaction and the context, several fluorescence-based methods are available, including solvatochromism, anisotropy, and fluorescence resonance energy transfer (FRET) (Fig. 4.2). This chapter focuses on the different applications of fluorescence technology to monitor specific events in biology for both fundamental and mechanistic issues.

## 1.2. Solvatochromism and resonance energy transfer

### 1.2.1 Solvatochromism

Most fluorescent molecules can be considered environmentally sensitive probes since there are several environmental parameters that can affect their fluorescent properties. These environmental factors include the solvent,

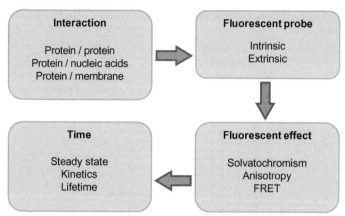

**Figure 4.2** Fluorescence technology to probe biological molecules. (For color version of this figure, the reader is referred to the online version of this chapter.)

inorganic and organic compounds, temperature, pH, and the concentration of the fluorescent molecules. The effects of these parameters vary widely from one fluorophore to another. In particular, the absorption and emission spectra, as well as the quantum yield, can be affected by the polarity of the environment. This phenomenon, more generally known as "solvatochromism," can be defined as the ability of a fluorescent molecule to undergo changes in position and/or intensity of absorption or emission bands according to solvent polarity variation.[5] The solvatochromic effect, hypsochromic (blue) or bathochromic (red) shifts, refers to a strong dependence of the absorption and emission spectra on solvent polarity (Fig. 4.3). Since the polarities of the ground and excited states of a fluorophore are different, a change in solvent polarity will induce different stabilization of the ground and excited states and thus a modification of the energy between these electronic states.[5] Consequently, variations in the position, intensity, and shape of the absorption and emission spectra can be used as a direct indicator of solvent changes and local environment modification.

In proteins, although the aromatic residues Trp, Phe, and Tyr can potentially be used as fluorescent probes, Trp is by far the most used intrinsic probe. Tryptophan is an important intrinsic solvatochromic probe that can be useful as a natural sensor of the conformational state of a protein or for assessment of the nature of its environment.[6–9] The indole group of the side chain confers maximum absorbance near 280 nm and maximum emission intensity located between 310 and 350 nm on the tryptophan residue depending on the environment. The use of denaturants, surfactants, or

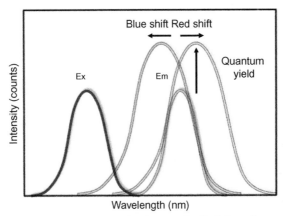

**Figure 4.3** Hypsochromic (blue) or bathochromic (red) shift with quantum yield increase. (For interpretation of the references to color in this figure legend, the reader is referred to the online version of this chapter.)

other amphiphilic molecules generally induces changes in the environment, providing information on the surrounding of tryptophan and thereby of proteins.[10] If a globular protein containing several tryptophans in its "hydrophobic" core is denatured, the environment of tryptophans is modified and a shift of the emission spectrum maximum to a longer wavelength (bathochromic/red shift) is observed. This effect arises from the exposure of tryptophan to an aqueous environment as opposed to a hydrophobic protein core. In contrast, the addition of a surfactant such as phospholipid vesicles to a protein that contains a tryptophan exposed to the aqueous solvent will induce a shift of the emission spectrum maximum to a shorter wavelength (hypsochromic/blue shift) if the tryptophan becomes embedded in the phospholipid vesicle. Thus the transfer of tryptophan residues from a polar to a less polar environment such as the membrane bilayer is usually associated with a blue shift. In addition, this blue shift may be accompanied by an increase in the quantum yield of tryptophan (Fig. 4.3).

An alternative approach to the intrinsic fluorescence of tryptophan is the covalent attachment of an extrinsic fluorophore to a single site on the target biomolecule. Indeed, there are a large number of fluorescent probes with photochemical properties that are more attractive than tryptophan and which are also sensitive to environmental changes.[5,11] By conjugating these to proteins, nucleic acids, or lipids, the sensitivity and quality of information provided by fluorescence spectrometry can be clearly improved. Synthesized for the first time in 1871 by Adolf von Baeyer, fluorescein was one of the first environment-sensitive probes, exhibiting pH–dependent excitation and emission wavelengths.[12,13] Although fluorescein and its chemical analogs are still used to label specific sites of target biomolecules, other probes have been developed with improved fluorescence stability, quantum yield, or solvatochromism. For example, Prodan, NBD (nitrobenzofurazan), Coumarin, Nile red, and their derivatives constitute well-known environment-sensitive dyes that are usually applied to sense changes in the environment proximal to the target.[11]

### 1.2.2 Resonance energy transfer

At the beginning of the twentieth century, Jean Perrin was the first scientist to consider the interaction through space between molecules.[14] He proposed that the excitation energy is transferred from one molecule to another through interactions between oscillating dipoles of closely spaced molecules. Based on this observation, Theodore Förster developed the theoretical basis of resonance energy transfer, that is, the "Förster resonance energy transfer" or FRET.[15,16] FRET generally occurs between a donor (D) molecule in the excited state and an acceptor (A) molecule in the ground state (Fig. 4.4A). The donor molecules typically emit at shorter wavelengths that overlap with the absorption spectrum of the acceptor (Fig. 4.4B). Energy transfer occurs without the appearance of a photon and is the result of long-range dipole–dipole interactions between the donor and the acceptor. When both molecules are fluorescent, the term "fluorescence resonance energy transfer" is used instead, although the energy is not transferred by fluorescence.[17] The rate of energy transfer depends on the spectral overlap of the emission spectrum of the donor with the excitation spectrum of the acceptor, the quantum yield of the donor, the relative orientation of the donor and acceptor transition dipoles, and the distance between the donor and acceptor molecules. The distance dependence of FRET allows for measurement of the distances between donors and acceptors.[17] Indeed, when the energy transfer efficiency ($E$) is measured, the distance ($r$) between the two fluorophores can be calculated, according the equation

$$E = R_0{}^6 / \left( R_0{}^6 + r^6 \right) \qquad [4.1]$$

where $R_0$ is the so-called Förster distance at which the efficiency of transfer equals 50%. This latter point is attractive with regard to the identification of molecular interactions *in vitro* or in a biological context. Thus the choice of a

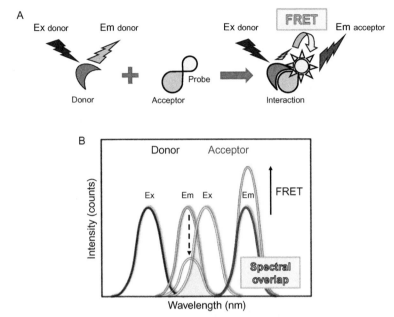

**Figure 4.4** Förster resonance energy transfer (FRET) principle. (A) When two molecules with appropriate donor and acceptor fluorophores interact, FRET occurs between the donor in the excited state and the acceptor in the ground state. (B) When the donor is excited at its absorption spectrum (blue), it typically emits at a shorter wavelength (green) that overlaps with the absorption spectrum of the donor (yellow). Then the resulting FRET enables the acceptor to emit fluorescence (red). (For interpretation of the references to color in this figure legend, the reader is referred to the online version of this chapter.)

specific acceptor/donor couple to associate with interacting partners can constitute a powerful tool to demonstrate both dynamic and *in situ* interactions as well as conformational changes.

### 1.2.3 Fluorescence polarization

Fluorescence Polarization was first described in 1926 by Perrin.[14] This specific type of polarization is based on the observation that fluorescent molecules in solution, excited with plane-polarized light, will emit light back in a fixed plane if the molecules remain stationary during the excitation of the fluorophore; the emitted light remains "polarized." However, molecules rotate and tumble, and the planes in which light is emitted can be very different from the plane used for initial excitation; the emitted light

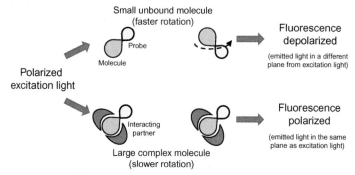

**Figure 4.5** Principle of fluorescence polarization. The interactions of a probe-associated molecule with interacting partners enable the formation of a large complex which induces polarization or increases the polarization value (fluorescence polarized), whereas no polarization variations occur for the unbound state of the probe-associated molecule (fluorescence depolarized). (For color version of this figure, the reader is referred to the online version of this chapter.)

is then "depolarized" (Fig. 4.5). With regard to intermolecular interactions, if a molecule is bound by a larger ligand, its effective molecular volume is increased and its rotation is slowed so that the emitted light is in the same plane as the plane-polarized light of excitation (Fig. 4.5). In this case, both the bound and unbound states of the molecule have an intrinsic polarization value: a high (polarized) value for the bound state and a low (depolarized) value for the unbound state.[17] Fluorescence polarization is a weighted average of the two values, providing a direct evaluation of the fraction of molecule/ligand binding. Thus fluorescence polarization measurements are also indicative of the formation of larger molecule/ligand complexes.

Fluorescence anisotropy is usually determined by the measurements of fluorescence emission in parallel and perpendicular planes. The degree of polarization ($p$) or anisotropy ($r$) is calculated according to the following equations:

$$p = \left(I_{\parallel} - I_{\top}\right)/\left(I_{\parallel} + I_{\top}\right) \text{ and } r = \left(I_{\parallel} - I_{\top}\right)/\left(I_{\parallel} + 2I_{\top}\right) \qquad [4.2]$$

where $I_{\parallel}$ is the fluorescence emission measured in the plane parallel to the plane of excitation and $I_{\top}$ is the fluorescence emission measured in the plane perpendicular to the plane of excitation. Anisotropy and fluorescence polarization approaches can be applied to study different biomolecular interactions.[18]

## 1.3. Fluorescence for biological molecules

Over the last two decades, fluorescence spectroscopy technologies have been extensively applied to understand specific molecular interactions in different biological and biochemical pathways, including monitoring major cellular events and enzyme mechanisms and sensing cellular response leading to aberrant behavior. Fluorescence approaches allow the characterization in a non-invasive manner, at the steady- and pre-steady-state levels of protein/protein, protein/nucleic acid, protein/substrate, and biomembrane/biomolecule interactions as well as enzymatic assays or competitive immunoassays, which constitutes a major advantage over other methods used for this purpose. Although fluorescence spectroscopy technologies provide valuable answers to numerous questions concerning the identity, location, conformation, and environment of any protein or nucleic acid, the sensitivity of the technologies also requires taking into account the limitations of the approach and keeping in mind that every small change in a parameter may induce variations.[19] Thus each specific approach has several advantages and limitations (Table 4.1).

**Table 4.1** Advantages and limitations of fluorescence techniques (adapted from Pope et al.)[19]

| Technique | Advantages | Disadvantages and limitations |
|---|---|---|
| Fluorescence intensity | – Simple<br>– Suitable for fluorigenic assays<br>– Readily miniaturized | – Little information for quality control<br>– Sensitive to inner-filter and autofluorescence interference |
| Fluorescence polarization/ anisotropy | – Simple and reasonably predictive<br>– Insensitive to inner-filter effects<br>– Ratiometric technique<br>– Improved well-level quality control<br>– Suitable for small ligands ($<15$ kDa) | – Local motion effects<br>– Suitability limited by lifetime of dye, ligand size, and molecular weight change<br>– Dynamic range limited<br>– Can suffer from autofluorescence |
| Fluorescence resonance energy transfer (FRET) | – Simple and reasonable predictable<br>– Suitable for short inter/ intramolecular distances ($<5$ nm)<br>– Range of available donors and acceptors | – Requires multiple lablels<br>– Sensitive to inner-filter and autofluorescence interference<br>– Limited to short distances to obtain high signal changes<br>– Most dyes monitor only donor quenching |

## 2. MONITORING PROTEIN/SUBSTRATE INTERACTIONS

Protein/partner interactions, such as enzyme/substrate, protein/ nucleic acid, and protein/nucleotide, play a crucial role in the regulation of biological pathways at the extracellular, cellular, viral, or drug delivery level. Development of fluorescence technologies using both natural and synthetic probes has provided new and appropriate tools to decipher the nature, strength, and impact on the life cycle of a wide range of molecular interactions. Protein/ligand complexes can be probed by intrinsic trypto-phan fluorescence at both steady-state and pre-steady-state levels. However, most proteins contain more than one tryptophan residue, which constitutes a limitation for the analysis of the specific interaction of one domain of the protein. Therefore, in most cases, extrinsic fluorescence constitutes an attractive alternative as a large panel of fluorescent dyes with well-suited spectral properties and more appropriate quantum yields for sensitive detec-tion and single-molecule assays.

### 2.1. Protein/nucleotide interactions

Nowadays, fluorescent- or caged-nucleotide analogs are widely used as probes for enzyme activities and as sensors for screening of inhibitors and/or activators.[20–22] Nucleotide chemistry has focused on the design of specific fluorescently labeled nucleotides to investigate either enzymatic parameters or interactions with partners (activator, inhibitor, etc.) of nucleotide binding proteins, such as nucleotide kinases, protein kinases, GTPases, ATPases, and so on.[23–26] Nucleotides can be modified on their base (benzo-, etheno-), on the phosphate (XTP-$\gamma$-naphthalene, $\gamma$-[(6-amino) hexyl]-, $\gamma$-(sulfo-1-naphthyl)amide, methylanthranylate (Mant)), and on the sugar (trinitrophenyl-, Mant-) moieties, depending on the nature/ structure of the nucleotide binding site of the enzyme (Fig. 4.6). New chemistry has also been proposed using the amino hexyl linker, allowing labeling with a large panel of available dyes with the amino group (Fig. 4.6). The binding of nucleotide analogs to proteins leads to large modifications of their extrinsic fluorescence, which can be used as a sensor for the determination of enzymatic parameters as well as for screening of inhibitors or natural substrates. However, it is important to keep in mind that the presence of the dye can modify the binding properties of the nucleotide, and displacement experiments of labeled nucleotides using unlabeled nucleotides are essential to validate the approach. For example,

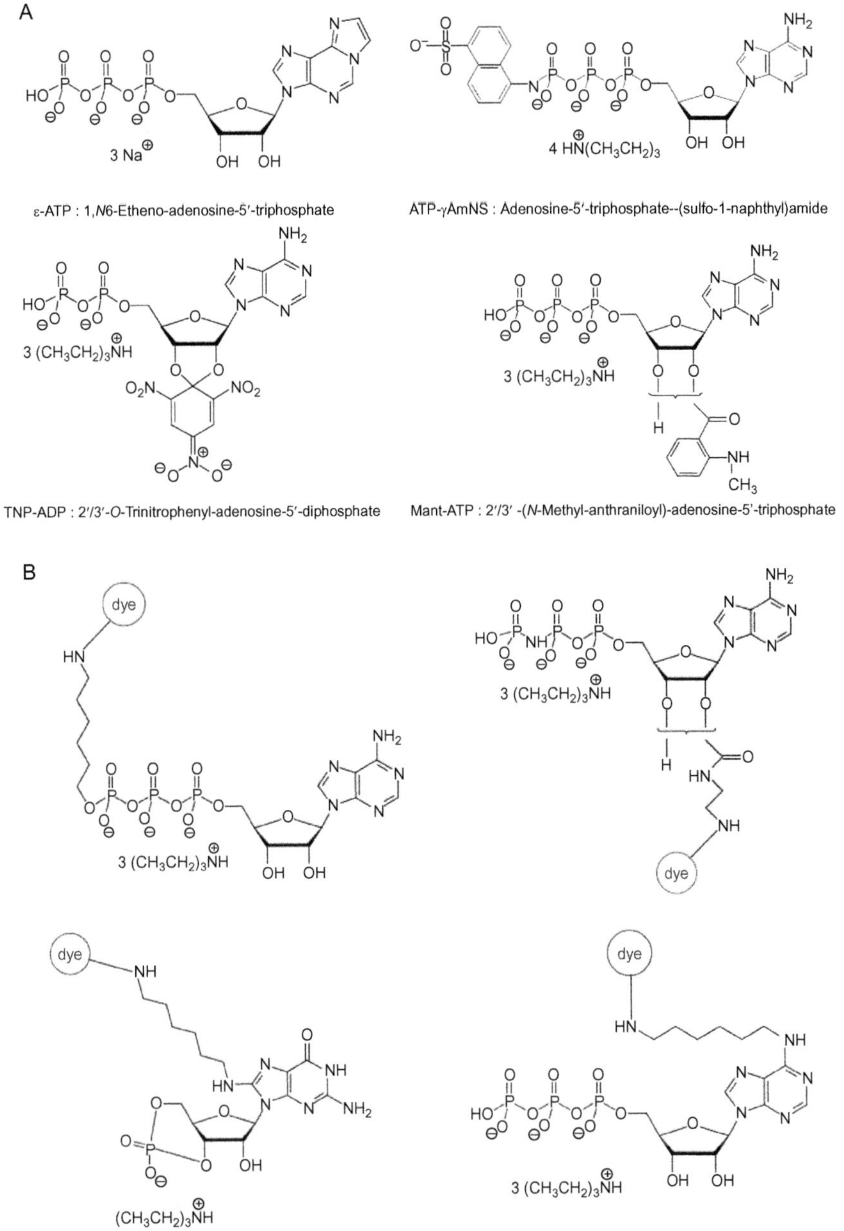

A

ε-ATP : 1,N6-Etheno-adenosine-5'-triphosphate

ATP-γAmNS : Adenosine-5'-triphosphate--(sulfo-1-naphthyl)amide

TNP-ADP : 2'/3'-O-Trinitrophenyl-adenosine-5'-diphosphate

Mant-ATP : 2'/3' -(N-Methyl-anthraniloyl)-adenosine-5'-triphosphate

B

**Figure 4.6** Fluorescently labeled nucleotide analog. (A) Most common dyes used on fluorescently labeled nucleotides on the ribose, the base, and the phosphate moieties. (B) Amino-hexyl-modified nucleotides for custom-made labeling with a large panel of dyes. (For color version of this figure, the reader is referred to the online version of this chapter.)

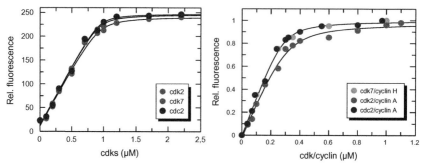

**Figure 4.7** Binding of Mant-ATP to cyclin-dependent kinase (Cdk) and Cdk/cyclin complex. Mant-ATP, the *N*-methyl-anthraniloyl moiety, is linked to 2′- and/or 3′-hydroxyl of the ribose. The fluorescence of the Mant-group is excited at 350 nm and emission monitored at 450 nm. The high sensitivity of the Mant-ATP was used to quantify the binding of nucleotide to different Cdks in the free form (A) or in complex with their Cyclin partner (B). Upon binding to Cdk and Cdk/cyclin complex, Mant fluorescence increased by 2.5- and 5-fold, respectively. In order to determine dissociation constant, the curves were fitted using quadratic equation (Eq. 4.3) (adapted from Morris *et al.*).[27] (For color version of this figure, the reader is referred to the online version of this chapter.)

the binding of Mant-ATP to the cell cycle protein kinase cyclin–dependent kinase (Cdk2) results in a 2.5-fold increase in fluorescence (Fig. 4.7) and allows for the determination of the dissociation constant at the steady-state level using either hyperbolic or quadratic equations, which take into account the concentration of the enzyme and probe:[27,28]

$$F = F_{ini} - (\Delta F)\left\{ (E_t + L + K_d) - \left[ (E_t + L + K_d)^2 - 4E_tL \right]^{1/2} \right\}/2E_t \quad [4.3]$$

where $F$ is the observed relative fluorescence intensity, $F_{ini}$ is the fluorescence intensity at the start of the titration, $\Delta F$ is the variation of the fluorescence intensity between the initial value and at a saturating concentration of substrate, $E_t$ is the total concentration of enzyme, $L$ is the total concentration of ligand, and $K_d$ is the dissociation constant of the enzyme–ligand complex. Bisubstrate nucleotides (Ap5A and Ap4A) have been designed to characterize enzymes harboring multiple nucleotide binding sites. An interesting system was described to measure the affinities of nucleotide kinases (NDP-, ADP-, AMP-, or TMP-kinases) for their substrates and inhibitors, based on a fluorescent analog of the bisubstrate inhibitor diadenosine pentaphosphate (AP5A): α,γ-di[(3′- or 2′-)-O-(*N*-methylanthraniloyl) adenosine-5′] pentaphosphate (mAP5Am).[29–31]

## 2.2. Protein/nucleic acid interactions

Several technologies have been developed to evaluate protein/nucleic acid interactions.[32] As for nucleotides, new chemistry has been proposed for labeling of nucleic acids, and a large panel of dyes have been attached to DNA or RNA molecules (Fluorescein, Cyanine, Alexa, ATTO, etc.). One of the major breakthroughs in the design of tools for measuring protein/nucleic acid interactions is the development of chemistry for accurate high-throughput DNA sequencing by a synthetic approach. DNA sequencing by synthesis is based on the extension of a primer hybridized to its target sequence by DNA polymerase with a reversible fluorescent chain terminator $2'$-deoxynucleotide.[33,34] Fluorescently labeled reversible chain terminator nucleotides are stable during the polymerase-mediated extension step, and their structure (geometry and size) and the location of the dye within the $2'$-deoxynucleotide moiety do not prevent their recognition by standard DNA polymerases.[35] Oligonucleotide can be easily labeled at the $5'$-position through the introduction of a primary amine ($NH_2$) at the $5'$-position to functionalize the corresponding terminus of the nucleotide for conjugation with an activated N-Hydroxysuccinimide (NHS) ester or isothiocyanate fluorescent label. Several hydrophobic spacers have been proposed of 3, 6, or 12 methylene ($CH_2$) groups between the terminal phosphate and the amino part. Amino-modified nucleotides have been used to incorporate a dye in post-coupling reactions on the base, with the main advantage of providing labeling anywhere in the oligonucleotide sequence without altering the $5'$-position for elongation. Finally, modifications on the deoxyribose have also been proposed to add more than one dye anywhere in the sequence or on either terminus after post-coupling reactions between the amine group and an activated label. In most cases, the dye is linked to the deoxyribose via a six-carbon-atom spacer, which reduces steric hindrance (Fig. 4.8). As most of the dyes and probes are largely hydrophobic, the risk of nonspecific association with the protein is present and it is essential to confirm interactions with displacement experiments using unlabeled oligonucleotides and other approaches to confirm the affinity values. As for fluorescently labeled nucleotides, it remains essential to take into account the impact of the probe on the interaction and binding parameters.

Although RNA and DNA polymerases are known to share the same general catalytic mechanism, they are all unique in their structural dynamics and constitute a major challenge for the enzymologist.[36] RNA and DNA

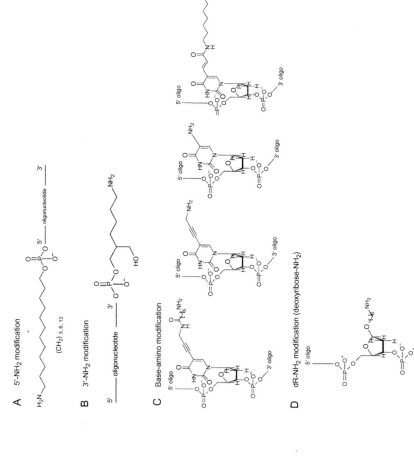

**Figure 4.8** Fluorescently labeled oligonucleotide. Oligonucleotide labeling can be performed by adding an amine group ($NH_2$) for post-coupling reactions with an activated dye. Labeling is performed at the 5′ (A) and 3′ (B) ends by the introduction of primary anime and variable spacers on the base with different chemistry and linkers (C) and on the deoxyribose, where the dye is attached to the deoxyribose via a six-carbon-atom spacer (D).

polymerases are very dynamic enzymes that need to control simultaneously parameters such as specificity, fidelity, and efficacy together with interactions with different types of substrates and partners. Fluorescently labeled oligonucleotides were used to investigate the mechanism of several DNA or RNA polymerases as well as to monitor the binding of regulatory proteins.[37] RNA or DNA polymerase activity has been probed by measuring changes in extrinsic fluorescence at the steady-state and kinetic levels as well as by fluorescence anisotropy as the size of the binder is significantly smaller than that of the protein.[18]

The reverse transcriptase (RT) of HIV (human immunodeficiency virus) is a key enzyme in the virus replication cycle and catalyzes a chain of reactions to convert the single-stranded HIV RNA genome into a double-stranded DNA for further integration in the cell host genome. This requires that RT is able to discriminate between different nucleic acids and to place them correctly in one of the three catalytic sites for RNA-dependent DNA synthesis, DNA-dependent DNA synthesis, and RNAse-H. RT structure and mechanism have been investigated in detail by combining steady-state transient kinetic, and FRET single-molecule assay using dyes attached to the enzyme or to its different partners: tRNA, viral-RNA, and DNA primer/template.[38–41] These investigations have shown that catalysis of RT is dependent on the binding orientation of the substrate, which adopts opposite conformations for DNA and RNA duplexes (Fig. 4.9). The large change in dye fluorescence upon binding of primer/template oligonucleotide to RT was used to follow this interaction at both the steady-state and pre-steady-state levels (Fig. 4.9). Stopped-flow technology has been developed for real-time fluorescence kinetic analysis.[42–44] A stopped-flow apparatus is a rapid mixing device used to study the chemical kinetics of a reaction in solution, and stopped-flow measurements involve the rapid mixing of two or more solutions (Fig. 4.9). The dead time, corresponding to the time between the end of mixing and the beginning of the observed kinetics of the reaction, ranges between 0.5 and 2 ms. In most cases, protein/partner interactions cannot be analyzed as a simple first-order or second-order reaction, as they often combine binding and conformational events that can be discriminated depending on the rate of each step. In the case of HIV RT, experimental use of fluorescently labeled double-stranded oligonucleotide for rapid kinetics studies has allowed for a detailed characterization of the mechanism and structure of the enzyme during initiation of replication. Kinetic analysis of the Fluorescein phosphoramidite (FAM) – labeled

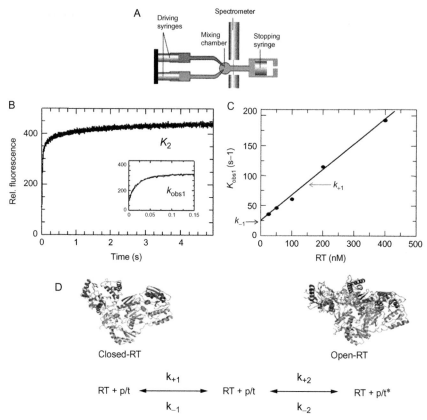

**Figure 4.9** Stopped-flow analysis of reverse transcriptase (RT)/primer:template binding. (A) Scheme or a stopped-flow apparatus with two driving syringes and a single mixing chamber. (B, C) Kinetics of binding of a fluorescently labeled p/t to RTs. Typical stopped-flow time course is shown, where 20 nM of 3'-FAM-labeled p/t was rapidly mixed with 100 nM of RT. Excitation was at 492 nm and emission was detected through a 530-nm cutoff filter. The kinetics were fitted using a two-exponential equation. (C) Secondary plot of the dependence of the fitted pseudo first-order rate constants for the first phase on RT concentrations. The on rate constant of the first phase ($k_{+1}$), corresponding to the formation of an RT–p/t collision complex, is dependent on the concentration of RT and was extrapolated from the slope of the secondary plot of $k_{obs1}$ versus RT concentration. The off rate $k_{-1}$ was estimated on the same plot from the intercept on the $k_{obs}$ axis. The kinetics of p/t binding onto RT followed a two-step mechanism (D) including a rapid diffusion-controlled second-order step leading to the formation of the RT–p/t collision complex, followed by one slow, concentration-independent conformational change from the closed to the open conformation of the polymerase (adapted from Divita et al.[38] and Agopian et al.).[41] (For color version of this figure, the reader is referred to the online version of this chapter.)

primer/template (p/t) binding to RT reveals two-exponential kinetics, corresponding to a two-step mechanism (Fig. 4.9). The first phase corresponds to the formation of a RT–p/t collisional complex inducing a conformational transition of RT from the closed- to the open-polymerase typical right-hand organization. The on rate $(k_{+1})$ of the first phase is dependent on the concentration of RT and is extrapolated from the slope of the secondary plot of $k_{obs1}$ versus enzyme concentration. The off rate $(k_{-1})$ can be estimated from the same plot from the intercept on the $k_{obs}$ axis. This first step is followed by conformational changes of the preformed RT–p/t complex, which correctly places the nucleic acid in the appropriate binding site for catalysis. The $k_2$ $(k_{+2}+k_{-2})$ rate constant of the second phase is independent of the enzyme concentration and corresponds to the second exponential term of the kinetics.[40,41,45]

By combining FRET and single-molecule fluorescence, Patel and colleagues have elucidated the initiation of transcription catalyzed by RNA polymerase[46,47] using two double-stranded DNA strands corresponding to the upstream promoters and downstream template DNA, labeled at different positions with two different dyes. The donor dye is associated to the upstream promoter and the acceptor to the template strand. Essential changes in DNA and polymerase conformation associated with initiation and abortive RNA synthesis have been identified, and a mechanism involving DNA scrunching following by its rotation has been proposed for early initiation catalyzed by T7 RNA polymerase. Similarly, combining single- and double-stranded labeled DNA offers a powerful sensing system to monitor the open/closed conformational transition of the RNA or DNA polymerase during initiation as well as nucleotide incorporation in real time and the walking of the polymerase along the DNA or RNA during elongation, as reported for mitochondria RNA polymerase[48] and for HVC polymerase (NS5B).[49]

## 2.3. Peptide/nucleic acids

During the last two decades, understanding peptide/nucleic acids interactions has become a major challenge in the field of drug delivery. Small peptides called cell-penetrating peptides (CPPs) have been developed to improve the cellular internalization of a wide range of nucleic acids (from plasmid DNA, siRNA, and single-stranded antisense molecule).[50] The main development of CPPs has focused on noncovalent delivery involving formation of mixed complexes between peptides and nucleic acids,[51,52] which improve the cellular internalization of oligonucleotides.[50,53] In this case, the ability of peptides to form stable interactions with nucleic acids constitutes a key parameter for the selection of potent carriers, and both extrinsic and intrinsic fluorescence have been used for screening peptides and for a better understanding of peptide/nucleic acid complex formation.

Analysis of peptide/nucleic acid interactions can be carried out by monitoring intrinsic peptide fluorescence as well as specific labeling of the cargo. It has been demonstrated for several CPPs that the fluorescence emission maximum of their tryptophan residues is affected by the presence of increasing amounts of nucleic acids and that the fluorescence emission maximum of a labeled oligonucleotide varies with increasing concentrations of the carrier peptide. Thus by monitoring both the intrinsic tryptophan fluorescence of peptides and the extrinsic fluorescence of a labeled cargo, it is possible to investigate the stability of CPP/nucleic acid interactions and thereby characterize complex formation.[54–57] The intrinsic fluorescence of tryptophans within peptides usually exhibits a strong quenching in the presence of nucleic acids.[54,55,58,59] This quenching might be attributed to both direct peptide/nucleic acid interactions and peptide/peptide interactions that occur when forming complexes with nucleic acids. Although positively charged residues of peptide are able to carry out electrostatic interactions with negative charges of the phosphate groups of nucleic acids, changes in tryptophan fluorescence intensity also suggest aromatic stacking effects. It has been demonstrated that short peptides or protein domains are able to undergo tryptophan fluorescence quenching when interacting with single-stranded nucleic acids as well as DNA duplex and triplex.[60–63] For example, strong binding of the KWGK peptide to a 21-mer duplex involves intercalation and stacking interactions of the tryptophan with GC regions of the oligonucleotide.[62] The quenching of fluorescence of the tryptophan has been ascribed to an electron transfer from indole of the tryptophan side chain (in the excited state) to purine and pyrimidine bases.[62]

With regard to extrinsic fluorescence of nucleic acids, the use of different fluorescent probes can show distinct behaviors with peptides. Although most of the peptide/nucleic acid interactions induce fluorescence

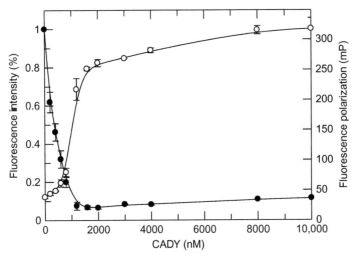

**Figure 4.10** Fluorescence analysis of peptide/nucleic acid interactions through the formation of CADY/siRNA complexes. While a strong quenching of the fluorescence intensity of a FITC-labeled siRNA is induced by the presence of CADY peptide (solid label), a net increase in the fluorescence polarization is also detected (opened label), suggesting both interactions and formation of a large complex between peptides and the siRNA (adapted from Deshayes *et al.*).[57]

quenching, enhancement of the quantum yield has also been observed.[54,58] However, one has to keep in mind that the hydrophobic part of the fluorescent probe confers a hydrophobic anchor for the peptide/cargo interactions: first through $\Pi$-stacking on the probe, then through electrostatic interactions between charged residues and the phosphates. Nevertheless, the extrinsic approach provides alternative information through fluorescence anisotropy/polarization measurements. Indeed, the probe possesses a degree of freedom that may vary in the presence of peptides. The variation of the steric environment of the probe induces a variation in its degree of freedom, resulting in a modification of fluorescence polarization. As shown in Fig. 4.10, in the case of CADY/ siRNA complexes, a net fluorescence quenching of an FITC (fluorescein isothiocyanate) labeled siRNA was also associated with a clear increase in fluorescence polarization.[55–57] These data tend to demonstrate that a strong reduction in the degree of freedom of the FITC conjugated to the siRNA occurs in parallel to fluorescence quenching. Thus both fluorescence intensity and polarization variations support the formation of mixed peptide/siRNA complexes.

## 3. MONITORING PROTEIN/PROTEIN AND POLYPEPTIDE INTERACTIONS

Protein/protein interactions control major pathways in biology and numerous technologies have been developed to better understand these processes.[32] Indeed technologies such as cross-linking, pull-downs, and mass spectrometry require cytosolic extraction by cellular lysis, which may induce a loss of the natural properties, especially *in situ* interactions. In addition, double or reverse hybrid strategies involve genetic constructs that may interfere with the real function of the targeted proteins and/or induce detection of false positives. In order to optimize sensitivity and specificity and to decrease the invasiveness of other methods, large numbers of both natural and synthetic fluorescent probes have been developed, yielding a toolbox of probes for the characterization of protein/protein interactions.[64] The grafting of fluorescent probes enables the detection and quantification of the interactions between protein and partners. A large number of fluorescence approaches have been developed to improve the knowledge of protein/protein interactions *in vitro*, from steady-state to kinetic investigations, and a combination of all of these approaches provides a better understanding of protein/protein and peptide/protein interactions and multiprotein complex formation.

## 3.1. Probing protein/protein interactions

An important requirement in setting up fluorescence experiments to investigate or monitor protein/protein interactions is the selection of an appropriate probe to follow the interaction/dissociation of complexes in a noninvasive manner. Ideally, one should monitor changes in intrinsic protein fluorescence, mainly related to tryptophan. Tryptophan residues constitute sensitive probes that are often located at protein/protein interfaces and/or involved in substrate or ligand binding domains. Extrinsic fluorescence has also been applied using solvatochromic dyes which are conjugated to proteins or dyes compatible for FRET between two partners (protein/protein or protein/ligand). In the latter, it is important to validate that the presence of the dye does not alter protein structure and/or function.

Protein complex association/dissociation and reversible unfolding can be followed by size exclusion chromatography, circular dichroïsm, analytical centrifugation, and fluorescence spectroscopy.[37] The sensitivity of fluorescence to its environment has been largely employed to understand transition states involved in protein complex formation/dissociation and protein/polypeptide reversible unfolding by both steady-state and rapid kinetics, and anisotropy analysis.[65] Investigation of protein/protein association by mixing the two purified partners in solution and following their association in a time-dependent manner is limited to only few examples.[37] Usually, solvent (acetonitrile, dimethyl formamide, isopropanol, etc.) or chaotropic agents (guandinium chloride, urea, etc.) are used to promote complex dissociation and protein unfolding. Fluorescence changes can be transformed to yield the relative fraction of unfolded/monomeric protein in order to determine the thermodynamic parameters of protein complex stability.[66] In most cases, fluorescence follows a sigmoidal transition according to a one-step/two-state model:

$$D^2 \leftrightarrow 2U/M_t \qquad [4.4]$$

The process can be described by the following equations in which the folded dimer (D) is at equilibrium with the unfolded monomer ($U/M_t$). The total concentration of monomers ($M_t$) at any concentration of solvent can be defined in terms of the fraction of monomeric protein ($M_m$) and $K_d$ can be expressed in terms of measurable values $M_t$ and $M_m$.

$$K_d = 2M_t(M_m)^2/(1 - M_m) \qquad [4.5]$$
$$K_d = [U]^2/[N_2] = 2P_t\left[f_u^2/(1 - f_u)\right] \qquad [4.6]$$

The free energy of unfolding/dissociation for a two-state model is defined as a linear function of the concentration of the unfolding/dissociating agent.

$$\Delta G_d = \Delta G^{H_2O} + m[\text{solvent}] = -RT\ln K_d \qquad [4.7]$$

where $m$ corresponds to the slope of the plot of $\Delta G_d$ versus [solvent]. [Solvent] is the concentration of solvent, and $R$ and $T$ are the gas constant and absolute temperature, respectively. $\Delta G_d$ was calculated via $K_d$ at the corresponding concentrations of solvent used. $P_t$ corresponds to the total protein concentration and $f_u$ is the fraction of unfolded/monomeric protein. $\Delta G^{H_2O}$ is the extrapolated free energy of unfolding in the absence of any unfolding agent.

Steady-state fluorescence measurements have been combined with lifetime anisotropy to sense protein complex dissociation,[67] as described for the small protein Dim2, a regulatory component of the splicesome machinery. Analysis of Dim2 dissociation using guadinium chloride by steady fluorescence and lifetime anisotropy (Fig. 4.11) has revealed that the small Dim2 protein exists in two states, monomer and dimer, with a dissociation constant in the nanomolar

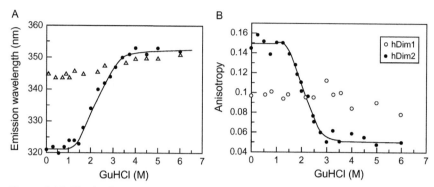

**Figure 4.11** Monitoring protein/protein interaction using chaotropic agent. Unfolding transition of hDim2 was monitored by fluorescence spectroscopy. (A) GuHCl-induced unfolding transition curves of hDim1 and hDim2 followed by a shift in the fluorescence emission maximum wavelength of hDim1 (triangle), hDim2 (filled circle). Experiments were performed at 25 °C using a concentration of 3 $\mu$M protein. (B) Steady-state anisotropy of hDim2 (closed circle) and hDim1 (triangle) was determined at an emission wavelength of 340 nM upon excitation at 290 nm in the presence of increasing concentrations of GuHCl (adapted from Simeoni et al.).[68]

range, in contrast to Dim1 homolog, and that Dim2 dimerization regulates its association with the splicesome machinery.[68]

Noncovalently linked external probes can also be applied to monitor changes in protein/protein interactions. 1-Anilinonaphthalene-8-sulfonate (ANS) or bis-ANS interacts with hydrophobic pockets that are accessible at the surface of proteins and has been used to follow transition states upon dissociation and unfolding processes. Moreover, spectral properties of ANS (emission 490 nm and excitation 340 nm) are compatible with tryptophan for FRET measurements. ANS and bis-ANS were used to monitor the dissociation of heterodimeric RT. The interface between the two subunits (p66 and p51) involves large hydrophobic patches and the dissociation of RT results in a large increase in the fluorescence of the probe due to noncovalent interactions of ANS to exposed surface hydrophobic motifs on the subunits, thereby providing a good signal for following RT dissociation in a time-dependent manner (Fig. 4.12).[69]

Another interesting alternative to monitor protein/protein interactions is the use of fluorescently labeled substrates, which, upon binding, reflect the formation of stable or/and active enzymatic complexes. The major advantage of this approach lies in the ability it provides to discriminate between active and inactive forms of the protein/enzyme. Two scenarios can be observed: either the binding site is located and formed by the protein/protein interface;

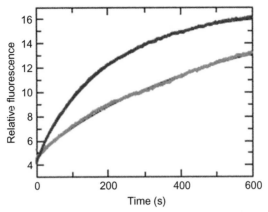

**Figure 4.12** Monitoring acetonitrile-mediated RT dissociation kinetics using a noncovalent external probe and the implication for inhibitor sensing. Heterodiomeric form or RT (0.5 μM) can be dissociated using acetonitrile and dissociation kinetics followed in the presence of 0.8 μM bis-ANS. The kinetics of dissociation was monitored by following the fluorescence resonance energy transfer between tryptophan of RT and bis-ANS. Tryptophan excitation was performed at 290 nm, and the increase of bis-ANS fluorescence emission at 490 nm was detected through a 420 nm cutoff filter, by adding 10% acetonitrile in the absence (black line) or presence of a dimerization inhibitor(gray line). Data were fitted according to a single-exponential equation (adapted from Agopian et al.).[41] (For color version of this figure, the reader is referred to the online version of this chapter.)

or the protein/protein interaction results in conformational changes within the substrate binding site located on one of the subunits. The substrate binding site is located at the interface[37] or formed by the association of the two subunits as in the case of the dimeric polymerase RT,[70] Mdm2/P53 interfaces,[71] or protease.[72] Peptide beacon/biosensor (HIV and caspase biosensors)[73] or fluorescently labeled nucleic acid[40] has been used to monitor protease or RNA or DNA polymerase activation as discussed in Section 3.

Monitoring protein/protein interactions does not necessarily require an interface substrate binding site. Several studies have shown that protein/protein interactions induce a marked conformational change in the closed environment of the catalytic site. As reported in Fig. 4.13, binding of cyclin A to Cdk2 results in an important fluorescence change in Mant–ATP bound to the Cdk moiety.[28] This change in fluorescence has been used to probe cyclin/Cdk interactions at the steady-state and pre-steady-state levels and to discriminate between the different Cyclin partners (Fig. 4.13).[27] Similar studies were also performed to understand the interaction of the ras oncogene with the GAP exchange factor.[74]

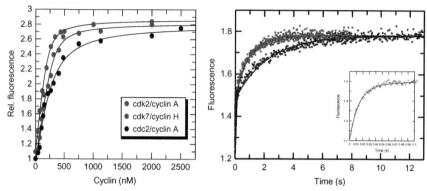

**Figure 4.13** Monitoring protein/protein interaction using fluorescently labeled substrates. (A) Binding titration of cyclin to their Cdk partners. The binding was followed by Mant-ATP fluorescence enhancement. A fixed concentration of protein kinase (Cdk2, Cdc2, or Cdk7) previously saturated with Mant-ATP was titrated with increasing amounts of cyclin. The enhancement of Mant-ATP fluorescence was monitored at 450 nm upon excitation at 340 nm. Dissociation constants were obtained by fitting the data with a quadratic equation. (B) Kinetics of binding of Cdk2 to cyclin A and cyclin H. kinetics of binding of Cdk2 (0.1 μM) to cyclins (0.4 μM). Cyclin A (red) and cyclin H (blue). Excitation was performed at 350 nm, and fluorescence emission of Mant-ATP was monitored through a cutoff filter (408 nm). Curves were fitted using a double-exponential term. (For interpretation of the references to color in this figure legend, the reader is referred to the online version of this chapter.)

## 3.2. Probing peptide/protein interactions

Protein/protein interaction studies provide a better understanding of the different parameters that rule the association of protein complexes. Analyses of these parameters enable the development of shorter peptides with similar interaction properties. These peptides may then associate with complexes in the same fashion as the protein partners and induce activation and/or inhibition of the whole complexes. This approach has been extensively developed to design novel inhibitors of specific protein/protein interactions as well as biosensors of protein complexes.[75–78] In parallel to the inhibitor, activator, or biosensor approaches, short peptides have also been developed for drug delivery. From the fluorescence point of view, as for protein/protein interactions, the intrinsic tryptophan fluorescence cannot be systematically used to monitor association since the cargo protein has to be devoid of tryptophan residues. In this case, the approach allows for the recording of intrinsic tryptophan fluorescence spectra of peptides in the presence of increasing amounts of cargo protein. Thus interactions may induce variations in the fluorescence intensity, which can be correlated with partner association.[79] However, if cargo proteins possess one or several tryptophan residues, the intrinsic fluorescence approach can no longer be applicable. In this case, an extrinsic probe is clearly required to determine peptide/protein affinity and characterize the formation of carrier/cargo complexes. Whatever be the location of the probe, on the cargo protein/peptide or on the carrier, its fluorescence emission can be monitored to gain insight into carrier/cargo interactions. As shown for several carrier peptides and various cargoes, this fluorescence approach allows for the comparison of the affinity between different carrier/cargo combinations.[80]

## 4. FLUORESCENCE FOR PROTEIN/MEMBRANE INTERACTIONS

Biological membranes constitute important components of living cells. Constituting a physical barrier between the cytosol and the extracellular environment, lipid membranes also compose the architecture of several intracellular compartments such as the Golgi, the endoplasmic reticulum, mitochondria, or the different types of endosomal/lysosomal vesicles involved in intracellular traffic. Membrane studies are often associated with analyses of specific proteins or peptides that are able to interact with, or are especially localized through, the lipid bilayer or the surface of membranes. From the role of membrane proteins to the translocating properties of some peptides, the protein/membrane interactions have been widely investigated. Among the different strategies for deciphering the nature and strength of protein/membrane interactions, the use of fluorescent probes as well as the involvement of new fluorescence technologies provides a better insight into protein/lipid affinity. In this section, the focus is on fluorescence approaches for protein or peptide/membrane interactions, from lipid bilayer insertion to direct phospholipid interactions.

### 4.1. Probing membrane interactions with tryptophan solvatochromism

Membrane insertion of a molecule may be easily correlated with its ability to be partially or fully embedded through fatty acyl chains that maintain phospholipid bilayer integrity. The development of membrane models has led to different systems that can mimic the natural membrane. Indeed, biological membranes are usually composed of numerous compounds. From phospholipids to membrane proteins, there are clearly too many components to study in a single fashion. Moreover, with regard to biological diversity, cellular membrane composition depends on the cellular type and can vary with the environment or with the physiological state.[81] Conception of a universal cellular membrane model is consequently too complicated. In order to circumvent this limitation, step-by-step approaches have been developed with artificial membranes. Liposomes, that is, small, large, and giant unilamellar vesicles (SUV, LUV, and GUV), have been developed to provide reliable models.[82] The combination of these models with fluorescent probes enables accurate monitoring of molecule/membrane interactions.

The first basic approach to investigate membrane insertion by fluorescence consists in monitoring the intrinsic tryptophan fluorescence of peptides or proteins in the presence of liposomes. The effect of membranes on the fluorescence emission of tryptophan provides information on the protein or peptide localization in the phospholipid bilayer by sensing environmental changes. Among the water-soluble membrane-active proteins, the channel-forming family of Colicins has been largely studied. These proteins are able to spontaneously insert into negatively charged membranes from aqueous media at low pH. A lack of change in the tryptophan fluorescence spectrum upon insertion indicates that their hydrophobic environment is preserved while becoming accessible to lipids, suggesting that membrane insertion only induces modifications in the relative positioning of the helices.[83] In contrast, monitoring fluorescence of four tryptophans of phospholipases indicates a change in the environment of one or more tryptophan residues, suggesting enzyme/membrane interactions.[84] However, the solvatochromism of tryptophan was mainly studied for membrane-active peptides.[59,85–88] For example, comparison of the membrane insertion of two analogous carrier peptides enabled the identification of slightly different behaviors (Fig. 4.14). The liposome effect on tryptophan fluorescence of the CPP MPG-α and MPG-β was compared. Depending on the nature of the phospholipid, headgroups, that is, neutral or negatively charged, different results were found. Addition of negatively charged phospholipids to a solution of MPG-β promotes a blue shift with maximum fluorescence (from 348 to 328 nm) associated with an enhancement of fluorescence intensity (about twofold). This behavior is characteristic of a tryptophan moving from a polar to a nonpolar environment. On the contrary, neutral lipids containing vesicles do not induce any modification of the fluorescence spectrum, suggesting no insertion of peptides into neutral phospholipid bilayers. For MPG-α, while negatively charged vesicles induce a similar blue shift, neutral lipids promote

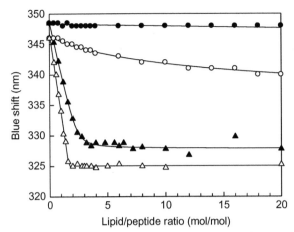

**Figure 4.14** Effect of membrane on the fluorescence emission of tryptophan residue of two cell-penetrating peptides. The wavelengths of the maximal tryptophan fluorescence emission of MPG-α (open label) and MPG-β peptides (solid label) were recorded in the presence of liposomes composed of neutral (POPC, circle) or negatively charged (POPG, triangle) phospholipids at different lipid/peptide molar ratios.

a small but significant shift, indicating that MPG-α also inserts in neutral bilayers with a positioning of the tryptophan closer to the lipid/water interface (Fig. 4.14). This kind of behavior was also observed for other peptides.[59,79,88–90] Further analyses also enabled the determination of the proportion of bound versus free peptides through the plot of binding isotherms.[86] Thus the insertion of a single tryptophan residue in the sequence of several carrier peptides led to comparative membrane insertion analyses.[87]

## 4.2. Probing membrane insertion by tryptophan quenchers

Although the solvatochromism approach can provide information on protein/membrane interactions, the possibility of insertion through the lipid bilayer or at the interface is usually ruled out by quenching measurements and FRET investigations.[91–93] Indeed, the combination of tryptophan fluorescence with both improvements in vesicle formation and phospholipid chemistry enabled the development of useful methods to determine the depth of insertion of molecules through a lipid bilayer. Acrylamide and potassium iodide (KI) are well known to interact with tryptophan and to induce a specific quenching of fluorescence intensity.[94–96] In fact, by combining these tryptophan quenchers with LUVs, it is possible to sense the accessibility of the indole in phospholipid bilayers and then the penetration of the protein or peptide in a membrane. Basically, the observed enhancement of intensity in the fluorescence of tryptophan in the presence of LUVs is usually titrated with increasing concentration of quenchers. The resulting data are then analyzed by using the Stern–Volmer equation:[17]

$$F_0/F = 1 + K_{SV}[Q] \qquad\qquad [48]$$

where $F_0$ and $F$ are the fluorescence intensities in the absence and in the presence of quencher, respectively, $[Q]$ is the molar concentration of quencher, and $K_{SV}$ is the Stern–Volmer quenching constant. A strong decrease of the $K_{SV}$ value in the presence of LUV is indicative of a loss of accessibility of the tryptophan to the quencher, suggesting thus a deeper insertion of molecule in the lipid bilayer.[87]

By specifically labeling fatty acyl chains or polar heads, quenching as well as enhancement of fluorescence can provide a more accurate mapping of the bilayer location of a protein. A large number of different labeled phospholipids have also been developed to enable direct FRET on the surface membrane or inside the bilayer. Although more difficult to use, this approach is more accurate. For example, brominated phospholipids were engineered by the addition of bromide at specific positions of the acyl chains of phospholipids: that is, position (6, 7), (9, 10), and (11, 12) of a stearoyl fatty acid chain. From the hydrophobic tail of the acyl chain to the polar head of phospholipids, the use of distinct positions of bromide enables the mapping of the insertion of peptide in a lipid bilayer by bromide quenching of the tryptophan intrinsic fluorescence.[97,98] Thus the resulting depth-dependent fluorescence quenching profiles enables comparison between different membrane-active peptides.[87,98]

## 4.3. Probing membrane interactions with specific probes

Based on the principle of solvatochromism, several different fluorescent probes were developed for specifically sensing membrane environments and membrane domains. Indeed, membranes cannot be considered simply as single phospholipid bilayers with a hydrophobic core and a polar surface. Involving different types of domains with various physical states, the "fluid mosaic model" of the structure of cell membranes is clearly in agreement with the dynamics of membrane components throughout the phospholipid bilayer.[99] Like the physical state of phospholipids, the polarity of the lipid bilayer, the membrane potential, and the hydration of membrane are parameters that contribute to the integrity of the membrane. Studies of protein/

membrane interactions generally require the use of specific probes.[11] In this context, several probes have been designed and synthesized in order to confer specific environmental sensitivity.[100,101]

In contrast to the monitoring of the intrinsic fluorescence of tryptophan, extrinsic labeling of proteins and peptides usually involves the covalent attachment of a fluorophore to a single site on the target molecule, generally via the thiol function of a cysteine residue. Indeed, cysteines are of low abundance in proteins and peptides, and their modification chemistry proceeds under conditions that do not compromise protein structure or function. There are many different probes that can be associated with cysteines. Among them, Acrylodan and Badan are two derivatives of the environment-sensitive dye Prodan, which can be used to label peptides or proteins to investigate their interactions with lipid membranes.[102] However, their poor solubility in water may induce their own burial in a lipid bilayer or in the hydrophobic pocket of proteins. Thus polarity-sensitive fluorophores that are more water soluble, such as NBD, are generally used.[103,104] NBD is one of the most famous membrane-sensitive probes and has been used as a reporter group in many studies, including investigations of cotranslational protein translocation and integration at the Endoplasmic reticulum (ER) membrane[105] and of toxin insertion into bilayers.[106]

Another approach consists in the use of specifically labeled phospholipids to measure FRET with membrane-active peptides and proteins. For example, Laurdan can act as a FRET acceptor of tryptophan emission and can be used to study the physical state of lipids within Förster distance from donor tryptophan residues in integral membrane protein.[107] However, some probes are more sensitive to the environment than others. Thus several technologies have been developed to insert specific donor/acceptor couples to induce membrane FRET through the lipid bilayer or at the surface of a model or natural membrane.[18,100] For example, NBD-labeled phosphatidylethanolamine can be combined with rhodamine-labeled proteins to investigate FRET between membranes and proteins.

## 5. CONCLUSIONS AND PERSPECTIVES

During the last 10 years, fluorescence technology has been used by scientists from many disciplines and for a large panel of applications. Fluorescence constitutes one of the most powerful methods to monitor interactions in a biological context and to answer a wide range of biological and chemical questions. This technology is very advantageous in understanding protein/protein, protein/ligand, and protein/membrane interactions in real time and in a noninvasive manner. The development of accurate equipment to measure FRET, single molecule, rapid kinetics, fluorescence anisotropy, and fluorescence polarization together with the access to new chemistry for fluorescent probes has opened new routes to identify critical protein conformational changes involved in the regulation of major signaling pathways, as well as aberrant cell behavior and pathological disorders such as cancer. Therefore, fluorescence technology constitutes a major piece of the puzzle for the development of future medicine both at the therapeutic and diagnostic levels.

## ACKNOWLEDGMENTS

This work was supported by the Centre National de la Recherche Scientifique (CNRS), the Agence Nationale de la Recherche (ANR- ANR-06-BLAN-0071), and the Institut National du Cancer (INCA-BIOSENSIMAG). We thank Dr. May C. Morris (CRBM-UMR5237-CNRS) for her critical reading of the chapter and all members of our laboratory for the fruitful discussions.

## REFERENCES

1. Herschel Sir JFW. On a case of superficial colour presented by a homogeneous liquid internally colourless. *Philos Trans Roy Soc (Lond)* 1845;**135**:143–5.
2. Stokes GG. On the change of refrangibility of light. *Philos Trans R Soc (Lond)* 1852;**142**:463–562.
3. Wikipedia: http://en.wikipedia.org/wiki/Fluorescence.
4. Jäblonski A. Uber den mechanisms des photolumineszenz von Farbstoffphosphore. *Z Phys* 1935;**94**:38–46.
5. Loving GS, Sainlos M, Imperiali B. Monitoring protein interactions and dynamics with solvatochromic fluorophores. *Trends Biotechnol* 2010;**28**(2):73–83.
6. Weber G. Fluorescence-polarization spectrum and electronic-energy transfer in tyrosine, tryptophan and related compounds. *Biochem J* 1960;**75**:335–45.
7. Longworth JW. Luminescence of polypeptides and proteins. In: Steiner RF, Weinryb I, editors. *Excited states of proteins and nucleic acids*. New York: Plenum; 1971. p. 319–483.
8. Spector AA. Fatty acid binding to plasma albumin. *J Lipid Res* 1975;**16**(3):165–79.
9. Vivian JT, Callis PR. Mechanisms of tryptophan fluorescence shifts in proteins. *Biophys J* 2001;**80**(5):2093–109.
10. Chen Y, Barkley MD. Toward understanding tryptophan fluorescence in proteins. *Biochemistry* 1998;**37**(28):9976–82.
11. Demchenko AP, Mély Y, Duportail G, Klymchenko AS. Monitoring biophysical properties of lipid membranes by environment-sensitive fluorescent probes. *Biophys J* 2009;**96**(9):3461–70.
12. Martin MM, Lindqvist L. The pH dependence of fluorescein fluorescence. *J Lumin* 1975;**10**:381–90.
13. Ma LY, Wang HY, Xie H, Xu LX. A long lifetime chemical sensor: study on fluorescence property of fluorescein isothiocyanate and preparation of pH chemical sensor. *Spectrochim Acta A Mol Biomol Spectrosc* 2004;**60**(8–9):1865–72.

14. Perrin F. Polarization de la lumière de fluorescence. Vie moyenne de molecules dans l'etat excite. *J Phys Radium* 1926;**7**:390–401.
15. Förster T. Energiewanderung und Fluoreszenz. *Naturwissenschaften* 1946;**33**:166–75.
16. Förster T. Zwischenmoleculare Energiewanderung und Fluoreszenz. *Ann Phys* 1948;**2**:55–75.
17. Lakowicz JR. *Principles of fluorescence spectroscopy.* 3rd ed.; New York, NY, USA: Springer; 2006.
18. Royer CA, Scarlata SF. Fluorescence approaches to quantifying biomolecular interactions. *Methods Enzymol* 2008;**450**:79–106.
19. Pope AJ, Haupts UM, Moore KJ. Homogeneous fluorescence readouts for miniaturized high-throughput screening: theory and practice. *Drug Discov Today* 1999;**4**(8):350–62.
20. Bagshaw C. ATP analogues at a glance. *J Cell Sci* 2001;**114**:459–60.
21. Jameson DM, Eccleston JF. Fluorescent nucleotide analogs: synthesis and applications. *Methods Enzymol* 1997;**278**:363–90.
22. Weisbrod SH, Marx A. Novel strategies for the site-specific covalent labelling of nucleic acids. *Chem Commun (Camb)* 2008;**44**:5675–85.
23. Goody RS, Frech M, Wittinghofer A. Affinity of guanine nucleotide binding proteins for their ligands: facts and artefacts. *Trends Biochem Sci* 1991;**16**(9):327–8.
24. Divita G, Goody RS, Gautheron DC, Di Pietro A. Structural mapping of catalytic site with respect to alpha-subunit and noncatalytic site in yeast mitochondrial F1-ATPase using fluorescence resonance energy transfer. *J Biol Chem* 1993;**268**(18):13178–86.
25. Rudolph MG, Veit TJ, Reinstein J. The novel fluorescent CDP-analogue (Pbeta) MABA-CDP is a specific probe for the NMP binding site of UMP/CMP kinase. *Protein Sci* 1999;**8**(12):2697–704.
26. John J, Sohmen R, Feuerstein J, Linke R, Wittinghofer A, Goody RS. Kinetics of interaction of nucleotides with nucleotide-free H-ras p21. *Biochemistry* 1990;**29**(25):6058–65.
27. Morris MC, Gondeau C, Tainer JA, Divita G. Kinetic mechanism of activation of the Cdk2/cyclin A complex. Key role of the C-lobe of the Cdk. *J Biol Chem* 2002;**277**(26):23847–53.
28. Heitz F, Morris MC, Fesquet D, Cavadore JC, Dorée M, Divita G. Interactions of cyclins with cyclin-dependent kinases: a common interactive mechanism. *Biochemistry* 1997;**36**(16):4995–5003.
29. Lavie A, Konrad M, Brundiers R, Goody RS, Schlichting I, Reinstein J. Crystal structure of yeast thymidylate kinase complexed with the bisubstrate inhibitor P1-(5′-adenosyl) P5-(5′-thymidyl) pentaphosphate (TP5A) at 2.0 Å resolution: implications for catalysis and AZT activation. *Biochemistry* 1998;**37**(11):3677–86.
30. Reinstein J, Vetter IR, Schlichting I, Rösch P, Wittinghofer A, Goody RS. Fluorescence and NMR investigations on the ligand binding properties of adenylate kinases. *Biochemistry* 1990;**29**(32):7440–50.
31. Shutes A, Der CJ. Real-time in vitro measurement of GTP hydrolysis. *Methods* 2005;**37**(2):183–9.
32. Gomelis E. *Protein-protein interactions: a molecular cloning manual.* 1st ed.; New York, NY, USA: Cold Spring Harbor Laboratory Press; 2005.
33. Metzker ML. Sequencing technologies—the next generation. *Nat Rev Genet* 2010;**11**(1):31–46.
34. Turcatti G, Romieu A, Fedurco M, Tairi AP. A new class of cleavable fluorescent nucleotides: synthesis and optimization as reversible terminators for DNA sequencing by synthesis. *Nucleic Acids Res* 2008;**36**(4):e25.
35. Guo J, Xu N, Li Z, Zhang S, Wu J, Kim DH, et al. Four-color DNA sequencing with 3′-O-modified nucleotide reversible terminators and chemically cleavable fluorescent dideoxynucleotides. *Proc Natl Acad Sci USA* 2008;**105**(27):9145–50.

36. Cameron CE, Moustafa IM, Arnold JJ. Dynamics: the missing link between structure and function of the viral RNA-dependent RNA polymerase? *Curr Opin Struct Biol* 2009;**19**(6):768–74.

37. Creighton TE. *The biophysical chemistry of nucleic acids and proteins.* Helvetian Press; 2010.

38. Divita G, Müller B, Immendörfer U, Gautel M, Rittinger K, Restle T, et al. Kinetics of interaction of HIV reverse transcriptase with primer/template. *Biochemistry* 1993;**32**(31):7966–71.

39. Liu S, Harada BT, Miller JT, Le Grice SF, Zhuang X. Initiation complex dynamics direct the transitions between distinct phases of early HIV reverse transcription. *Nat Struct Mol Biol* 2010;**17**:1453–60.

40. Abbondanzieri EA, Bokinsky G, Rausch JW, Zhang JX, Le Grice SF, Zhuang X. Dynamic binding orientations direct activity of HIV reverse transcriptase. *Nature* 2008;**453**(7192):184–9.

41. Agopian A, Depollier J, Lionne C, Divita G. p66 Trp24 and Phe61 are essential for accurate association of HIV-1 reverse transcriptase with primer/template. *J Mol Biol* 2007;**373**(1):127–40.

42. Johnson KA. Advances in transient-state kinetics. *Curr Opin Biotechnol* 1998;**9**(1):87–9.

43. Gutfreund H. Rapid-flow techniques and their contributions to enzymology. *Trends Biochem Sci* 1999;**24**(11):457–60.

44. Barman TE, Bellamy SR, Gutfreund H, Halford SE, Lionne C. The identification of chemical intermediates in enzyme catalysis by the rapid quench-flow technique. *Cell Mol Life Sci* 2006;**63**(22):2571–83 [Review].

45. Rittinger K, Divita G, Goody RS. Human immunodeficiency virus reverse transcriptase substrate-induced conformational changes and the mechanism of inhibition by nonnucleoside inhibitors. *Proc Natl Acad Sci USA* 1995;**92**(17):8046–9.

46. Patel SS, Pandey M, Nandakumar D. Fluorescence-based assay to measure the real-time kinetics of nucleotide incorporation during transcription elongation. *Curr Opin Chem Biol* 2011;**15**(5):595–605.

47. Tang GQ, Roy R, Ha T, Patel SS. Mechanism of transcription initiation by the yeast mitochondrial RNA polymerase. *Mol Cell* 2008;**30**(5):567–77.

48. Kim H, Tang GQ, Patel SS, Ha T. Dynamic coupling between the motors of DNA replication: hexameric helicase, DNA polymerase, and primase. *Nucleic Acids Res* 2012;**40**(1):371–80.

49. Fourar M, Divita G. Fluorescence-based methods to monitor the real-time kinetics of nucleotide incorporation by NS5B, RNA polymerase (submitted).

50. Lehto T, Ezzat K, Langel U. Peptide nanoparticles for oligonucleotide delivery. *Prog Mol Biol Transl Sci* 2011;**104**:397–426.

51. Crombez L, Morris MC, Deshayes S, Heitz F, Divita G. Peptide-based nanoparticle for ex vivo and in vivo drug delivery. *Curr Pharm Des* 2008;**14**(34):3656–65.

52. Heitz F, Morris MC, Divita G. Twenty years of cell-penetrating peptides: from molecular mechanisms to therapeutics. *Br J Pharmacol* 2009;**157**(2):195–206.

53. Morris MC, Deshayes S, Heitz F, Divita G. Cell-penetrating peptides: from molecular mechanisms to therapeutics. *Biol Cell* 2008;**100**(4):201–17.

54. Morris MC, Vidal P, Chaloin L, Heitz F, Divita G. A new peptide vector for efficient delivery of oligonucleotides into mammalian cells. *Nucleic Acids Res* 1997;**25**(14):2730–6.

55. Crombez L, Aldrian-Herrada G, Konate K, Nguyen QN, McMaster GK, Brasseur R, et al. A new potent secondary amphipathic cell-penetrating peptide for siRNA delivery into mammalian cells. *Mol Ther* 2009;**17**(1):95–103.

56. Konate K, Crombez L, Deshayes S, Decaffmeyer M, Thomas A, Brasseur R, et al. Insight into the cellular uptake mechanism of a secondary amphipathic cell-penetrating peptide for siRNA delivery. *Biochemistry* 2010;**49**(16):3393–402.

57. Deshayes S, Konate K, Aldrian G, Crombez L, Heitz F, Divita G. Structural polymorphism of non-covalent peptide-based delivery systems: highway to cellular uptake. *Biochim Biophys Acta* 2010;**1798**(12):2304–14.

58. Morris MC, Chaloin L, Méry J, Heitz F, Divita G. A novel potent strategy for gene delivery using a single peptide vector as a carrier. *Nucleic Acids Res* 1999;**27**(17):3510–7.

59. Deshayes S, Gerbal-Chaloin S, Morris MC, Aldrian-Herrada G, Charnet P, Divita G, et al. On the mechanism of non-endosomial peptide-mediated cellular delivery of nucleic acids. *Biochim Biophys Acta* 2004;**1667**(2).141–7.

60. Helene C, Dimicoli JL. Interaction of oligopeptides containing aromatic amino acids with nucleic acids. Fluorescence and proton magnetic resonance studies. *FEBS Lett* 1972;**26**(1):6–10.

61. Toulmé JJ, Hélène C. Specific recognition of single-stranded nucleic acids. Interaction of tryptophan-containing peptides with native, denatured, and ultraviolet-irradiated DNA. *J Biol Chem* 1977;**252**(1):244–9.

62. Akanchha S, Jain A, Rajeswari MR. Binding studies on peptide oligonucleotide complex: intercalation of tryptophan in GC-rich region of c-myc gene. *Biochim Biophys Acta* 2003;**1622**:73–81.

63. Jain A, Akanchha S, Rajeswari MR. Stabilization of purine motif DNA triplex by a tetrapeptide from the binding domain of HMGBI protein. *Biochimie* 2005;**87**(8):781–90.

64. Giepmans BN, Adams SR, Ellisman MH, Tsien RY. The fluorescent toolbox for assessing protein location and function. *Science* 2006;**312**(5771):217–24.

65. Thomson JA, Shirley BA, Grimsley GR, Pace CN. Conformational stability and mechanism of folding of ribonuclease T1. *J Biol Chem* 1989;**264**(20):11614–20.

66. Pace CN. Evaluating contribution of hydrogen bonding and hydrophobic bonding to protein folding. *Methods Enzymol* 1986;**131**:266–80.

67. Bujalowski WM, Jezewska MJ. Fluorescence intensity, anisotropy, and transient dynamic quenching stopped-flow kinetics. *Methods Mol Biol* 2012;**875**:105–33.

68. Simeoni F, Arvai A, Bello P, Gondeau C, Hopfner KP, Neyroz P, et al. Biochemical characterization and crystal structure of a Dim1 family associated protein: Dim2. *Biochemistry* 2005;**44**(36):11997–2008.

69. Agopian A, Gros E, Aldrian-Herrada G, Bosquet N, Clayette P, Divita G. A new generation of peptide-based inhibitors targeting HIV-1 reverse transcriptase conformational flexibility. *J Biol Chem* 2009;**284**(1):254–64.

70. Jacobo-Molina A, Clark Jr. AD, Williams RL, Nanni RG, Clark P, Ferris AL, et al. Crystals of a ternary complex of human immunodeficiency virus type 1 reverse transcriptase with a monoclonal antibody Fab fragment and double-stranded DNA diffract x-rays to 3.5-A resolution. *Proc Natl Acad Sci USA* 1991;**88**(23):10895–9.

71. Nicholson J, Hupp TR. The molecular dynamics of MDM2. *Cell Cycle* 2010;**9**(10):1878–81.

72. Chène P. Inhibition of the p53-MDM2 interaction: targeting a protein-protein interface. *Mol Cancer Res* 2004;**2**(1):20–8 Review.

73. Bardet PL, Kolahgar G, Mynett A, Miguel-Aliaga I, Briscoe J, Meier P, et al. A fluorescent reporter of caspase activity for live imaging. *Proc Natl Acad Sci USA* 2008;**105**(37):13901–5.

74. Nassar N, Horn G, Herrmann C, Scherer A, McCormick F, Wittinghofer A. The 2.2 A crystal structure of the Ras-binding domain of the serine/threonine kinase c-Raf1 in complex with Rap1A and a GTP analogue. *Nature* 1995;**375**(6532):554–60.

75. Kurzawa L, Morris MC. Cell-cycle markers and biosensors. *Chembiochem* 2010;**11**(8):1037–47.

76. Kurzawa L, Pellerano M, Coppolani JB, Morris MC. Fluorescent peptide biosensor for probing the relative abundance of cyclin-dependent kinases in living cells. *PLoS One* 2011;**6**(10):e26555.
77. Gondeau C, Gerbal-Chaloin S, Bello P, Aldrian-Herrada G, Morris MC, Divita G. Design of a novel class of peptide inhibitors of cyclin-dependent kinase/cyclin activation. *J Biol Chem* 2005;**280**(14):13793–800.
78. Pommier Y, Cherfils J. Interfacial inhibition of macromolecular interactions: nature's paradigm for drug discovery. *Trends Pharmacol Sci* 2005;**26**(3):138–45 Review.
79. Deshayes S, Heitz A, Morris MC, Charnet P, Divita G, Heitz F. Insight into the mechanism of internalization of the cell-penetrating carrier peptide Pep-1 through conformational analysis. *Biochemistry* 2004;**43**(6):1449–57.
80. Kurzawa L, Pellerano M, Morris MC. PEP and CADY-mediated delivery of fluorescent peptides and proteins into living cells. *Biochim Biophys Acta* 2010;**1798**(12): 2274–85.
81. Spector AA, Yorek MA. Membrane lipid composition and cellular function. *J Lipid Res* 1985;**26**(9):1015–35.
82. Szoka Jr. F, Papahadjopoulos D. Comparative properties and methods of preparation of lipid vesicles (liposomes). *Annu Rev Biophys Bioeng* 1980;**9**:467–508.
83. Lakey JH, Massotte D, Heitz F, Dasseux JL, Faucon JF, Parker MW, et al. Membrane insertion of the pore-forming domain of colicin A. A spectroscopic study. *Eur J Biochem* 1991;**196**(3):599–607.
84. Mosmuller EW, Pap EH, Visser AJ, Engbersen JF. Steady-state fluorescence studies on lipase-vesicle interactions. *Biochim Biophys Acta* 1994;**1189**(1):45–51.
85. Magzoub M, Eriksson LE, Gräslund A. Conformational states of the cell-penetrating peptide penetratin when interacting with phospholipid vesicles: effects of surface charge and peptide concentration. *Biochim Biophys Acta* 2002;**1563**(1–2):53–63.
86. Persson D, Thorén PE, Herner M, Lincoln P, Nordén B. Application of a novel analysis to measure the binding of the membrane-translocating peptide penetratin to negatively charged liposomes. *Biochemistry* 2003;**42**(2):421–9.
87. Thorén PE, Persson D, Esbjörner EK, Goksör M, Lincoln P, Nordén B. Membrane binding and translocation of cell-penetrating peptides. *Biochemistry* 2004;**43**(12): 3471–89.
88. Mano M, Henriques A, Paiva A, Prieto M, Gavilanes F, Simões S, et al. Cellular uptake of S413-PV peptide occurs upon conformational changes induced by peptide-membrane interactions. *Biochim Biophys Acta* 2006;**1758**(3):336–46.
89. Christiaens B, Symoens S, Verheyden S, Engelborghs Y, Joliot A, Prochiantz A, et al. Tryptophan fluorescence study of the interaction of penetratin peptides with model membranes. *Eur J Biochem* 2002;**269**(12):2918–26.
90. Christiaens B, Grooten J, Reusens M, Joliot A, Goethals M, Vandekerckhove J, et al. Membrane interaction and cellular internalization of penetratin peptides. *Eur J Biochem* 2004;**271**(6):1187–97.
91. Torrent M, Cuyás E, Carreras E, Navarro S, López O, de la Maza A, et al. Topography studies on the membrane interaction mechanism of the eosinophil cationic protein. *Biochemistry* 2007;**46**(3):720–33.
92. Coïc YM, Vincent M, Gallay J, Baleux F, Mousson F, Beswick V, et al. Single-spanning membrane protein insertion in membrane mimetic systems: role and localization of aromatic residues. *Eur Biophys J* 2005;**35**(1):27–39 [Epub 2005 Jul 15].
93. Yin M, Ochs RS. A mechanism for the partial insertion of protein kinase C into membranes. *Biochem Biophys Res Commun* 2001;**281**(5):1277–82.
94. Eftink MR, Ghiron CA. Exposure of tryptophanyl residues in proteins. Quantitative determination by fluorescence quenching studies. *Biochemistry* 1976;**15**(3):672–80.

95. De Kroon AI, Soekarjo MW, De Gier J, De Kruijff B. The role of charge and hydrophobicity in peptide-lipid interaction: a comparative study based on tryptophan fluorescence measurements combined with the use of aqueous and hydrophobic quenchers. *Biochemistry* 1990;**29**(36):8229–40.

96. Kaszycki P, Wasylewski Z. Fluorescence-quenching-resolved spectra of melittin in lipid bilayers. *Biochim Biophys Acta* 1990;**1040**(3):337–45.

97. Bolen EJ, Holloway PW. Quenching of tryptophan fluorescence by brominated phospholipid. *Biochemistry* 1990;**29**(41):9638–43.

98. Ladokhin AS. Distribution analysis of depth-dependent fluorescence quenching in membranes: a practical guide. *Methods Enzymol* 1997;**278**:462–73.

99. Singer SJ, Nicolson GL. The fluid mosaic model of the structure of cell membranes. *Science* 1972;**175**(4023):720–31.

100. Maier O, Oberle V, Hoekstra D. Fluorescent lipid probes: some properties and applications. *Chem Phys Lipids* 2002;**116**(1–2):3–18.

101. Klymchenko AS, Mely Y. Fluorescent environment-sensitive dyes as reporters of biomolecular interactions. *Prog Mol Biol Transl Sci* 2012;**113**:35–58.

102. Weber G, Farris FJ. Synthesis and spectral properties of a hydrophobic fluorescent probe: 6-propionyl-2-(dimethylamino)naphthalene. *Biochemistry* 1979;**18**(14):3075–8.

103. Kenner RA, Aboderin AA. A new fluorescent probe for protein and nucleoprotein conformation. Binding of 7-(p-methoxybenzylamino)-4-nitrobenzoxadiazole to bovine trypsinogen and bacterial ribosomes. *Biochemistry* 1971;**10**(24):4433–40.

104. Crowley KS, Reinhart GD, Johnson AE. The signal sequence moves through a ribosomal tunnel into a noncytoplasmic aqueous environment at the ER membrane early in translocation. *Cell* 1993;**73**(6):1101–15.

105. Liao S, Lin J, Do H, Johnson AE. Both lumenal and cytosolic gating of the aqueous ER translocon pore are regulated from inside the ribosome during membrane protein integration. *Cell* 1997;**90**(1):31–41.

106. Heuck AP, Hotze EM, Tweten RK, Johnson AE. Mechanism of membrane insertion of a multimeric beta-barrel protein: perfringolysin O creates a pore using ordered and coupled conformational changes. *Mol Cell* 2000;**6**(5):1233–42.

107. Antollini SS, Barrantes FJ. Laurdan studies of membrane lipid-nicotinic acetylcholine receptor protein interactions. *Methods Mol Biol* 2007;**400**:531–42.

# CHAPTER FIVE

# From FRET Imaging to Practical Methodology for Kinase Activity Sensing in Living Cells

François Sipieter*,¶,‡ and Pauline Vandame¶,‡, Corentin Spriet*,‡, Aymeric Leray*,‡, Pierre Vincent‡,§, Dave Trinel*,‡, Jean-François Bodart¶,‡, Franck B. Riquet¶,‡,1 and Laurent Héliot*,‡,1

*Biophotonic Team (BCF), IRI CNRS, University Lille1–University Lille2 USR 3078, Villeneuve d'Ascq, France
‡Groupement de Recherche Microscopie Fonctionnelle du vivant, GDR 2588-CNRS, Villeneuve d'Ascq, France
§Neurobiologie des processus adaptatifs—UMR 7102, UPMC, Paris, France
¶Laboratoire de Régulation des Signaux de division EA4479 University Lille1, Villeneuve d'Ascq, France
1Corresponding authors: Emails: franck.riquet@iri.univ-lille1.fr; laurent.heliot@iri.univ-lille1.fr

## Contents

*Progress in Molecular Biology and Translational Science*, Volume 113
ISSN 1877-1173
http://dx.doi.org/10.1016/B978-0-12-386932-6.00005-3

145

# ABBREVIATIONS

**A** acceptor
**AKAR** A-kinase activity reporter
**Akt** protein kinase B
**AMDI** adaptive Monte Carlo data inflation
**ATP** adenosine triphosphate
**BFP** blue fluorescent protein
**cAMP** cyclic adenosine monophosphate
**CCD** charge-coupled device
**cDNA** complementary DNA
**CFP** cyan fluorescent protein
**CKAR** C-kinase activity reporter
**Cp** Circularly permuted
**CyPET** optimized eCFP for FRET
**D** donor
**DC** direct current
**DCR** dual channel ratio
**DNA** deoxyribonucleic acid
**DsRed** *Discosoma* sp. red fluorescent protein
**DVR** dual view ratio
**eGFP** enhanced green fluorescent protein
**EKAR** extracellular regulated kinase activity reporter
**Epac** exchange protein activated by cAMP
**ERK** extracellular regulated kinase
**FD** frequency domain
**FHA** forkhead associated
**FLIM** fluorescence lifetime imaging microscopy
**FP** fluorescent protein
**FRET** Förster resonance energy transfer
**GFP** green fluorescent protein
**GTPase** hydrolase enzyme that can bind and hydrolyze guanosine triphosphate (GTP)
**IRF** instrumental response function

**JNK**  c-Jun N-terminal kinase
**KAR**  kinase activity reporter
**LSM**  least squares method
**MAPK**  mitogen-activated protein kinase
**MLE**  maximum likelihood estimation
**MPK**  mitogen-activated protein kinase phosphatase
**MRE**  molecular recognition element
**PAABD**  phosphoamino acid-binding domain
**PKA**  protein kinase A
**PKC**  protein kinase C
**PKG**  protein kinase G
**PMT**  photomultiplier tube
**PTB**  phosphotyrosine binding
**R**  ratio
**Rac**  subfamily of the Rho family of GTPases
**RFP**  red fluorescent protein
**RNA**  ribonucleic acid
**siRNA**  small interfering RNA
**SR**  sequential ratio
**TCSPC**  time-correlated single-photon counting
**TD**  time domain
**U2OS**  osteosarcoma cell line
**YFP**  yellow fluorescent protein
**YPet**  optimized YFP for FRET

## Foreword

Biological processes are intrinsically dynamic. Although traditional methods provide valuable insights for the understanding of many biological phenomena, the possibility of measuring, quantifying, and localizing proteins within a cell, a tissue, and even an embryo has revolutionized our train of thoughts and has encouraged scientists to develop molecular tools for the assessment of protein or protein complex dynamics within their physiological context. These ongoing efforts rest on the emergence of bio-photonic techniques and the continuous improvement of fluorescent probes, allowing precise and reliable measurements of dynamic cellular functions. The march of the "in vivo biochemistry" has begun, already yielding breathtaking results.

## 1. INTRODUCTION

How cells sense external and internal signals and how these signals are processed to drive specific responses in a multiscale context are major questions in biology.

Proten phosphorylation plays a significant role in a wide range of cellular processes such as cell proliferation, differentiation, and cellular death. In eukaryotes, phosphorylation occurs on serine, threonine, tyrosine, and histidine residues. Protein phosphorylation can alter the activity of many proteins, causing a chain reaction leading to the phosphorylation of many proteins involved in a particular cellular process. Conventional analytical methods have identified and characterized various posttranslational modifications, but the major drawback is that these methods provide only a snapshot of the cell. In fact, in order to assess activity of protein kinases, immunoblotting and immunocytochemistry with phospho antibodies toward specific residues described to report on kinase activation, are global and indirect/static approaches. They are limited to the time resolution and the quality of the cell fractionation assay under analysis. In addition, antibodies toward specific phospho residues do not really reflect activity of the kinase of interest.

To go beyond the snapshot, tools have been developed to answer the new challenges in today's biology quest: protein localization, interaction, and activity, when applicable. Concerning the latter, fluorescent biosensors have now become the biologist's toolbox to visualize, especially, the spatiotemporal dynamics of kinase signaling in living systems. They provide high sensitivity and versatility while only minimally perturbing cell physiology.

The word "biosensor" is a rather generic term that is used to define a wide array of systems that enable the sensing of various analytes. Typically, a biosensor is composed of two parts having distinct functions. The first, referred to as the "bioreceptor," recognizes the analyte and is responsible for the selectivity and the sensitivity of the whole biosensor. The second, named "transducer," is in charge of conveying the signal from the recognition part toward the adapted instrument (Fig. 5.1).

This definition is, of course, applicable to molecular biosensors dedicated to sensing biological events in living cells. Indeed, the analyte is represented by ions such as $Ca^{2+}$, second messengers such as cAMP (cyclic adenosine monophosphate), an enzymatic activity such as kinase activity, or an active enzyme conformation. The bioreceptor or molecular recognition element (MRE) is thus materialized by calmodulin/M13 or TroponinC/M13, Epac (exchange protein activated by cAMP), and the substrate/phosphoamino acid-binding domain (PAABD) respectively. The transducer element is represented in these cases by fluorescent proteins, which, when their distance allows it, will generate a change in their fluorescent signal or spectral properties. Thus, the instrument dedicated to signal detection is an "optical fluorescent microscope" or a "spectrofluorimeter."

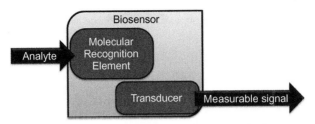

**Figure 5.1** Schematic representation of the basic principle of the biosensor.

The first of a long line of such biosensors was developed in R. Tsien's lab and reported in 1996.[1] While of simple design, it was the first to demonstrate that fluorescent protein variants could be used for biosensing. This first genetically encoded biosensor was composed of a blue fluorescent protein (BFP) and a green fluorescent protein (GFP) encompassing a trypsin cleavable linker. Thereafter, many different types of biosensors have been developed with various molecular structures (Fig. 5.2).

All genetically encoded fluorescent protein (FP)-based biosensors are classified into groups depending on their structure.[2,3] Kinase activity reporters (KARs) are included in the group based on intramolecular Förster resonance energy transfer (FRET). The archetypal structure of such biosensors consists of two FPs flanking an MRE (Fig. 5.3). The interaction of an active kinase with its specific MRE (substrate + PAABD) leads to a change in the molecular conformation of the MRE. This change alters the distance and/or relative orientation between the two FPs and, consequently, the FRET signal. This chapter focuses on genetically encoded biosensors, and specifically those dedicated to kinase activity measurements. Although kinase activity reporters (KARs) vary in specificity depending on the choice of the substrate, the design strategy of these reporters remains universal.[4]

Genetically encoded FRET biosensors can be used to analyze molecular events in single living cells and tissue, and even expressed in animals (e.g., aquatic animals, transgenic mice). They are expressed in the native context of a living cell to report on dynamic events. However, as described by Frommer et al.,[2] these FRET biosensors actively sense ("active reporters") cellular microenvironments and even the subcellular microenvironment and cannot be regarded as "passive reporters" such as for instance, regular FP-fused protein.

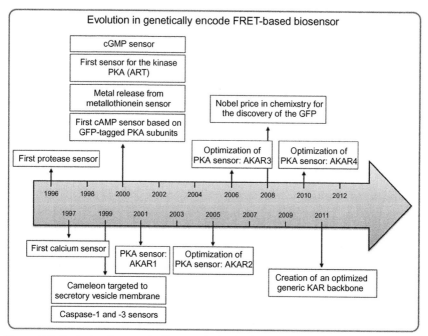

**Figure 5.2** Evolution in genetically encoded FRET-based biosensors.

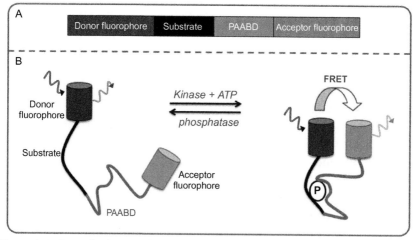

**Figure 5.3** General scheme of the functional domains of FRET-based kinase activity reporter (KAR). The schematic representation shows the biosensor conformational changes upon phosphorylation by the specific kinase leading to an increase in FRET between donor and acceptor fluorophores. (For color version of this figure, the reader is referred to the online version of this chapter.)

## 2. FLUORESCENCE GENERALITIES

### 2.1. Introduction

The relaxation of a fluorophore from an excited state to its ground state after the absorption of electromagnetic radiation may result in the emission of photons, which is called "luminescence". If this transition occurs for an electron in the excited singlet state (with a spin opposite to that of a paired electron in the ground state), this emission is called "fluorescence". The fluorescent molecule is often promoted to the $S_1$ excited state. In most cases, a rapid relaxation subsequently occurs from the lowest vibrational level of the first excited state $S_1$; this is the internal conversion process that usually occurs within approximately $10^{-12}$ s and results in some energy loss from the system, which is responsible for the energy difference (Stokes' shift) between the absorption and emission spectra (see the Jablonski diagram in Fig. 5.4). The energy of the emitted photon is dependent upon the ground state toward which the transition occurs.

### 2.2. The absorption process

The energy of the excited photon must be equal to, or greater than, the energy difference ($E_0$ and $E_1$) between the ground state ($S_0$) and the excited state ($S_1$). The frequency of this photon is $\nu = (E_1 - E_0)/h$, where $h$ is the Plank constant. When a photon is absorbed, its energy is transferred to the valence electron and this electron is promoted to a higher electronic orbit, thus putting the molecule in the excited state. This absorption is very fast, since it occurs within $10^{-15}$ s.

Experimentally, the efficiency of light absorption at a wavelength $\lambda$ is characterized by the *absorbance* $A(\lambda)$ related to the *transmittance* $T(\lambda)$ by

$$A(\lambda) = \log\left(\frac{I_0}{I}\right) = -\log T(\lambda) \qquad [5.1]$$

where $I_0$ is the intensity of a monochromatic incident light of wavelength $\lambda$ passing through an isotropic sample containing absorbing molecules at a concentration $c$ (mol$^{-1}$), $I$ is the light intensity leaving the absorbing medium, and $l$ (cm) is the absorption path length (sample thickness) of the sample (Fig. 5.5).

The absorbance follows the Beer–Lambert law

$$A(\lambda) = \varepsilon(\lambda) l c \qquad [5.2]$$

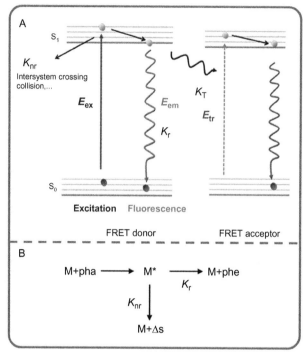

**Figure 5.4** (A) Jablonski diagram. $S_0$ is the fundamental energy level of the electron, $S_1$ is the first exited level, and the small lines represent the vibrational energy levels. $E_{ex}$ is the energy of excitation, $E_{em}$ is the energy of fluorescence emission, $K_r$ is the radiative de-excitation rate, and $K_{nr}$ is the nonradiative de-excitation rate. (B) Scheme of the molecular equilibrium occurring between the absorption and emission processes where M is the fluorophore (molecule), M* is the excited state of the molecule, pha is the photon absorbed, phe is the photon emitted, $\Delta S$ is the entropy, $k_r$ is the radiative de-excitation rate, and $k_{nr}$ is the nonradiative de-excitation rate. (For color version of this figure, the reader is referred to the online version of this chapter.)

where $\varepsilon(\lambda)$ is the molar absorption coefficient (L mol$^{-1}$ cm$^{-1}$) and $l$ (cm) is the absorption path length (sample thickness) of the sample (Fig. 5.5).

The absorbance reflects the probability of a population of fluorophores to jump to an excited state under the effect of an incident photon at the wavelength $\lambda$.

The *absorption coefficient* $\alpha(\lambda)$ is the absorbance divided by the optical path length ($l$) in the medium:

$$\alpha(\lambda) = \frac{A(\lambda)}{l} = \frac{1}{l}\log\left(\frac{I_0}{I}\right), \quad I = I_0 e^{-\alpha(\lambda)l} \qquad [5.3]$$

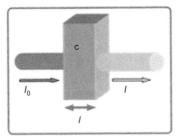

**Figure 5.5** Physical parameters implicated in the absorbance measurements. $I_0$ is the intensity of the incident light. I correspond to residual intensity after absorption by the sample. l is the path length of the sample. (For color version of this figure, the reader is referred to the online version of this chapter.)

## 2.3. The emission process

When a molecule has been promoted to an excited state upon the absorption of electromagnetic radiation, it necessarily returns to the ground state through competition between radiative ($K_r$) and nonradiative ($K_{nr}$) pathways. The radiative pathways involve photon emission, and nonradiative pathways include energy transfer through collisions, resonance energy transfer through near-field dipole–dipole interactions (such as FRET detailed in the next section), and photochemical decomposition. A change in the vibrational and rotational states of the molecule can also cause a loss of energy via a nonradiative process.[5]

The Jablonski diagram shown in Fig. 5.4 illustrates the balance of energy through the excitation–relaxation cycle.

The difference in energy (or wavelength) between the absorbed and the emitted photons is known as the "Stokes shift" shown in Fig. 5.6. This phenomenon was first described by Sir G. G. Stokes in 1852. A large Stokes shift is often highly desirable for simplifying the wavelength separation between the fluorescence emission and the excitation.[6]

There is competition between the different de-excitation processes previously discussed ($K_r$ and $K_{nr}$). The quantum yield ($\Phi$) is the ratio of the number of photons emitted to the number of photons absorbed. It can also be described using the rates of radiative ($K_r$) and nonradiative ($K_{nr}$) processes of de-excitation.

$$\Phi = \frac{K_r}{K_r + K_{nr}} \qquad [5.4]$$

The quantum yield can vary from 0 to 1, where 0 corresponds to non-fluorescent materials and 1 corresponds to highly fluorescent materials in which each photon absorbed results in an emitted photon.

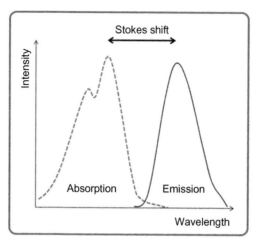

**Figure 5.6** Illustration of the Stokes shift between the absorption and emission spectra. (For color version of this figure, the reader is referred to the online version of this chapter.)

The excited molecules ($M^*$) could be de-excited by radiative ($K_r$) or nonradiative ($K_{nr}$) processes whose intersystem transfer as FRET ($K_T$). In classical kinetics, the rate of disappearance of excited molecules is expressed by the differential equation

$$-\frac{d[M^*]}{dt} = (K_r + K_{nr})[M^*] \qquad [5.5]$$

Integration of this equation yields the time evolution of the concentration of the excited molecules $[M^*]$ (see Fig. 5.4).

$$[M^*] = [M^*]_0 e^{(-t/\tau)} \qquad [5.6]$$

$$\text{With } \tau = \frac{1}{K_r + K_{nr}} \qquad [5.7]$$

$[M^*]_0$ is the concentration of the excited molecules at time 0, which results from light excitation. The *fluorescence lifetime* $\tau$ is in the range of $10^{-9}$ s (0.5–20 ns for commonly used fluorescence transitions). $\tau$ describes the average time for which a molecule stays in its excited state before emitting a photon.[7] The resonance transfer or FRET ($K_T$) is included in nonradiative deactivation pathways ($K_{nr}$). It can be defined as the difference between the rates $K_{nr}$ of the donor only and $K_{nr}$ of the donor in the presence of the acceptor. The fluorescence intensity begins to decrease when molecules are in their excited states. This decrease depends upon the rate of electron de-excitation and it can be deduced from Eq. (5.7):

$$I_t = I_0 e^{-t/\tau} \qquad [5.8]$$

$$\text{With } \tau = \frac{1}{K_\gamma + K_{nr} + K_T} \qquad [5.9]$$

where $K_{nr}$ is the nonradiative rate of the donor only. The fluorescence lifetime is in fact the inverse of the slope of the curve measuring the fluorescence as a function of time in a semilogarithmic representation (Fig. 5.7).

The fluorescence impulse response function $I(t)$ is often represented by a multiexponential decay model

$$I(t) = \sum_i \alpha_i e^{-t/\tau_i} \qquad [5.10]$$

where $\tau_i$ are the decay times and $\alpha_i$ are the amplitudes of the components. The values of $\alpha_i$ and $\tau_i$ may have a direct or an indirect molecular significance. For a mixture of fluorophores, if each component has a single decay time, $\tau_i$ are their decay times (Fig. 5.7). The parameters $\alpha_i$ and $\tau_i$ cannot always be attributed to molecular features of the sample. Alternatively, the measured intensity decay can be fitted with Eq. (5.10). The values of $\alpha_i$ and $\tau_i$ can be used to calculate the *fractional contribution* $f_i$ of each decay time $\tau_i$ to the steady-state intensity:

$$f_i = \frac{\alpha_i \tau_i}{\sum_i \alpha_i \tau_i} \qquad [5.11]$$

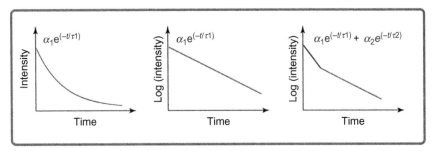

**Figure 5.7** Fluorescence lifetime decay profiles: $\tau_2 < \tau_1$ (adapted from the Nikon website). (For color version of this figure, the reader is referred to the online version of this chapter.)

The resolution of multiple $\tau_i$ becomes increasingly difficult as they get more closely spaced. The statistically significant resolution of close $\tau_i$ requires a high signal-to-noise ratio and a large number of collected photons. The calculation of the energy transfer efficiency from the donor fluorescence lifetime in the presence and absence of the acceptor (Eq. 5.11) assumes that, under the experimental conditions, the donor decays according to a single-exponential model. If the donor displays a multi-exponential decay, $\tau_i$ values can be used to calculate the average lifetime $\langle \tau \rangle$. It is defined as the average time that the fluorophore remains in the excited state and is defined as

$$\langle \tau \rangle = \frac{\int_0^\infty t I(t) dt}{\int_0^\infty I(t) dt} = \sum_i f_i \tau_i \qquad [5.12]$$

In order to simplify data representation, mean lifetime $\tau_{mean}$ has been largely used in FRET experiments.[8–11] It is given by

$$\tau_{mean} = \sum_i \alpha_i \tau_i = \int I(t) dt \qquad [5.13]$$

## 3. FRET MEASUREMENT

### 3.1. Introduction

Among the different previously described nonradiative deactivation processes $(K_{nr})$, the energy transfer between dipoles was first described by Förster in 1926 and Perrin in 1932. Förster resonance energy transfer or FRET is a physical process in which the energy of a chromophore (called "donor" [D]) in its excited state is transferred nonradiatively to a neighboring chromophore (called "acceptor" [A]) while in its ground state.[5] This physical process has been often applied experimentally for investigating molecular interactions at distances beyond the diffraction-limited resolution (for review see Refs. 5,7). For instance, FRET measurements (generally used in spectroscopy and microscopy) allow the investigation of the formation of protein complexes in living cells and tissues, as well as the conformational changes of single proteins such as biosensors (for review see Refs. 6,7,12).

## 3.2. FRET basis

FRET is one possible pathway for the relaxation of the excited state molecules. This phenomenon occurs only under appropriate conditions of proximity and orientation between two fluorophores (donor and acceptor), which are as follows:

1. This form of energy transfer occurs in the near field of the donor. In other words, the distance ($r$) between D and A must be less than 10 nm ($r < 10$ nm) so that $K_T \neq 0$.

2. The energy transfer is achieved between molecules with resonant oscillation dipole moments (overlapping wave functions). This requires an overlap between the D emission spectrum and the A excitation spectrum.

3. The orientation of the emission dipole moment of D with respect to the excitation dipole moment of D must be such as for FRET to occur.

The theoretical concept for FRET was developed following both the classical model by Perrin in 1925 followed by the quantum mechanical model in 1932 and by Förster in 1946–1949[13–15] (for review see Refs. 5,6).

If one considers a single donor and acceptor separated by a distance $r$, the rate of energy transfer $K_T(r)$ can be calculated as a probability of the transfer of an energy quantum from D to A per time unit, given by the fundamental equation

$$K_T(r) = \frac{\Phi_D \kappa^2}{\tau_D r^6} \left( \frac{9000 \ln(10)}{128 \pi^5 n^4 N_A} \right) J(\lambda) \int_0^\infty F_D(\lambda) d\lambda \qquad [5.14]$$

where $\Phi_D$ is the donor quantum yield (as previously described in this chapter) in the absence of acceptor, $\tau_D$ the donor lifetime in the absence of acceptor, $n$ is the refractive index of the medium, $N_A$ is the Avogadro's number, $J(\lambda)$ is the overlap integral, $F_D$ is the normalized fluorescence intensity of the donor, and $\kappa^2$ is a dimensionless orientation factor describing the relative spatial orientation of the donor and acceptor transition moments. Note that $\varepsilon_A$, $\Phi_D$, and $n$ are fixed by the choice of FRET pairs and the medium. Therefore, $K_T(r)$ variation is mainly dependent on $r$ and $\kappa$.

Equation (5.14) is not easy to use for the design of biochemical experiments[6]. This is why the Förster distance $R_0$ was introduced by Förster in 1948. When the transfer rate $K_T(r)$ is equal to the decay rate of the donor in absence of an acceptor, one-half of the donor molecules decay by the energy transfer process. Once the value of $R_0$ is known, the rate of energy transfer $K_T$ can be easily calculated as

$$K_T(r) = \frac{1}{\tau D}\left(\frac{R_0}{r}\right)^6 \qquad [5.15]$$

for $r = R_0$

$$K_T(r) = 1/\tau_D \qquad [5.16]$$

One obtains (from Eq. 5.14)

$$R_0^6 = \Phi_D \kappa^2 \left(\frac{9000\ln(10)}{128\pi^5 n^4 N_A}\right) \int_0^\infty F_D(\lambda)\varepsilon_A(\lambda)\lambda^4 d\lambda \qquad [5.17]$$

$R_0$ is given in Angström and may be simplified as

$$R_0^6 = 8.79 \times 10^{-5}\left[\Phi_D \kappa^2 J(\lambda) n^{-4}\right] \qquad [5.18]$$

This expression allows the Förster distance to be calculated from the spectral properties of the donor and the acceptor and from the donor quantum yield $\Phi_D$.[13]

FRET efficiency is dependent on the inverse sixth power of the intermolecular separation ($r$) (discussed in the next section).

Additionally, Förster distance is usually reported for an assumed value of $\kappa^2$ of 2/3 characterizing free FRET pairs. The question of dipole–dipole orientation is discussed later.

Directly, if the transfer rate is much faster than the decay rate, then energy transfer will be efficient; otherwise, FRET will be inefficient.

A crucial step in the practical implementation of FRET is the knowledge of several major parameters:
- the quantum yield $\Phi_D$ of the fluorophore donor only;
- the overlap integral $J(\lambda)$ between the donor and acceptor fluorophores;
- the orientation factor $\kappa^2$ between two fluorophore dipoles.

## 3.3. Overlap integral $J(\lambda)$

$J(\lambda)$ is the *overlap integral* between the donor emission and the acceptor absorption spectra (expressed as $M^{-1} cm^{-1} nm^4$) and is defined as

$$J(\lambda) = \frac{\displaystyle\int_0^\infty F_D(\lambda)\varepsilon_A(\lambda)\lambda^4 d\lambda}{\displaystyle\int_0^\infty F_D(\lambda)d\lambda} \qquad [5.19]$$

where $F_D(\lambda)$ is the corrected fluorescence donor (dimensionless) and $\varepsilon_A$ is the extinction coefficient of the acceptor at $\lambda$ (expressed in $M^{-1}\,cm^{-1}$).

## 3.4. Orientation factor $\kappa^2$

$\kappa^2$ can vary from 0 to 4 according to the following equation:

$$\begin{aligned}\kappa^2 &= (\cos\theta_T - 3\cos\theta_D\cos\theta_A)^2 \\ &= (\sin\theta_D\sin\theta_A\cos\Phi - 2\cos\theta_D\cos\theta_A)^2\end{aligned} \quad [5.20]$$

where $\theta_A$, $\theta_D$, $\theta_T$, and $\Phi$ are as shown in in Fig. 5.8.

Additionally, the fluctuation of *refractive index* $n$ can induce errors in the calculated distance $r$ but this effect is not usually considered.

The orientation factor ($\kappa^2$) characterizes the statistical average of the relative fluorophore orientations, and determines both how well the fluorophore dipoles are coupled and how efficiently energy is transferred (Fig. 5.8). However, if the dipoles are perpendicular, $\kappa^2$ becomes zero, which would result in serious errors in the calculated distance. This question has been discussed in detail.[16–18] In general, the variation of $\kappa^2$ does not induce major errors in the calculated distance; however, for intramolecular FRET in biosensors this question is of importance. In fact, in a rigid molecule (such as a polypeptide with four to nine amino acid residues) with isotropic orientational distribution (statistically randomly distributed) of the donor and the acceptor transition moments but with no rotation during the lifetime of excited state (frozen), the value of $\kappa^2 = 0.476$ can be used. The optimal value of $\kappa^2$ at room temperature is $2/3$.[19] For a fluorophore bound to macromolecules (i.e., fluorescent

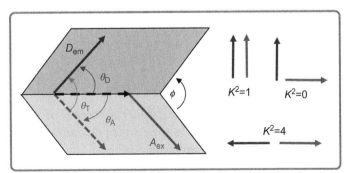

**Figure 5.8** Parameters involved in the calculation of the orientation factor $\kappa^2$. (For color version of this figure, the reader is referred to the online version of this chapter.)

proteins), segmental motions of the donor and acceptor tend to randomize the orientations and $\kappa^2 = 2/3$ is classically used.[6,20] Computational simulations showed that $\kappa^2$ converges to 2/3 in a FRET sensor where the D/A pair is presumed to be freely mobile.[21] Finally, the evaluation of errors in the distance ($r$) due to approximation on $\kappa^2$ has been reported, but they are no more than 10%.[20]

## 3.5. Energy transfer efficiency

The *efficiency of energy transfer* ($E$) can be defined as the ratio of the relaxation rate due to energy transfer to the sum of all relaxation rates.

$$E = \frac{K_T(r)}{\tau_D^{-1} + K_T(r)} \qquad [5.21]$$

The rate of energy transfer is often defined as a function of inverse sixth power of the distance between the two molecules.

$$E = \frac{R_0^6}{R_0^6 + r^6} \qquad [5.22]$$

The first factor that affects the FRET signal is the distance ($r$) between the fluorophores. The most sensitive range of $r$ is 0.7–1.4 $R_0$, corresponding to 90–10% FRET efficiency (Fig. 5.9). $R_0$ usually ranges from 4 to 7 nm; hence protein conformational change in this range is ideal for the largest FRET dynamic in biosensors.

The transfer efficiency is typically measured using the relative fluorescence intensity of the donor in the absence ($F_D$) or presence ($F_{DA}$) of acceptor.

$$E = 1 - \frac{F_{DA}}{F_D} = 1 - \frac{\int I_{DA}(t)dt}{\int I_D(t)dt} \qquad [5.23]$$

where $F_{DA}$ is the intensity of donor fluorescence emission in presence of acceptor, $F_D$ is the intensity of donor fluorescence in absence of acceptor, and $I_D$ and $I_{DA}$ are, respectively, the intensity decays of the donor alone and the donor in the presence of the acceptor.

The transfer efficiency can also be calculated from the lifetimes under these respective conditions ($\tau_D$ and $\tau_{DA}$):

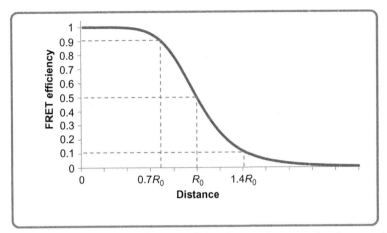

**Figure 5.9** FRET efficiency as a function of the distance between the donor and the acceptor. The FRET efficiency is 50% when the distance is equal to the Förster distance ($R_0$). FRET efficiencies ranging from 10% to 90% correspond to distances between fluorophores of $1.4R_0$ and $0.7R_0$, respectively.

$$E = 1 - \frac{\tau_{DA}}{\tau_D} \qquad [5.24]$$

The FRET kinetic measurement can be performed by the calculation of the ratio ($R$) between the two stationary states' kinetics. $R$ is classically used for measurements by FRET-based biosensors (see next section).[22]

Both Eqs. (5.23) and (5.24) are applicable only to donor/acceptor pairs that are separated by a fixed distance. However, single-exponential decays are rare in biology.

The mean lifetime $\tau_{mean}$ defined by Eq. (5.15) has been largely used in FRET experiments.[8–11] This mean lifetime is then equivalent to the area of the fluorescence intensity decay, which is related to the FRET efficiency $E$. However, the mean lifetime does not correspond to the correct average lifetime, which is defined by Eq. (5.12).

## 3.6. FRET measurements of molecular populations

In FRET analysis, particularly for biosensors, two elements must be usually considered: the interacting fluorophore population and the FRET efficiency.

The distance distribution density function $P(r)$ describes the probability of finding the specific donor/acceptor pair separation.

The apparent transfer efficiency *(E)* can be written as

$$E = \int_0^{\infty} \frac{P(r)R_0^6}{R_0^6 + r^6} dr \qquad [5.25]$$

Bulk measurements of FRET efficiency by intensity-based methods cannot distinguish between an increase in FRET efficiency (i.e., coupling efficiency) and an increase in FRET population (concentration of FRET species) since both parameters are not resolved. FRET measurements based on the analysis of the donor fluorescence lifetime may resolve this problem with multiexponential decay models.[7,23] The assumption that interacting and noninteracting populations are present allows the determination of both the efficiency of interaction and the fractional population of the interacting molecules. In the first instance, the presence/absence of FRET is determined by fitting the experimental data to a single-exponential decay.[6] Sufficient reduction in the measured lifetime indicates the existence of FRET. Additional analysis is subsequently applied to determine the source of lifetime reduction. In this case, a biexponential fluorescence decay model applied to the data allows the determination of the fluorescence lifetimes of noninteracting and interacting subpopulations or two distinct levels of interaction in case of many biosensors. In time-domain measurements (see next section), data may be fitted by iterative convolution with

$$I(t) = IRF(t) \otimes \left\{ Offset + \sum_i \alpha_i e^{-t/\tau_i} \right\} \qquad [5.26]$$

where $IRF(t)$ is the instrumental response and Offset is the baseline, $\tau_i$ is the lifetime of interacting or noninteracting populations, and $\alpha_i$ is the the pre-exponential factor relating to the absolute species concentration (Fig. 5.10).[24]

## 3.7. Conclusion on FRET principles and applications

It is difficult to obtain quantitative determination of labeled interacting molecules from steady-state images. The fluorescence intensity does not only depend on the FRET efficiency but also on the unknown local concentration of dyes. Up to eight measurements at different excitation wavelengths and in different emission wavelength bands can be used to obtain calibrated FRET results from steady-state data.[7] In lifetime data, however, FRET shows up as a dramatic decrease of the donor lifetime.[6] Initially, qualitative FRET results can be obtained by fitting decay curves with a single-exponential approximation.

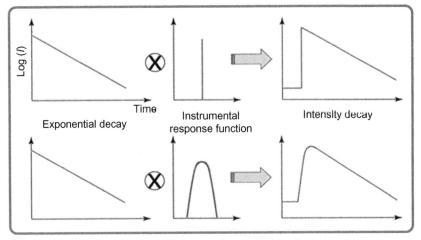

**Figure 5.10** Illustration of the convolution product. When the instrumental response function (IRF) of the FLIM system is temporally short, the measured intensity decay is almost identical to the sample decay. If the IRF is large, the collected decay becomes very much different of the sample decay, and it becomes thus necessary to take into account this IRF to obtain correct lifetime estimations. (For color version of this figure, the reader is referred to the online version of this chapter.)

Quantitative measurements require multiexponential decay analysis.[6] The proportion of donor molecules involved in energy transfer is given by the ratio of the two exponential decay amplitudes, $\alpha_2/\alpha_1$, while the average coupling efficiency of the FRET pairs is given by the lifetime ratio $\tau_2/\tau_1$.

A major complication of many *in vivo* FRET experiments is the large and differential contribution of cellular autofluorescence to the measured donor and acceptor fluorescence. Autofluorescence contributions can be sometimes corrected by acquiring an additional image,[25] or they can be minimized by the use of spectral lifetime image microscopy.[26]

When autofluorescence is substantial, determination of FRET efficiency by fluorescence lifetime measurement might be advantageous because autofluorescence lifetime is usually very short and can be included in the fit of lifetime data. Finally, one has to consider the limitations imposed by the available instrumentation. For example, lifetime measurements require relatively sophisticated instrumentation, which is not yet widely available.

## 3.8. Few considerations on fluorescent proteins

Green fluorescence proteins from the jellyfish *Aequorea victoria* are well characterized and have provided myriad applications in cellular biology.[27,28] GFP engineering has generated a large range of fluorescent proteins of different colors,[29] enabling scientists to consider dynamic localization of several proteins of interest in living cells.[1]

The properties of these FPs reveal that they are excellent candidates for FRET-based biological applications (see Table 5.1). The choice of a particular FP as a donor or an acceptor is very important and is mainly based on the analysis of its respective excitation and emission spectra. FPs must meet certain criteria to form a FRET pair: (i) an effective overlap between the emission spectrum of the fluorophore donor and the excitation spectrum of the acceptor; (ii) a large extinction coefficient at the region of excitation; (iii) a high quantum yield (ratio of photons emitted/photons absorbed); (iv) a separation between the excitation and emission spectra of the donor and of the acceptor; (v) good photostability; (vi) high brightness; (vii) minimal perturbation to the environment by FPs (toxicity); (viii) minimum sensitivity to the cellular environment (pH, chloride); and (ix) inability or at least limited capacity to dimerize and/or oligomerize. The last requirement is very important because the use of oligomerizing FPs may compromise the interpretation of the FRET signal. In 2006, Dunn et al.[43] elegantly demonstrated that the use of monomeric FPs significantly increased the FRET efficiency of a KAR. However, the FPs' expression level must be sufficiently high to provide enough signal, but not too high, it should become cytotoxic. In contrast, some tetrameric FPs can be toxic to bacteria when produced in large quantities, but this is not the case with monomeric FPs. All these requirements need to be considered and therefore a compromise in the choice of a particular FP in a specific context is suggested. Fortunately, FPs are continually being subjected to molecular engineering to improve their intrinsic properties and to increase the number of variants.[44,45] Generally, for biosensing approaches, FRET can be evaluated by ratiometric methods by measuring the fluorescence emitted by the acceptor in response to the excitation of the donor fluorophore. However, when considering protein interactions studies by FRET measurement, FLIM is a preferred method. Thus, other approaches have emerged in recent years with fluorescent protein engineering that rely on donor FLIM for biosensing studies.

## 4. FRET MEASUREMENTS: METHODS AND INSTRUMENTATION

As previously described, FRET induces modification of several properties of the emitted fluorescence. Different techniques thus arise from these modified measurements,[46,47] such as monitoring the fluorescence emission

**Table 5.1** Properties of frequently used FPs and their implementation as FRET pair for ratiometric and fluorescence lifetime imaging

| Protein | Color[a] | Organism | Excitation peak (nm)[b] | Emission peak (nm)[b] | Brightness | Photostability | Oligomerization | References |
|---|---|---|---|---|---|---|---|---|
| | | | | Main fluorescent proteins | | | | |
| eCFP | Cyan | *Aequorea victoria* | 433/445 | 475/503 | ++ | +++ | Weak dimer | Cubitt et al.[32] |
| meCFP | Cyan | *Aequorea victoria* | 433/452 | 475/505 | ++ | +++ | Monomer | Zacharias et al.[33] |
| Cerulean | Cyan | *Aequorea victoria* | 433/445 | 475/503 | +++ | ++ | Weak Dimer | Rizzo et al.[34] |
| mCerulean | Cyan | *Aequorea victoria* | 433/445 | 475/503 | +++ | ++ | Monomer | Rizzo et al.[34] |
| eGFP | Green | *Aequorea victoria* | 488 | 507 | +++ | ++++ | Weak dimer | Tsien[29] |
| meGFP | Green | *Aequorea victoria* | 488 | 507 | +++ | ++++ | Monomer | Heim et al.[35] |
| eYFP | Yellow | *Aequorea victoria* | 514 | 527 | ++++ | ++ | Weak dimer | Miyawaki et al.[36] |
| Venus | Yellow | *Aequorea victoria* | 515 | 528 | ++++ | + | Weak dimer | Nagai et al.[37] |
| mVenus | Yellow | *Aequorea victoria* | 515 | 528 | ++++ | + | Monomer | Nagai et al.[37] |
| DsRed | Red | *Discosoma* sp. | 558 | 583 | ++++ | ++++ | Tetramer | Matz et al.[38] |
| mRFP1 | Red | *Discosoma* sp. | 584 | 607 | ++ | + | Monomer | Campbell et al.[39] |
| tdTomato | Red | *Discosoma* sp. | 554 | 581 | ++++ | +++ | Tandem dimer | Shaner et al.[40] |
| mCherry | Red | *Discosoma* sp. | 587 | 610 | ++ | +++ | Monomer | Shaner et al.[40] |

**Table 5.1** Properties of frequently used FPs and their implementation as FRET pair for ratiometric and fluorescence lifetime imaging—cont'd

| | | | Example of FRET pair fluorescent proteins | | | | |
|---|---|---|---|---|---|---|---|
| FRET pair | Recommended donors | Recommended acceptors | Donor excitation (nm) | Acceptor emission (nm) | Laser | $R_0^c$ | References |
| Cyan–yellow | meCFP/ mCerulean | eYFP/mVenus | 433–452/ 433–445 | 527/528 | Violet | $4.92 \pm 0.10$ | Allen and Zhang[41] |
| Green–red | meGFP | mRFP1/mCherry/ tdTomato | 488 | 607/610/581 | Argon | $4.73 \pm 0.09$ | Harvey et al.[42] |

[a]The colors represent the emission fluorescence of the corresponding fluorescent protein.
[b]Maximum wavelengths of the excitation and emission spectra.
[c]$R_0$: Förster distances $r_0$ for FRET pairs of fluorescent proteins. $R_0$ values are given in nanometers.
References 29–42.
*Adapted from Refs. 30,31.*

spectrum, lifetime, anisotropy, and so on. In the particular case of biosensor application, the most popular technique is called the "sensitized emission," which consists in acquiring the fluorescence emitted by the donor only and the acceptor only, after donor excitation. However, several other techniques offer very interesting alternatives for quantifying the molecular activity, for example, (i) spectral imaging, which consists of excitation at one wavelength and measurement of the whole emission spectrum, and (ii) FLIM for measuring the lifetime changes of donor fluorescent proteins.

Each of these techniques has its own advantages and drawbacks. Using two extreme examples, sensitized emission is the simplest method and can be performed on nearly all conventional microscopes, but it requires great care with regard to biological references; FLIM, on the other hand, needs tricky instrumentation but can yield unambiguous measurements of FRET efficiency. In the following section, systems needed to perform reliable biosensor imaging experiments are therefore described.

## 4.1. Intensity-based approaches

The most intuitive and easy methods to perform FRET measurements are based on fluorescence intensity. The technique discussed now consists in imaging the sensitized emission, that is, the fraction of acceptor emission induced by the nonradiative energy transfer from the donor molecule. These measurements can be based on either ratiometric or spectral imaging.

### 4.1.1 Ratiometric approach

FRET measurements using biosensors are usually performed by acquiring the fluorescence emitted by the donor and the acceptor. The resulting data are usually represented by the ratio of these fluorescence measurements after appropriate corrections. The classical ratiometric approach, which is based on Equation (5.14), consists in measuring at least three channels: (i) excitation and observation of the donor ($I_{donor}$); (ii) excitation of the donor and observation of the acceptor ($I_{FRET}$); and (iii) excitation and observation of the acceptor ($I_{acceptor}$). $I_{acceptor}$ is needed to compensate for donor and acceptor concentration differences, thus accounting for bleed-through and nonspecific excitation.[48] In the case of biosensor measurements performed on living cells, the donor and acceptor amounts are identical and the last channel acquisition is useless. In this particular case, only (i) and (ii) are required and, as will be seen in Section 5, correction by acceptor channel will even decrease the signal-to-noise ratio of the measured signal. This experimental condition is formally similar to the well-described conditions used to

**Table 5.2** Example of the adapted filter set for the CFP/YFP FRET pair[48]

| Channel | Excitation filter | Excitation dichroic | Emission filter |
|---------|-------------------|---------------------|-----------------|
| $I_{donor}$ | 420/20 | 450 | 475/40 |
| $I_{FRET}$ | 420/20 | 450 | 535/25 |

image calcium with ratiometric indicators, except that calcium probes such as fura-2 show a ratiometric change for two different excitation wavelengths while biosensors show a ratiometric change of the fluorescence emission for one excitation wavelength.

One major advantage of this technique is that it requires only two wide-field or confocal images, allowing for high-speed acquisitions of fluorescence images with systems equipped with appropriate filter sets available in most imaging core facilities (Table 5.2).

Special care must, however, be taken to limit the delay between the acquisition of $I_{donor}$ and $I_{FRET}$ to avoid artifacts due to cell movement between two image acquisitions (see Section 7 of this chapter). The traditional system needs a filter cube change and thus a delay of several hundreds of milliseconds between image acquisitions is possible, which is not negligible compared to the movement times of cells or organelles. Instead of changing the fluorescence cube, which may worsen image registration, a faster solution is the use of a filter wheel between the microscope and the camera, which allows changes to be made in the emission filter in less than 100 ms. Solutions have also been developed to perform simultaneous acquisitions of both channels, as described in Fig. 5.11.[49–51]

These wide-field configurations allow fast acquisition of both channels but they lack optical sectioning capability. Confocal microscopes can also be used in the same way, with excitation of the donor and simultaneous collection of the fluorescence emitted by donor and acceptor molecules. This allows for a three-dimensional localization of the biosensor response, and video-rate confocal microscopes can be used when a high spatial resolution is needed, as shown by the early use of biochemical biosensors.[52] The use of confocal microscopes also opens the way to more complex acquisition procedures such as spectral imaging.

### 4.1.2 Spectral imaging

Most of the latest confocal microscopes are designed for spectral imaging either in the sequential mode or, most interestingly, for biosensor imaging in the simultaneous mode.[53] Indeed, for FRET applications, spectral imaging is

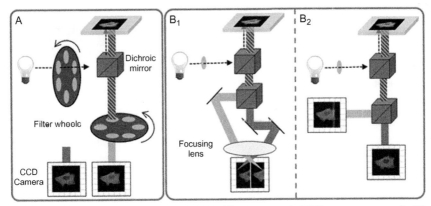

**Figure 5.11** Examples of the setup optimized for ratiometric FRET measurements. In (A), images are acquired sequentially. A fast filter wheel allows fast switching between both acquisitions (10–30 ms). In (B), images are acquired simultaneously. A dichroic mirror is used to separate the emission from donor and acceptor molecules. A focusing lens can then be used for directing the light of both channels on each half of the same camera (B1), or two different cameras can be used for light collection (B2). (For color version of this figure, the reader is referred to the online version of this chapter.)

a slower but more precise configuration for measuring sensitized emission. It consists in measuring for each pixel of an image the overall emission spectrum and is not limited to only two bandwidths using filters (see Fig. 5.12 for more details about the setup).

Gathering fluorescence emission spectra then allows donor and acceptor emission spectra to be separated according to the distinct shape of both spectra after spectral unmixing. The fluorescence signal is then analyzed in the same way as traditional ratiometric images are. It, however, allows for distinguishing the real FRET signal from other elements that may alter this measurement, such as autofluorescence or the presence of multiple fluorophores in the sample, and it can then be used in more complex biological environments.

### 4.1.3 Lifetime-based approaches

Fluorescence lifetime is inherently quantitative and is most of the time independent of the concentration of the fluorophore. Furthermore, FRET-FLIM experiments need measurements of the donor fluorescence lifetime only (cf. Eqs. 5.16, 5.18, and 5.29), which makes it extremely valuable for simultaneous multibiosensor measurements. Lifetime measurement, however, requires dedicated and more sophisticated instruments than for ratiometric imaging. The measurement can be done in either the time domain (TD FLIM) or the frequency domain (FD FLIM) (Fig. 5.13). In

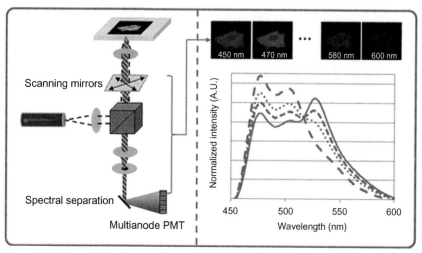

**Figure 5.12** Simplified scheme of a confocal setup allowing spectral imaging in simultaneous mode. Briefly, a laser beam scans the specimen using scanning mirrors. The pinhole conjugated to the focal plane rejects light emitted by objects outside this focal plane, which results in optical sectioning. Spectral separation is then achieved by a grating combined to recycling systems, allowing minimal loss of light. The light is then collected using a multianode PMT. The fluorescence spectrum is recorded for each pixel. (For color version of this figure, the reader is referred to the online version of this chapter.)

the following sections, the two most representative techniques, time–correlated single-photon counting (TCSPC) for TD FLIM and phase and modulation measurements for FD FLIM are discussed in detail.

### 4.1.4 Time domain: TCSPC

Most TCSPC systems are implemented on a confocal microscope equipped with

- a pulsed laser source. The source must produce short laser pulses (from several hundred femtoseconds to picoseconds' width) with a frequency usually ranging from 10 to 80 MHz. It is interesting to note that the Ti:Sa laser matches these specifications, which can be of great help in deep and noninvasive biosensor imaging;

- detectors with a fast instrumental response. Optimal instrumental response function can be obtained using a multichannel plate or the latest generation of avalanche photodiodes ($<50$ ps full-width at half-maximum), but they are extremely fragile and require careful handling. TCSPC manufacturers therefore provide more robust detectors, that is, optimized photomultiplier tubes, with an IRF of around 250 ps adapted

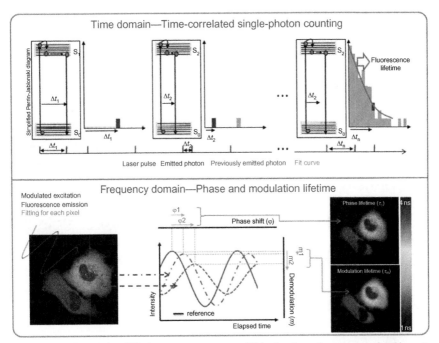

**Figure 5.13** Principle of lifetime measurements. (A) In the time domain, a pulsed laser is used to excite fluorophores. The time between photon excitation and emission is measured and accumulated to get an histogram of photon emission time. Fluorescence lifetime is then estimated from the slope of this exponential decay. (B) In the frequency domain, a modulated excitation is used to excite the sample. Monitoring of the fluorescence phase and modulation shift compared to a reference with known fluorescence lifetime is used to calculate phase and modulation lifetimes. (See Color Insert.)

for TCSPC experiments, which, however, necessitates particular attention during photon decay curves analysis;

- a photon counting card. All the systems rely on the same principle, which is based on a time amplitude converter. It consists in a linear voltage ramp started by the arrival of a photon and stopped by the next laser pulse. The output voltage will thus be proportional to the photon arrival time. However, the ramp is triggered only by a photon arrival followed by a laser pulse, which means that if two photons are acquired between two laser pulses, only the first photon will be measured. This is the "pulse pile-up" effect (Fig. 5.14A). To avoid this statistic selection of fastest photons, one has to limit the acquisition frequency to one-hundredth of the excitation frequency (giving rise to an error every 10,000 photons), which explains the longer acquisition time of this technique.

An example of such a setup[23] is presented in Fig. 5.14B.

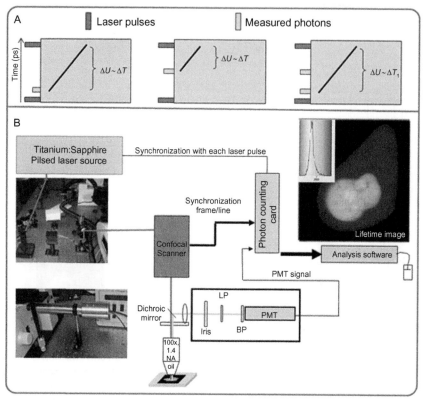

**Figure 5.14** (A) Principle of the time amplitude conversion. On the left, a fast photon is measured and starts the linear tension ramp resulting on a large $\Delta U$ corresponding to the difference in time between a photon emission and the following laser pulse. While the excitation is at constant frequency, the measured $\Delta U$ allows the retrieval of the photon emission time. The same explanation is also valid for a slow photon (middle scheme). However, if two photons are emitted between two laser pulses, only the first one is measured. This effect, called "pulse pile-up," induces an artifactual decrease in the measured fluorescence lifetime. (B) Scheme of a typical TCSPC acquisition setup with a laser source allowing two-photon excitation. Pictures on the left show the injection of the infrared laser in a confocal scan head (upper panel) and the detection module adapted on the descanned position of the confocal microscope. (See Color Insert.)

### 4.1.5 Frequency domain: Phase and modulation

In many experimental FD FLIM systems described in the literature, the modulated excitation light source is composed of a laser (diode, solid-state, gas, or dye lasers) combined with an external modulator (either an acousto-optic or an electro-optic modulator).[54–56] The advent of commercially available LEDs (light-emitting diodes), which can be directly modulated, has contributed to the simplification of the instrumentation and reduced the cost of FD FLIM systems.[57–59]

Measurement of the phase and modulation quantities can be performed with two approaches: the heterodyne and the homodyne methods. The heterodyne method is the preferred approach for accurately estimating fluorescence lifetime components in cuvette experiments. It has also been successfully applied in FLIM experiments in combination with scanning-mode and single-channel detectors (usually with a gain-modulated photomultiplier[54]). This method is thus compatible with laser scanning microscopes (such as confocal and multiphoton microscopes), which offer high three-dimensional spatial resolution and good signal-to-noise ratios.

The homodyne method, which consists in modulating the excitation light and the detector at the same frequency, is routinely used in many biological and biophysical laboratories because it can be performed with wide-field detectors (such as a modulated intensified CCD camera). This approach has been implemented in classical wide-field fluorescence microscopy, which enables rapid FLIM image acquisition.[57–60] However, one limitation inherent in this system is the nonconfocality of the excitation and, consequently, of the fluorescence emission. To improve the axial discrimination of the FD FLIM system, it can be simply combined with either a single-plane illumination strategy[61] or a spinning disk module[62,63] (an example of such an implementation is presented in Fig. 5.15).

Nowadays, several companies (ISS, Intelligent Imaging Innovations, Lambert Instruments) have introduced commercial FD FLIM systems that can be fully integrated with all commercial multiphoton, confocal, or wide-field microscopes.

# 5. DATA ANALYSIS

## 5.1. Ratiometric imaging

Data processing methods for reaching quantitative FRET measurements between two fluorescent molecules with ratiometric imaging have been exhaustively covered.[64–66] Briefly, in order to correct for both varying, unknown concentrations of the donor and acceptor and instrumental artifacts (e.g., spectral bleed-through), a large number of methods have been proposed with different numbers of images and thus different levels of complexity.[64,65,67] Among these, the most robust and widely used method requires the acquisition of three fluorescence images. In this case, the ratio $R$ is defined by[68]

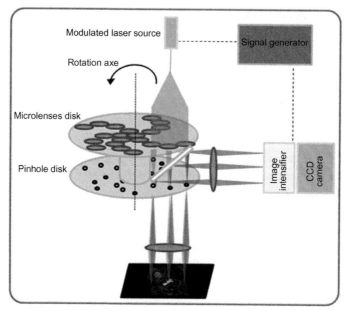

**Figure 5.15** Scheme of a phase and modulation acquisition system adapted on a spinning disk system to ensure fast fluorescence lifetime measurement with optical sectioning capability. (See Color Insert.)

$$R = \frac{I_{FRET} - \alpha \times I_{Donor} - \beta \times I_{Acceptor}}{I_{Donor}} \qquad [5.27]$$

where $I_{Acceptor}$ is the fluorescence intensity measured with an acceptor emission filter after acceptor excitation, $I_{Donor}$ and $I_{FRET}$ are fluorescence intensities measured, respectively, with the spectral bandpass of the donor and acceptor after donor excitation, and $\alpha$ and $\beta$ are the correction factors for, respectively, the bleed-through of the donor into the acceptor emission filter after donor excitation and the bleed-through of the acceptor into the acceptor emission filter after donor excitation.

In the particular case of a single-chain biosensor, the expression (27) can be simplified. Indeed, donor and acceptor concentrations are identical, which implies that bleed-through correction factors $\alpha$ and $\beta$ are linearly dependent and that fluorescence intensities $I_{Acceptor}$ and $I_{Donor}$ are proportional. Taking into account these simplifications, a straightforward calculation leads to

$$R = \frac{I_{FRET}}{I_{Donor}} - c \qquad [5.28]$$

where $c$ is a constant that is dependent on the fluorophore's properties. This constant, which is just an offset modification of the ratio $R$, does not give any supplementary information.

For a single-chain biosensor, the ratio $R$ is then fully described by the ratio of the fluorescence intensities, $I_{FRET}/I_{Donor}$. Consequently, the third fluorescence image $I_{Acceptor}$ (see Eq. 5.27) is not useful; this additional image acquisition will just increase the complexity of instrumentation, the time of acquisition (and thus the related movement artifacts), and the noise propagation in the ratio calculation.

Even though the expression of the ratio is extremely simple, the fact must be emphasized that obtaining reliable $R$ values requires special care and numerous corrections, which have been exhaustively detailed elsewhere.[68,69] Briefly, we need to compensate for instrumental artifacts (camera offset subtraction, shading or flat-field correction for the nonuniformity of the illumination, correction for misalignments between two fluorescence images due to chromatic aberration) and for chemical and biological artifacts (photobleaching correction for taking into account the different donor and acceptor photobleach rates[70], optional autofluorescence subtraction, correction for cell movement, or deformation between acquisition time).

Calculation of the corrected ratio $R$ from acquired fluorescence images can be performed instantaneously pixel by pixel, and the calculated $R$ values can be displayed with pseudocolor superimposed on the fluorescence intensity image in real time, making it possible to follow the biosensor activity[71–73] (see Fig. 5.16). Ideally, for a sensor binding to a single molecule (such as fura-2), the ligand concentration [L] is directly related to the ratio as[74]

$$[L] \propto \frac{R - R_{min}}{R_{max} - R} \qquad [5.29]$$

where $R_{min}$ and $R_{max}$ are, respectively, the minimum and maximum ratios. However, regarding single-chain biosensors, it is usually not possible to have access to these ratios. In this case, the normalized ratio $\Delta R/R_{t_0}$ can be used to facilitate the comparison between all cell responses (cf. Fig. 5.16). This normalized ratio is defined by

$$\frac{\Delta R}{R_{t_0}} = \frac{|R - R_{t_0}|}{R_{t_0}} \qquad [5.30]$$

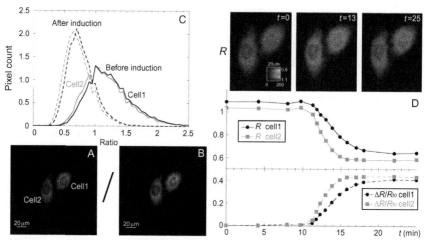

**Figure 5.16** Data analysis for FRET imaging with a standard wide-field epi-fluorescence microscope. The fluorescence intensity images of living cells transfected with $^T$Epac$^{VV}$ acquired with two distinct filters sets: donor emission and acceptor emission (after donor excitation) are shown, respectively, in (A) and (B). The resulting ratio $R$ can be calculated pixel by pixel (after instrumental correction) and the ratio distributions of both cells are indicated in (C) before ($t = 0$) and after induction ($t = 25$ min). The mean ratio is extracted from these distributions for each cell and plotted as a function of time in (D). The corresponding ratio images are also represented in the upper part of the panel. Finally, also indicated in (D) is the normalized ratio $\Delta R/R_{t_0}$ which simplifies the comparison between cell responses. (See Color Insert.)

where $R_{t_0}$ is the initial ratio (measured at time $t = 0$).

The fact that this normalized ratio (and the ratio $R$) gives just an indication of the relative changes between the biosensor "ON" and "OFF" states must be emphasized. In other words, determination of individual concentrations of both biosensor states with ratiometric imaging is not feasible in the cellular environment,[75] since all biosensors expressed are not optically active. In fact, the maturation and the photodegradation (photobleaching) of the two fluorescent proteins (donor and acceptor) are processed at different rates.[70,76] This leads to a population of biosensors whose fluorescence emission is no more related to the physiological state of the cell (unresponsive donor-only and acceptor-only single-chain biosensors).

## 5.2. Fluorescence lifetime imaging

Unlike ratiometric imaging, which is based on fluorescence intensity being directly related to the fluorophore concentration (of both the donor and the acceptor), lifetime measurements are generally independent of these concentrations. It allows determining quantitatively the proportion of interacting donor and the FRET efficiency in living cells.

### 5.2.1 Frequency-domain lifetime imaging

5.2.1.1 Recovering phase shift and modulation depth in single frequency experiments

The first step of FD FLIM data analysis consists in retrieving the phase and modulation values ($\varphi$ and $m$) from fluorescent images acquired with either the heterodyne or the homodyne method (see Section 4).

With the heterodyne method, the detector is modulated at a frequency $\omega + \Delta\omega$ (where $\Delta\omega$ is a low frequency in the kilohertz range), which is slightly different from the modulation frequency $\omega$ of the excitation source. The resulting signal collected by the FLIM system ($S_{\text{heterodyne}}$) is thus modulated in time at the low frequency $\Delta\omega$:

$$S_{\text{heterodyne}} \propto 1 + \frac{m_{\text{ex}} m_{\text{de}} m}{2} \cos\left(\Delta\omega t + \Delta\phi - \varphi\right) \qquad [5.31]$$

where $\Delta\phi$ is the phase shift between excitation and detection, $m_{\text{ex}}$ and $m_{\text{de}}$ are, respectively, the modulation amplitude of excitation and detection. By recording this signal $S_{\text{heterodyne}}$ at various time delays $t$ and fitting it as a function of time with a cosine function of the low cross-correlation frequency $\Delta\omega$, we can extract the phase shift $\varphi$ and modulation depth $m$ for each pixel of the FLIM image.

In the homodyne implementation (when the frequencies of excitation and detection are identical), the collected signal, which is no longer modulated, is a direct current (DC) component defined by

$$S_{\text{homodyne}} \propto 1 + \frac{m_{\text{ex}} m_{\text{de}} m}{2} \cos\left(\Delta\phi - \varphi\right) \qquad [5.32]$$

with $m_{\text{ex}}$ and $m_{\text{de}}$, respectively, the modulation amplitude of excitation and detection. The phase shift $\Delta\phi$ between excitation and detection varies from 0 to 360° ($2\pi$) with $K$ equally spaced intervals. For each phase shift $\Delta\phi$, the DC collected signal is recorded for each pixel of the FLIM image. By fitting this collected signal $S_{\text{homodyne}}$ with a cosine function of $\Delta\phi$, the resulting phase $\varphi$ and modulation $m$ are calculated pixel by pixel (cf. Fig. 5.17).

5.2.1.2 Calculation of the fluorescence lifetime and data representation

Once the phase $\varphi$ and modulation $m$ have been determined for each pixel of the image, these $m$ and $\varphi$ values are further manipulated for evaluating the fluorescence lifetime of the sample. In order to obtain correct lifetime values

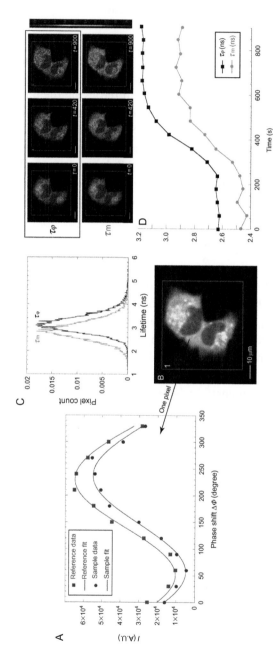

**Figure 5.17** Data analysis of fluorescence lifetime measurements acquired with homodyne method. (A) Experimental reference and sample fluorescence intensities (dots) as a function of the phase shift $\Delta\phi$ between excitation and detection for one pixel; the corresponding fits from which the phase $\varphi$ and modulation $m$ are deduced are indicated with lines. (B) Fluorescence intensity image of living cells transfected with $^{T}Epac^{VV}$. Phase and modulation lifetime ($\tau_m$ and $\tau_\varphi$) distributions for the selected area are represented in (C). Mean values are deduced from these distributions and their evolutions in time are shown in (D). Corresponding phase and modulation lifetime images are also displayed for three distinct times. (See Color Insert.)

with both methods (heterodyne and homodyne method), calibration of the FLIM system with a reference of known fluorescence lifetime $\tau_{ref}$ is indispensable for taking into account the phase shift and modulation introduced by the electronics and optics (cf. Eqs. 5.31 and 5.32). The fluorescence lifetime of an unknown sample is then defined by[77]

$$\tau_m = \frac{1}{\omega}\sqrt{\frac{m_{ref}^2}{m^2}\left(1+\omega^2\tau_{ref}^2\right)-1} \qquad [5.33]$$

$$\tau_\varphi = \frac{1}{\omega}\tan\left(\varphi - \varphi_{ref} + \tan^{-1}\left(\omega\tau_{ref}\right)\right) \qquad [5.34]$$

where $\varphi_{ref}$ and $m_{ref}$ are, respectively, the phase and modulation values estimated from the reference. The fact that these fluorescence lifetimes ($\tau_m$ and $\tau_\varphi$) are calculated for each pixel must be emphasized. The resulting FLIM image, which is displayed using a color scale, is usually superimposed on the intensity image in order to highlight the brightest regions (see Fig. 5.17).

When using the FD FLIM system to record biosensor activity in living cells, evolutions in time of the phase and modulation lifetimes ($\tau_\varphi$ and $\tau_m$) are sometimes represented.[73] However, this method gives just an indication of the relative changes between the two states of the biosensor. In fact, these phase and modulation lifetimes acquired during a single-frequency experiment do not correspond to the true lifetimes of the biosensor. To recover these true lifetime components, it is necessary to acquire multiple frequency FLIM images and to fit pixel by pixel the experimental phase $\varphi$ and modulation $m$ for each frequency $\omega$ with a function of $\omega$.[59] Experimental data are usually adjusted to the theoretical values by minimizing an error function (using a Levenberg–Marquardt algorithm) without any *a priori* information on the lifetime components.[59,78,79] If it can be assumed that the lifetime information is the same for all pixels of the FLIM image, data can be globally analyzed.[80,81] In both cases, obtaining reliable lifetime components is time consuming and it is hardly accessible to the nonexpert.

### 5.2.2 Time-domain lifetime imaging

#### 5.2.2.1 Extracting lifetime components

In time-domain methods, fluorescent samples are repeatedly excited by short pulses of light, and the resulting fluorescence intensity decay histograms can be recorded for each pixel of the FLIM image with different detectors (see Section 4). Regardless of the technique employed, experimental lifetime components are usually deduced by adjusting the experimental decay histograms with the theory. In TD FLIM, the theoretical detected intensity profile $F(t)$, which is dependent on both the fluorescence sample and the instrumentation, is defined by Eq. (5.26).

This fitting procedure is usually performed with the least squares method (LSM), which consists in minimizing an error function $\chi^2$ defined as the total squared differences between experimental data points $x_i$ and theoretical values $s_i$ deduced from Eq. (5.26):

$$\chi^2_{\text{LSM}} = \frac{1}{N_d - p} \sum_{i=1}^{N_d} \frac{(s_i - x_i)^2}{x_i} \qquad [5.35]$$

where $N_d$ is the number of data points and $p$ is the number of fitting parameters. The minimization of this error function is generally performed with the Levenberg–Marquardt algorithm, which has been implemented in most of the commercially available FLIM analysis software. This commonly used FLIM image analysis strategy is robust and gives reliable results when the number of photons is large[8,11,23,82] (cf. Fig. 5.18).

However, if the total number of photons is low, the LS method becomes inaccurate because its error function assumes that the noise is described with a Gaussian instead of a Poisson distribution, which is incorrect. In the case of a small number of photons, a solution named "adaptive Monte Carlo data inflation" (AMDI) algorithm is used, which consists in inflating statistically the number of photons for being compatible with the LS method.[11] Another possibility, which is called the "maximum likelihood estimation" (MLE) method,[82,83] is to modify the error function to take into account the Poisson noise distribution. This error function is now defined as[83]

$$\chi^2_{\text{MLE}} = \frac{2}{N_d - p} \sum_{i=1}^{N_d} (s_i - x_i) - \frac{2}{N_d - p} \sum_{i=1}^{N_d} x_i \ln\left(\frac{s_i}{x_i}\right) \qquad [5.36]$$

It was demonstrated that both solutions (AMDI and MLE methods) give an accurate lifetime component estimation of multiexponential intensity decays with a reduced number of photons, in comparison with the LS method.[11,82] Finally, it has been recently demonstrated that an alternative method based on Bayesian analysis enables the correct estimation of the lifetime of monoexponential decays with a few photons.[84] However, this method was never applied with multiexponential decays, which limits its application in the context of biosensor activity measurements.

### 5.2.2.2 Exploiting data

As already mentioned, the fluorescence intensity decay of a single-chain biosensor is described with multiexponential terms (cf. Eq. 5.10).

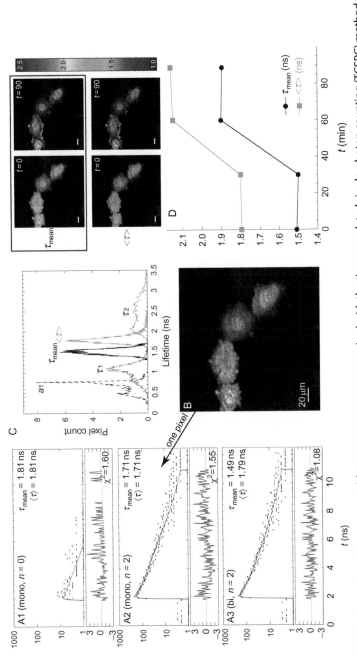

**Figure 5.18** Data analysis of fluorescence lifetime measurements acquired with the time-correlated single-photon counting (TCSPC) method. (A1) Typical experimental intensity decay (dots) acquired in a few milliseconds per pixel (corresponding to an acquisition time of 300 s for the complete image represented in (B); the fit obtained with standard least squares method (LSM) is indicated with a line. Because of the small photon count $N$, the error function ($\chi^2$) is nonflat and elevated, indicating that the adjustment is not perfect. (A2) One possibility to increase $N$ consists in applying a spatial binning of factor $n$ (sum all pixels comprised in a $(2n+1)^2$ squared region). In this case, the fit is slightly improved but $\chi^2$ is still nonflat because the monoexponential model is not adapted. (A3) When the biexponential model is used, the fit is satisfactory, which means that the sample proportion and lifetime can be correctly estimated. (B) Fluorescence intensity image of living cells transfected with $^T$Epac$^{VV}$. The distributions of all fitting parameters ($\alpha_1$, $\tau_1$, and $\tau_2$) for the complete image are shown in (C). Also represented are the mean lifetime $\tau_m$ and the average lifetime $\langle\tau\rangle$ distributions. Their evolutions as a function of time are plotted in (D). The corresponding mean and average lifetime images are also displayed in the upper part of the panel. (See Color Insert.)

Regardless of the fitting method employed (and as previously described), the final purpose of the FLIM data analysis is to correctly extract both the proportions and the lifetime components for each pixel of the image. This can be achieved preferentially without constraining any fitting parameters (cf. Fig. 5.18). However, for sample-emitted fluorescence whose intensity decay is multiexponential, the correct estimation of all parameters with a standard fitting method requires a large number of photons ($N > 100,000$[85]). For decreasing this number and consequently the acquisition time and phototoxicity, we have already detailed in previous paragraphs two solutions based on algorithm implementation (MLE and AMDI). An alternative solution consists in reducing the number of fitting parameters by constraining one of them, for example, the donor lifetime, but it makes it necessary that the donor lifetime is monoexponential and that it has been precisely determined in a previous experiment (which requires a modified biosensor without acceptor). Another possibility is to analyze the data globally, but it is valid only if one can assume the lifetime information to be identical for all pixels of the FLIM image (or a selected area).[86,87] In all cases, it requires expertise and computation time for obtaining reliable lifetime values with fitting methods.

Once the lifetime components have been correctly estimated, the resulting lifetime images are displayed using a color scale and are usually overlaid on the intensity image for weighting the lifetime value with the fluorescence intensity (cf. Fig. 5.18). However, due to the large number of parameters (proportion and lifetime components), it is not possible to represent the evolution of all of them. In order to simplify data representation, the mean lifetime $\tau_{mean}$ defined by Eq. (5.13) has been largely used in FRET experiments.[8–11] However, the mean lifetime does not correspond to the correct average lifetime, which is defined by Eq. (5.12). This average lifetime has been recently used for biosensor activity recording.[88] However, the reader should be informed that this quantity is not monotonous as a function of the donor lifetime in the presence of the acceptor. In other words, one average lifetime value can be found with two distinct donor lifetimes in the presence of the acceptor for a fixed proportion of the interacting donor. Consequently, the knowledge of both the average lifetime and the fraction of the interacting donor is not enough to entirely characterize the fluorescent sample. Due to the nonuniqueness of the average lifetime, it is also necessary to combine it with another parameter (e.g., FRET efficiency).

### 5.2.3 Nonfitting-based approaches

In the previous section, it was mentioned that the correct analysis of the acquired fluorescent signal with fitting methods is time consuming and necessitates a high level of expertise. In order to simplify the analysis of FLIM images and to make it accessible to the nonexpert user, novel methods based on nonfitting approaches have been developed recently.[89–92] This section limits the review to two approaches that are applicable to a large range of lifetime acquisition techniques: the minimal fraction of interacting donor $(mf_D)$[90] and the polar approach or phasor.[89,91] The rapid lifetime determination, which is restricted to time-gated FLIM images with two temporal channels, is voluntarily omitted.[93,94]

The minimal fraction of interacting donor $mf_D$, which was introduced by Padilla-Parra et al.,[90] can be applied with all TD FLIM techniques (TCSPC, time-gated, etc.). It is defined by

$$mf_D = \frac{1 - \langle \tau \rangle / \tau_2}{\left( \langle \tau \rangle / (2\tau_2) - 1 \right)^2} \qquad [5.37]$$

where $\tau_2$ is the lifetime of the donor alone and $\langle \tau \rangle$ is the mean lifetime defined by Eq. (5.12) (see previous section). The computation of $\langle \tau \rangle$ is straightforward and can be performed pixel by pixel and recovered online during FLIM acquisition. When the lifetime of the donor alone $\tau_2$ is monoexponential and known (implying that it has been measured in a previous experiment which is typically not performed in a biosensor experiment), it can be deduced from Eq. (5.37) that the computation of the $mf_D$ is a simple calculation that can be performed online on a standard computer. This is a major advantage when following the biosensor activity over time in live cells, since the user has immediate access to quantitative information, which is the minimal fraction of the interacting donor. This indicator varies between 0 when there is no interaction and 1 when all the donors interact. It is a robust indicator, but the reader is reminded that it gives information on the minimum of the interacting proportion alone and not on the exact quantity $\alpha_1$.

The polar approach or phasor is another nonfitting method that has been successfully applied in FLIM experiments. This method was initially described by Jameson et al.[95] and then successively improved by different groups.[96–98] It consists in converting the lifetime image into a new two-dimensional histogram called "polar" or "phasor." In this polar representation, each point that is defined with $[u; v]$ coordinates corresponds to one pixel of the FLIM image and vice versa. Consequently, one FLIM image is transformed into a scatter diagram whose position gives an indication on the number of exponentials present in the intensity decay. For example, when the fluorescence emitted by the sample decays monoexponentially, the scatter diagram is localized on the semicircle centered at [0.5, 0] with a radius of 0.5 (see Fig. 5.19). Short fluorescence lifetimes are close to the coordinates [1, 0], whereas long lifetimes approach the origin ([0, 0]). If multiple lifetime components are present in the sample, the scatter histogram is located inside the semicircle, which can be helpful for identifying a mixture of several molecular species or a FRET phenomenon.[89,96] This approach can be applied with both FD[96]

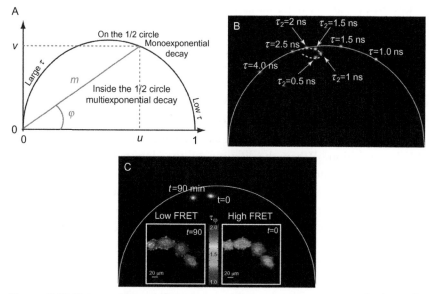

**Figure 5.19** Data analysis of fluorescence lifetime measurements with the polar approach or phasor. (A) Illustration of the polar representation. Lifetime measurements of fluorescent sample with single-exponential decay are localized on the semicircle. If multiple lifetime components are present in the sample, the FLIM acquisitions are located inside the semicircle. (B) Example of the polar representation with simulated FLIM images. As expected, spots corresponding to FLIM images simulated with mono-exponential decays are positioned on the semicircle whereas spots issues from simulated data with biexponential decays are inside. Also represented with dashed line is the FRET trajectory for a proportion of interacting donor $\alpha_1 = 0.5$. (C) Application of the polar representation to biosensor experiment realized in living cells transfected with $^T$Epac$^{VV}$. We can clearly distinguish two spots corresponding to both conformations of the biosensor (high FRET and low FRET). We have also displayed resulting phase lifetime images. (See Color Insert.)

and TD[89,92] FLIM acquisition systems and it has been recently used with biosensors.[75,99]

In TD FLIM experiments, the $u$ and $v$ coordinates are, respectively, the cosine and sine transforms of the fluorescence intensity decay $I(t)$ and they are defined by

$$u = \frac{\int I(t)\cos(\omega t)\mathrm{d}t}{\int I(t)\mathrm{d}t} \qquad [5.38]$$

$$v = \frac{\int I(t)\sin(\omega t)\mathrm{d}t}{\int I(t)\mathrm{d}t} \qquad [5.39]$$

where $\omega$ is the laser repetition angular frequency.

For FD FLIM experiments, $\omega$ is the angular frequency of light modulation, and the $u$ and $v$ coordinates, which are directly related to the phase $\varphi$ and modulation $m$ measured with the FD FLIM system, are given by

$$u = m\cos(\varphi) \qquad [5.40]$$
$$v = m\sin(\varphi) \qquad [5.41]$$

In both cases (TD and FD FLIM experiments), determination of the $[u, v]$ coordinates can be performed instantaneously. However, the polar representation gives just visual information on the biosensor state and does not allow for obtaining really quantitative information on FRET parameters.

Recently, Leray and coworkers have demonstrated that it is possible to retrieve quantitatively the FRET parameters from the polar coordinates.[91] In fact, the fraction of interacting donors $\alpha_1$ and the fluorescence lifetime of the donor in the presence of the acceptor $\tau_1$ can be analytically expressed as[91]

$$\tau_1 = \frac{1 - u - v\tau_2\omega}{\omega(v - u\tau_2\omega)} \qquad [5.42]$$

$$\alpha_1 = \frac{\tau_2\left(1 + \tau_1^2\omega^2\right)\left(1 - u - u\tau_2^2\omega^2\right)}{(\tau_1 - \tau_2)\left(-1 + u + u\tau_1^2\omega^2 + \tau_1\tau_2\omega^2 + u\tau_2^2\omega^2 + u\tau_1^2\tau_2^2\omega^4\right)} \qquad [5.43]$$

If the lifetime of the donor alone, $\tau_2$, is monoexponential and has been measured in a previous experiment (requiring a modified biosensor without acceptor), from Eqs. (5.42) and (5.43) both the lifetime of the donor in the presence of the acceptor and the fraction of the interacting donor can be easily deduced. We emphasize the fact that these calculations can be performed during the time-lapse acquisition of the biosensor activity, which makes it possible to follow over time the evolution of both the FRET efficiency and the proportion of the interacting donor $\alpha_1$.

## 6. DESIGN AND OPTIMIZATION OF GENETICALLY ENCODED KARs

The design of a FRET biosensor for imaging biochemistry in living cells is based on the development of a single polypeptide capable of generating a conformational change that modulates FRET efficiency in response to a biochemical event, such as phosphorylation in this context.

A genetically encoded KAR is composed of the following key elements: an MRE consisting of a substrate peptide for the kinase of interest, a phosphoamino acid-binding domain to detect the target activity, and a reporting element consisting of a fluorescent protein-based FRET couple flanking the sensing element. These functional parts are joined together with linkers whose optimization is needed as they readily affect the overall dynamic range of the biosensor (Fig. 5.20).

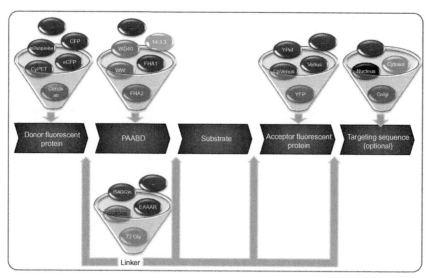

**Figure 5.20** Necessary ingredients for making KARs. (For color version of this figure, the reader is referred to the online version of this chapter.)

The identification and/or optimization of the different components of a KAR are reviewed in this section. A range of KARs have been designed on this model, which allow activity monitoring of many tyrosine and serine/threonine kinases.

## 6.1. Substrate peptide identification

Many methodological approaches have proven to be useful to identify and characterize that a peptide sequence acts as a specific substrate for a kinase of interest.

The first approach identifies a specific substrate *in silico* by using databases of known substrate sequences of kinases such as UniProtKB/Swiss-Prot, which provide reliable protein sequences associated with a very high level of annotation and a high level of integration with additional databases. However, peptides selected by some knowledge-based libraries (such as UniProt, for instance) can have multiple phospho-acceptor sites. Other databases such as KinasePhos can predict phosphorylation sites within given proteins and provide information on the exact positioning of phosphorylation sites with a link to the corresponding catalytic protein kinases involved.[4]

Another approach relies on the use of the kinase target sequence from a protein known to be phosphorylated by the kinase of interest. Since many kinases have multiple isoforms, substrate sequence could be modified to be more specific toward a protein kinase isoform. The protein kinase C family, with its 10 members, is a good example.[100,101] The first genetically encoded PKC (protein kinase C)-FRET-based reporter (CKAR, C-kinase activity reporter) is an effective reporter for all PKC isoforms,[102,103] but each isoform has its own activity signature.[104] Kajimoto *et al.* have designed a new genetically encoded reporter based on the first CKAR but with an ultraspecific substrate to measure only PKCδ activity in different cell compartments. In order to make CKAR more selective for PKCδ, they selected 11 known substrates of PKCδ threonine as phospho-acceptor residues, and kept isoleucine at the position +3 to facilitate binding of the PAABD (see Section 6.2). Sensors with candidate substrate sequences were characterized *in vitro* and *in cellulo* (see Section 6.6) for specificity and selectivity. In spite of being time consuming, this kind of approach can provide highly specific and selective KARs.

The last approach enables the identification of protein kinase substrates through large-scale analysis using high-density peptide microarrays from various biological samples. The use of high-throughput technologies has become an essential step to create and/or optimize a KAR especially when specific substrates of the kinase of interest are not yet known (for review see Ref. 105).

To this extent, two libraries have been established for substrate identification with peptide microarrays. First, knowledge-based libraries contain many small peptide sequences isolated from known proteins. The second library contains *de novo* synthesized peptides using either randomly generated peptides or a combinatorial approach. The first type of library allows the determination of phosphorylation sites using overlapping peptide scans from a known protein sequence. Furthermore, combinatorial and randomly generated peptide libraries are useful in detecting specific kinase substrates. Combinatorial libraries define one or more amino acids at fixed positions (e.g., isoleucine at $+3$ position after the unique phospho-acceptor residue), while other amino acids (except serine, threonine, tyrosine) are placed at randomized positions. The number of generated peptides is very high compared to that in knowledge-based libraries. Combinatorial libraries have successfully identified PKA (protein kinase A)[106,107] and PKG (protein kinase G) substrates.[107,108]

In some cases, a phosphorylation site cannot solely determine substrate specificity of protein kinases. Substrate specificity is also determined by short sequences named "docking sites," which are specifically recognized by a kinase. Indeed, many studies report that docking sites dramatically increase the efficiency with which a substrate is phosphorylated by a kinase *in vitro* and *in cellulo*.[109–111] The case of the kinase MAPK (mitogen-activated protein kinase)/ERK is a good example because MAPK family members such as ERK, p38, and JNK have similar phosphorylation sites. Distinct docking sites for ERK, p38, and JNK have been identified. It ensures substrate targeting and can be used to design a KAR with high sensitivity and selectivity.

Erkus was the first genetically encoded FRET biosensor of ERK activity in different compartments of single living cells.[112] Erkus has a docking site (D domain) connected to the C-terminal of the sensor. This D domain is a common docking site contained in most known endogenous substrates of ERK and increases the probability of substrate phosphorylation by ERK.[113,114] However, Harvey *et al.*[42] have developed a new genetically encoded FRET-based biosensor extracellular regulated kinase activity reporter (EKAR) (EKARc and EKARn target cytoplasm and nucleus, respectively). As an improvement, EKAR exhibits a new ERK-specific docking site (FQFP) adjacent to the substrate, resulting in a FRET signal three times larger than that of Erkus, while other potential docking sites tested in this context greatly reduced the FRET signal of the sensor. This shows that docking sites are very important in the design of new specific and selective KARs (see Ref. 115 for review), which was recently highlighted in a sensor for the M-phase promoting factor.[116]

## 6.2. Phosphoamino acid-binding domain

Upon specific phosphorylation, the PAABD recognizes and binds to the phosphorylated substrate, resulting in a conformational change of the polypeptide that somehow alters in the opposite way the fluorescence emission of the two fluorophores. The choice of a good PAABD is a key element of KAR design.[117] Protein phosphorylation may lead to the formation of molecular signaling complexes through interactions between specific phosphorylated residues and binding domains. Several binding domains named "modular domains" and "adaptors molecules" have been well characterized and are able to activate a specific pathway. Activation of transmembrane receptors triggers phosphorylation on tyrosine residues, which results in the recruitment and binding of adapter molecules such as Src-homology 2/3 (SH2 SH3) and phosphotyrosine binding (PTB) domains. Similarly, signal transduction involves phosphorylation on serine/threonine residues and thus constitutes consensus sequences recognized by other adapter molecules including 14.3.3 proteins, forkhead-associated (FHA), domains WW domains, and WD40 domains. The phospho-acceptor residue within the substrate sequence then first guides the choice of the PAABD.

Several factors should be considered as they can affect the efficiency and the reversibility of the PAABD binding to the phosphorylated substrate. For example, in order to measure compartmentalized PKA activity in single living cells, Zhang et al.[118] developed a genetically encoded A-kinase activity reporter (AKAR). This first-generation PKA sensor (AKAR1) used 14-3-3 as a PAABD. Because of the high binding affinity of this PAABD toward the phosphorylated substrate, once phosphorylated, the sensor was blocked in closed conformation, making it weakly sensitive to cellular phosphatase activity and thus irreversible and incompatible with dynamic measurements of PKA activity. To circumvent this problem, 14-3-3 was replaced by an FHA domain in subsequent versions of AKAR,[119] yielding AKAR2. When tested in cells, it showed a better dynamic response and a reversible behavior. Similarly, an FHA domain was also used in the first Erk biosensor Erkus.[112] Shortly afterward, the development of EKAR[42] saw the FHA domain replaced by a WW domain. Although both sensors followed the same design principle, comparative analysis of EKAR and Erkus revealed that the dynamic response of EKAR was greatly improved; the dynamic range of EKAR was higher than that of Erkus.[42]

Details regarding FHA and WW domains can be found in Refs. 120,121, respectively.

## 6.3. Optimal linker combinations in FRET-based biosensors

Manipulating the linkers (i.e., swapping the linker from one sensor to another) might prove to be an easier way and should be considered in the first instance. Making highly sensitive FRET biosensors remains difficult and requires fine-tuning. In spite of being time consuming, linker optimization is a crucial step in biosensor design. In fact, FRET efficiency depends essentially on the distance and the orientation of the two fluorophores (see Section 2),[122] which places linkers at the heart of the dynamic range of biosensors. Linkers are mainly composed of amino acids such as glycine, proline, and alanine, which gives them full flexibility. The classical flexible linker consists of (GGSGGS)$n$ which keeps fluorophores at a "safe" distance from one another,[123] and the rigid linker is composed of (EAAAR)$n$ where FPs are held in "fixed" distance and orientation.

In order to optimize and accelerate the development of FRET-based biosensors, Ibraheem et al.[124] used a reliable high-throughput method by undertaking the optimization of a methylation-sensitive H3K27 sensor—H3K27-MetBio (trimethylation of lysine 27 of histone 3)—mainly based on changes in the length of the linkers. They focused on the optimization of the linker between the PAABD and the substrate. Screening of biosensor variants was performed in colonies of Escherichia coli through the generation of many hundreds of different linker combinations using several screening libraries. The efficiency of H3K27-MetBio was improved, with a FRET signal efficiency 2.3 times larger than the original sensor[124].

In a recent study, Piljic et al. focused on the optimization of linkers flanking the FPs.[125] They developed a reliable and rapid method to generate multiple genetically encoded FRET sensor variants and tested them in reversely transfected mammalian cells. Long linkers improve the flexibility of the sensor and favor orientation of the donor toward the acceptor fluorophore, even though, sometimes, certain short and rigid linkers produce a greater FRET signal.[125] Improvements of biosensors could also be achieved by varying the linker composition in amino acids using combinatorial and randomly generated linker libraries for detecting the most effective linker composition.

Finally, Komatsu *et al.* established highly sensitive FRET biosensor backbones. Their strategy completely abolished the dependence of the sensors on the orientation of the two fluorophores. Indeed, prediction of the exact orientation of the donor and the acceptor in an optimized sensor is rather difficult.[103,126] Instead, they optimized biosensor backbones relying entirely on the distance between the fluorophores by modifying the linker between the PAABD and the substrate.[127] Regular repetitions of the motif (SAGG)*n* (where *n* is the number of repetitions (13-61)) and the 72 polyglycine linker[128] were utilized and compared.[127] All generated backbones highlighted that long and flexible linkers reduce the proportion of biosensors folding in the basal state, thereby improving the dynamic range of biosensors.

## 6.4. Choosing a FRET pair

Historically, the first FRET measurements were performed with the BFP and the enhanced green fluorescent protein (eGFP) as the FRET pair combination.[1,129] Although these FPs meet the requirements for FRET measurements, BFP has an unfortunate tendency to bleach much faster than eGFP. The cyan fluorescent protein (CFP) and the yellow fluorescent protein (YFP) FRET pair quickly replaced the BFP–eGFP FRET pair to monitor $Ca^{2+}$ variations in individual live cells.[130]

The continuous improvement of fluorescent proteins, while beneficial for biosensor optimization, could become overwhelming, as the possible combinations seem endless. Nowadays, cloning procedures are much easier, thus easing up the process. The choice of a FRET pair should be guided by (1) up-to-date/optimized fluorophores, (2) which are red-shifted to minimize phototoxicity, and (3) have monoexponential lifetimes when considering FRET-FLIM measurement methods. The idea here is to tap into the distance ($R_0$) and eventually the dipole–dipole orientation ($\kappa^2$) in order to maximize FRET efficiency (see Section 2)

### 6.4.1 Blue/yellow FRET pairs

This "original FRET pair" is still used in most biosensors today because of its good spectral properties. Many variants derived from these FPs have rapidly emerged and significantly increased FRET efficiency. The monomeric (m) Cerulean[34] and the mTurquoise2[131] are the preferred variants of the CFP, while monomeric Citrine and Venus[37] have proved to be the most popular variants of the YFP. Cerulean has a better quantum yield, a higher extinction coefficient, a fluorescence lifetime with a single-exponential decay, and an increased brightness than the conventional CFP (see Table 5.1). mCitrine shows better tolerance to pH variations and quenching by chloride ions, while Venus is characterized by a quicker maturation within cells.

Along these lines, a "sticky" FRET pair, known as CyPET (optimized eCFP for FRET) and YPet (optimized YFP for FRET), has been optimized from a cyan–yellow FRET pair by Daugherty's lab in 2005, using a quantitative evolutionary strategy.[132] The resulting FRET pairs showed a 20-fold ratiometric FRET signal change instead of the 3-fold for the parental pair, in a context where fluorophores were separated by a caspase 3 cleavable substrate. While this FRET pair provides substantial improvement in sensitivity and dynamic range for some biosensors,[132] the poor folding and expression of CyPET at 37 °C limits its usefulness in living cell studies.[31] Nevertheless, YPet remains one of the brightest YFPs and displays superior FRET efficiency when paired with a cyan donor fluorophore such as eCFP, alleviating any ill effects stemming from CyPET.[127,133]

Lately, circularly permuted (cp) fluorophores have appeared as a good means to enhance the dynamic range of biosensors. These cp fluorophores have flourished in scientific reports as a solid solution for biosensor optimization.[41,71] Technically, they consist of fluorescent proteins in which the N- and C-termini are fused together through a flexible linker. This protein engineering has allowed the emergence of other types of biosensors resting solely on the fluctuation of fluorescence properties upon binding of the analyte to a molecular switch positioned in the linker region.[134] Note that such fluorophores are more sensitive to their photochemical environment such as pH and temperature in this configuration/protein structure. The first biosensor successfully improved using this strategy is the $Ca^{2+}$ indicator (yellow cameleon), exhibiting nearly sixfold increase in dynamic range of FRET efficiency upon $Ca^{2+}$ binding to the biosensor.[135] Since then, cpFPs, in particular cpVenus, have been implemented in other biosensors, including AKAR3,[41] AKAR4,[71] and recently $^{T}Epac^{VV}$.[73] The simple swapping of the CFP for the Cerulean as donor between AKAR3 and AKAR4 in combination with the cpVenus improves the dynamic range of the AKAR series. Indeed, AKAR4 shows a slight average difference in fluorescence lifetime in control experiments when compared to AKAR3 (Fig. 5.21).

### 6.4.2 Green/red FRET pairs

Since 2004, many other variants of fluorescent proteins have emerged with the discovery and characterization of a new red fluorescent protein (RFP) named *Discosoma* sp. red fluorescent protein (DsRed) isolated from the coral

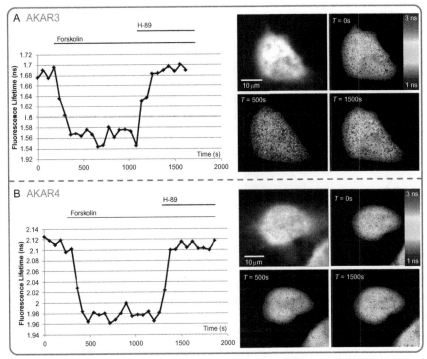

**Figure 5.21** U2OS cells transfected with AKAR3 (A) or AKAR4 (B). PKA was activated using an adenylate cyclase activator (forskolin) and then inhibited with H-89. Lifetime measurements were performed in the frequency domain, as described in Section 4. Graphs represent the average phase fluorescence lifetime measured for the entire cell as a function of time. Images represent intensity (top left) and phase lifetime (others) at specific times. (See Color Insert.)

of *Dicosoma* genus in 1999.[38] Although red-shifted FP variants would undoubtedly result in lower phototoxicity upon biological sample illumination (less energy), they are not commonly employed for sensing application. The only example is the ERK biosensor EKAR[42] in its green–red version which does not show a large dynamic range (unpublished results). However, a green–red FRET pair has proven its usefulness in the detection of molecular interaction by FRET–FLIM. Besides, even in their optimized versions, the intrinsic fluorescence properties of RFPs (oligomerization, quantum yield, brightness, etc.) do not stand a chance when compared with their yellow counterparts. Indeed, the historical FRET pair is cyan–yellow based and has received much attention directed toward its optimization. Ongoing efforts in developing red-shifted variants will most likely yield adequate acceptance and will prove themselves useful in multisensing approaches.

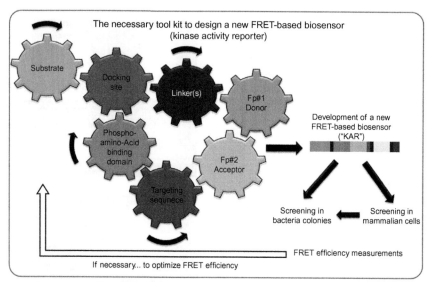

**Figure 5.22** Overall methodology for creation and/or optimization of KAR biosensors. (For color version of this figure, the reader is referred to the online version of this chapter.)

In the light of opportunities offered by newly engineered FPs, many studies suggest alternatives to conventional FRET pairs. Combining the YFP or GFP as a donor with orange or red fluorescent proteins as acceptor is now possible. Although these FRET pairs have mostly been used in protein interactions studies, they have recently emerged as an alternative to the CFP/YFP FRET pair in biosensor design.[42]

Figure 5.22 illustrates the process necessary for the creation and/or the optimization of KAR biosensors.

## 6.5. Targeting FRET biosensors to subcellular compartments

Biosensors can be directed at distinct subcellular compartments in order to gain information on the compartmentalized activity of a kinase of interest. This is achieved by adding an additional amino acid sequence at the N-terminal or C-terminal extremity of the sensor. Several subcellular targeting sequences have been isolated and are now well characterized (Table 5.3).

## 6.6. *In vitro* and *in cellulo* characterization of new FRET biosensors

Two paths can be exploited to test newly built and optimized biosensors: *in vitro* and *in cellulo*. Note that results from one approach may not necessarily be similar to those generated with the other one. However, characterization

**Table 5.3** Listing of validated targeting sequences for biosenseur subcellular compartimentation

| Subcellular targeted location | Targeting sequences | References |
|---|---|---|
| Cytosol | LALKLAGLDI (at C-terminal) | Gallegos et al.[102] |
| Nucleus | PKKRKVEDA (at C-terminal) | Gallegos et al.[102] Ananthanarayanan et al.[136] |
| Golgi | 33 amino residues of eNOS (at N-terminal) | Sasaki et al.[137] Gallegos et al.[102] |
| Endoplasmic reticulum | MLLPVLLLGLLGAAAD (at N-terminal) + KDEL (at C-terminal) | Palmer et al.[138] |
| Plasma membrane | MGCIKSK (at N-terminal) | Violin et al.[103] Kunkel et al.[126] Gallegos et al.[102] |

References 102,103,126,136–138.
*Adapted from Refs. 4.*

remains the bottleneck through which every biosensor has to pass. Characterization is a crucial step, as it determines the efficiency of new FRET-based biosensors and thus requires several considerations. The results should provide evidence that the sensor generated is specific to the kinase of interest and should assess the dynamic range of the biosensor in the presence of the activator and/or inhibitor. This step usually relies on pharmacological agents, the availability and specificity of which can be limited.

### 6.6.1 In vitro *characterization*

Upon expression of biosensors, in a bacterial system, for instance, lysates are subjected to stringent purification procedures to retrieve only full-length protein from the truncated protein produced as a result of proteolytic degradation. Now, in solution (see Chapter 4) with the purified protein kinase of interest and ATP, in a 96-well plate screening format or in a spectrophotometer cuvette, the properties of the biosensor are evaluated. Ratiometric measurement methods (see Section 4.1.1 for details) are then employed under different experimental conditions: before and after addition of ATP and purified kinase, and control wells for baseline ratio determination.

In addition, phosphorylation levels of the biosensor expressed in mammalian cells can be estimated by immunoprecipitation and detected by

PhosphoTag western blotting. The overall amounts of expressed biosensors are then revealed by an anti–GFP antibody.[127]

### 6.6.2 In cellulo characterization

*In cellulo* characterization consists of expressing the biosensor in a mammalian cell model and assessing the ratiometric signal in the presence of specific activators and inhibitors of the kinase of interest. Biosensor control constructs (earlier version or/and non-phosphorylatable biosensor) should undergo similar experimental conditions. Transient transfection of plasmid DNA encoding biosensors is an accessible and inexpensive approach, which is sufficient to produce biosensors using cellular transcriptional machinery. Kinase activity measurement in a cellular context can also evaluate the specificity of a biosensor candidate: kinase blockers should prevent the ratiometric response to activators, demonstrating the involvement of one specific kinase family in the process being studied. This strategy is more often seen as a complementary rather than an alternative method to evaluate FRET efficiency of candidate sensors.

In *vitro* and in *cellulo* characterizations are often used to evaluate quantitatively the effect of several kinase inhibitors to determine the most effective inhibitor of the kinase of interest. Indeed, they can serve as high-throughput screening to evaluate rapidly the effects of several pharmacological treatments on the activation or extinction of a particular pathway. For such assays, it is essential that the biosensor exhibits a robust and reproducible signal. This is generally assessed by the $Z$-factor, which is a statistical parameter that compares the dynamic range to data variation.[123] A schematic representation of such an experimental design is shown in Fig. 5.23.

## 7. CONSIDERATIONS FOR KAR MEASUREMENTS

### 7.1. Controlling acquisition systems

As previously described, several techniques allow quantifying FRET for KAR imaging. Each of them requires a dedicated setup and thus the corresponding characterization procedures. The most advanced acquisition techniques (spectral acquisition, TD and FD FLIM, etc.) mainly require rigorous characterization of the systems and optimized acquisition conditions as described in the literature.[23,26,59,139] They can then be directly used for KAR imaging and yield unambiguous quantification of FRET efficiencies. Ratiometric imaging, despite being an easier acquisition procedure, can results in erroneous interpretation of FRET efficiency due to the classical drawbacks of quantitative fluorescence imaging. The

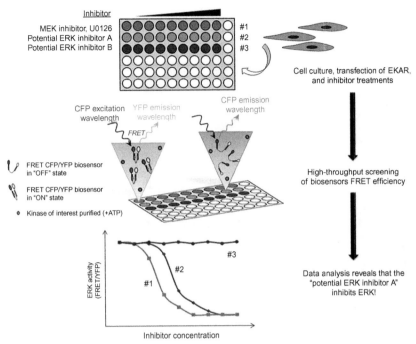

**Figure 5.23** Schematic representation of the method for the evaluation of kinase inhibitor with the FRET-based biosensor EKAR by high-throughput screening. (See Color Insert.)

apparent easier acquisition procedure makes it available to researchers who are unaware of the factors that might complicate fluorescence quantification. We will thus focus here on some key elements responsible for these misinterpretations.

Fluorescence microscopy is mainly used for qualitative imaging of protein, lipid, or nuclear acid distribution in cells. Indeed, quantifying the distribution of these molecules is highly complicated by various optical, physical, and biological parameters.[140] Many of them can be circumvented by ratiometric imaging and are not an issue in the case of KAR imaging. However, several parameters remain critical. For example, the properties of the illumination sources, the optics of the microscope, or the sensitivity and signal-to-noise ratio of the detectors can affect both fluorescence and ratio measurements. Some of these parameters can be easily optimized by a careful choice of the elements available on the acquisition systems. For instance, using apochromatic objectives with corrected chromatic aberration to avoid focusing the excitation wavelength at two different positions will prevent strong border effects. Others are more difficult to avoid (Fig. 5.24 gives several examples)

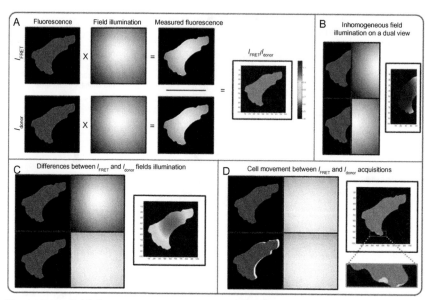

**Figure 5.24** Field illumination and cell movement between $I_{FRET}$ and $I_{donor}$ acquisition channels effects on ratio imaging. For each panel, the cell exhibiting an equal $I_{FRET}$ and $I_{donor}$ is imaged through a nonhomogeneous field illumination and ratio image is calculated. Procedure is detailed in (A). The expected ratio value is thus equal to one for the whole image. In (B)–(D), we can see erroneous ratio images due to different sources of artifacts. See text for more details. (See Color Insert.)

and will strongly depend upon the acquisition technique used for measuring the ratios (sequential ratio, SR; dual view ratio, DVR; dual channel ratio, DCR).

### 7.1.1 Field illumination

Nonhomogeneous field illumination and light collection is a classical problem encountered both in wide-field and in confocal microscopy and can yield to significant differences in intensity measurements from the center to the border of an image. We encourage users to verify the field homogeneity of the employed system using a reference sample. In ratio imaging, this may not be a major issue if the field nonhomogeneity is the same for the sequential ratio (this is most of the time the case for SR, Fig. 5.24A). Indeed, if each pixel of the emitted $I_{FRET}$ and $I_{donor}$ images is multiplied by a same factor, the ratio measurements will stay accurate despite nonhomogeneities in the field. On the contrary, for DVR or DCR, this may be a major issue because both images are measured on either two distinct detector parts or two different cameras. Detector variability is thus spatially uncorrelated and may strongly impact the ratio values (Fig. 5.24B and C). It is thus absolutely necessary to suppress this field illumination inhomogeneity by acquiring reference images using fluorescent solutions. When differences in field illumination persist, corrections can be carried out by normalizing pixel intensities using reference images. This procedure is called "shading correction".[141]

Besides correcting for spatial field inhomogeneities, shading correction also intrinsically calibrates the absolute ratio value. Shading images are acquired with a uniform sample of a fluorescent chromophore chosen to emit fluorescence at both wavelengths of interest. In the particular case of CFP-like and YFP-like chromophores, a suitable fluorescent dye is Coumarin 343, which emits at both wavelengths. When each acquired image is divided by its corresponding shading image, by definition the ratio of the Coumarin 343 dye is equal to 1.[74,142] This method thus defines ratio values with respect to a reference chromophore and therefore defines the "balance" between the two channels. This procedure allows easier comparison between setups that may have different spectral sensitivities.

### 7.1.2 Image misregistration between acquisition channels

In the sequential ratio setup, the images acquired at two wavelengths may not be perfectly registered. This may be due to wedge effects caused by the nonparallel faces of the emission filter. This is easily corrected by the translation of one of the images by a few pixels, and algorithms exist that determine the optimal subpixel registration correction.[143] In addition, if the objective suffers from chromatic aberration, sequential ratio allows a focus correction by a fixed value which can be accurately determined using fluorescent beads. When cellular movement occurs at the timescale of the sequential acquisition, no correction can be applied and the system is just not fast enough for measuring the biological event of interest, and simultaneous detection with dual view or dual channels may be needed. In these configurations, in addition to the linear translation of one image with respect to the other, nonlinear deformation may occur as a result of different optical paths followed by both wavelengths, and it will be necessary to apply more elaborate algorithms that correct for image skewing. Misregistration effects are nonetheless a common cause of data misinterpretation and the experimenter should look carefully at the images for border effects which usually show a typical rainbow appearance (Fig. 5.24D)

## 7.2. Controlling biological samples

Parameters such as temperature or pH can directly affect the chromophore properties. In addition, these physical parameters plus other biologically relevant parameters (osmolarity, ion and metabolite concentrations, etc.) may alter the activity of signaling cascades (including kinases) and hence the measurements obtained with KAR biosensors. Experimental conditions must therefore be controlled in order to minimize the impact of such parameters.

### 7.2.1 Temperature

Concerning the temperature, experiments must be carried out at a stable temperature (usually 37 °C) inside thermostated space surrounding the whole microscope. These incubators require hours to warm up and reach a stable temperature and should always be turned on to ensure optimal stability. The temperature probe should be positioned as close as possible to the sample. The thermal inertia due to contact with both the objective and the microscope stage might need direct control with thermostatic controllers. While temperature probes are most of the time positioned in air, there can be a large difference between the average enclosure temperature and the sample temperature. Calibration of this difference must thus be performed by positioning the temperature probe in the medium.[144]

### 7.2.2 Imaging media and perfusion

In long-term experiments, drying of the imaging medium modifies ionic concentrations and pH. A buffered medium can be used for short-term experiments (from 2–4 h depending on cell type). For longer measurements, the use of a $CO_2$ incubator and a perfusion system to renew the culture medium is recommended. In all cases, media containing phenol red should be avoided, as this molecule is slightly fluorescent and leads to an unwanted fluorescent signal. A phenol red-free medium must therefore be preferred. Details on the preparation of dedicated imaging media can be found in Refs. 42,51.

Perfusion systems are very useful to add drugs to the sample. Alternatively, drugs in a preheated medium can be applied manually in the dish with a Pasteur pipette. This can, however, lead to motion artifacts because of shear stress, and some practice is necessary.

### 7.2.3 Biosensor expression level

Transfection agents or electroporation is often chosen to transfect cells with biosensor plasmids. Viruses (e.g., adenoviruses or Sinbis) can be considered for kinase activity measurement in cells resistant to the classic transfection method[145] or in tissue.[146] However, the biosensor expression level can induce differences in the amplitude response. Indeed, the FRET probes destabilize the reaction equilibrium between a kinase and its substrates. We therefore recommend performing measurements on cells with equivalent KAR expression levels. To avoid the variations caused by the sensor expression levels, the creation of a stable cell line might be an alternative.

### 7.2.4 Photoinduced effects

Photoinduced effects are another source of artifacts. Among these effects, the most obvious is cell death, which is easily noticeable by changes in cell morphology. Photostress is more tricky to observe and can result in cell autofluorescence. While autofluorescence spectra are broad and cannot be corrected by filters, and autofluorescence lifetime is very short and can be confused with a FRET event, both spectral and lifetime properties of autofluorescence can create artifacts in measurements with both intensity ratio- and lifetime-based techniques. The last photoinduced effect is photobleaching, which can also falsify measurements with all techniques, thereby leading to unacceptable errors in data interpretation. Extensive controls of photoinduced effect on cells expressing a KAR negative reference are thus indispensable (see Section 7.2.5).

These photoinduced effects depend on the laser power and the exposure time. Thus, it will be less critical for techniques that require a lower illumination power ($P_{average}$ measured using a power meter). For example, ratiometric measurements will be less affected than FD FLIM experiments, and TSCPC is the most stressful method because of the time necessary to scan an entire cell. Since most biosensors are based on the CFP/YFP FRET pair (or their variants such as Cerulean, Turquoise, CyPET, Venus, circularly permutated Venus, YPet, etc.), we will focus on these fluorophores to exemplify the acquisition time and power required for each technique (performed on our systems).

In ratiometric measurements, after exciting the donor, the exposure time is set to $\sim$500 and $\sim$200 ms ($P_{average} < 1$ mW) for CFP and YFP, respectively, with our microscopy system.[48] As described in Section 5, acquiring a channel with excitation and observation of YFP is a prerequisite in intermolecular FRET. It is, however, not recommended in the case of sensor imaging, because this optional correction step will increase both photoinduced effects and noise, without improving results. Acceptors excitation can, however, be useful in another way. Indeed, acceptor photobleaching can be a good negative control to recover a basal level after activation of the kinase of interest. The photoinduced effect arising from this high-energy laser excitation must, however, be controlled.[71]

For FLIM, the acquisition time strongly depends on the technique used. For FD FLIM measurements using our setup, the exposure time of the donor is between 100 and 300 ms with a laser power $P_{average} = 4$ mW sent to a spinning disk system equipped with a $63 \times$ oil objective (NA = 1.4). To achieve optimal compromise between the number of phase images and the accuracy of lifetime determination, 12 phase-shifted images are chosen. Moreover, achieving a good signal-to-noise ratio often requires averaging each images three times. From these constraints, the overall exposure time for one lifetime image can be between 3 s to almost 10 s. For time-resolved lifetime measurements with our system, a field scan takes about 5 min (still with an average laser power of a few milliwatts) and will require data binning to achieve a precise lifetime map whereas the measurements per pixel only takes few tens of milliseconds.

The choice of imaging technique thus depends on (i) the time resolution, (ii) the number of images needed, and (iii) the spatial resolution. For example, TCSPC can be used to obtain a precise image of the subcellular localization of kinase activity before and after the activation of the biological signal, while ratiometric measurement or phase and modulation technique will be preferred to achieve a precise estimate of kinase activity kinetics over time. The use of a biosensor targeted to subcellular compartment measured by ratio imaging can also be a solution to achieve precise activity localization without compromising acquisition speed.

In conclusion, we suggest avoiding any unnecessary exposure of fluorophores, especially the donor, in FLIM experiments; moreover, photobleaching of the CFP will occur faster than for YFP variants. For example (and if it is possible), find the focus using transmitted light rather than a mercury lamp. Because of the numerous possible photoinduced artifacts, negative controls are indispensable.

### 7.2.5 Negative control

Sensing applications require the creation of a control biosensor. Concerning KAR, the phosphorylable residue within the substrate (serine, threonine, or tyrosine) may be mutated to a nonphosphorylable amino acid such as glycine or alanine. Expressed in cells, it should not produce a FRET signal in response to the stimulation of the kinase of interest. It will therefore serve as negative control and provide the basal level of FRET signal in living cells. An example of such an experiment is presented in Fig. 5.25. There is,

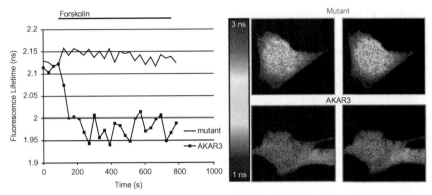

**Figure 5.25** U2OS cells transfected with AKAR3 or AKAR3 inactive mutant. PKA was activated using an adenylate cyclase activator (forskolin, 12.5 μM). Lifetime measurements were performed in the frequency domain as described in Section 4. Graphs represent the average phase fluorescence lifetime measured for the entire cell as a function of time. Experiments with either biosensor were performed separately, but are represented on the same graph for clarity. Images represent phase lifetime before (left panel) and after (right panel) induction. (See Color Insert.)

unfortunately, no positive control for FRET measurements in living cells (all sensors are in folded position), but a strong induction using activating agents can be envisaged as an alternative. Molecular strategies for such a positive control should be envisioned in the near future, as both control biosensors could provide a complete sensing range.

## 7.3. Biosensor dynamic range and reproducibility

The dynamic range of a biosensor is a critical parameter to consider, and increasing this range is often the purpose of biosensor optimization. The dynamic range depends directly on the FRET efficiency and can be increased by switching fluorophores, improving linkers, and enhancing kinase binding to the biosensor (with the insertion of a docking domain, for example), as described earlier (see Section 6). Possessing the appropriate pharmacology allowing maximum stimulation and inhibition is essential for estimating this dynamic range. More than just control, the revealed amplitude (as well as other criteria such as sensitivity) directs the choice of the biosensor to use.

As another example, the EKAR[42] shows a relatively low dynamic range, both for ratiometric or lifetime-based measures, compared to a cAMP biosensor ($^{T}Epac^{VV}$) recently published.[73] Indeed, cells expressing EKAR exhibit a 20% average $\Delta R/R$ and $120 \pm 20$ ps average $\Delta \tau$, while cells expressing $^{T}Epac^{VV}$ give a mean $\Delta R/R$ of more than 80% and a mean $\Delta \tau$

of $700 \pm 50$ ps, both after maximum stimulation and inhibition (10 cells per conditions, unpublished results). While these biosensors differ in their composition, the fluorophores of $^{T}Epac^{VV}$ were optimized to improve FRET efficiency whatever the technique employed. Indeed, the donor fluorophore of $^{T}Epac^{VV}$ (mTurquoise) is characterized by a phase lifetime of 3.7 ns[147] instead of 2.3 ns for Cerulean[148] in EKAR, which improved the dynamic range in FLIM. As an acceptor, $^{T}Epac^{VV}$ uses a dimer of CpVenus-Venus instead of a single Venus in EKAR, in order to improve detection of the sensitized emission (see Section 6).

After all these considerations related to fluorescence, FRET measurements, data analysis, creation/optimization of biosensor, and methodologies for kinase activity measurements, we would like to enlighten the reader about the way data are being employed today in systems biology approaches.

## 8. TOWARD QUANTITATIVE APPROACHES IN BIOLOGICAL PROCESSES

The molecular toolkit for kinases activity reporters is called to address the role of kinases in the orchestration of signal transmission within the cells. It remains challenging to understand how a signal pattern is organized and provides accurate integration and appropriate cellular decision while mobilizing common effectors in response to different inputs. Indeed, kinases such as Erk, PKA, and Akt are considered as common effectors and their spatiotemporal dynamics drives cellular functions. Receptors often act as organizers of signal integration, giving shape to the signal: transitory, sustained, all-or-none.

Regarding Erk, MAPK/Erk network exhibits plasticity, and different dynamical properties may arise from connections within this network. For instance, rewiring of the Erk pathway architecture in the context of NGF (nerve growth factor) or EGF (epidermal growth factor) stimulation in PC12 cells (model for neuron differentiation) is believed to build the dynamical response observed for these growth factors: transient or sustained MAPK/Erk cascade activation following EGF or NGF stimulation, respectively.[149,150]

Similar dynamical properties of the MAPK/Erk pathway have been suspected to affect the signaling cascade involved in cell death.[151] Actually, subcellular localization of the MAPK/Erk pathway components and dynamical spatiotemporal changes are believed to determine cell fate between proliferation and death.[152] Sustained Erk activation, which is sufficient to induce cell death, is often associated with cell death mediated by ROS (reactive oxygen species)[153] because these radicals can suppress protein-phosphatase activities of MKP (MAPK-specific phosphatases), exerting key roles in the regulation of MAPK/Erk dynamics.[154] Inhibition of MAPK/Erk activation using inhibitors against components of the MAPK cascade has a survival effect on cells induced to die following chemical or physical challenges. Forced MAPK/Erk cytosolic localization prevents the survival or a mitogenic response but potentiates activities of specific pro-apoptotic proteins such as DAPK (death-associated protein kinase),[155] Bik (bcl-2-interacting killer),[156] and PEA-15 (astrocytic phosphoprotein-15)[157] via the activation of the PI3K/Akt pathway. Depending on cell type and stimulus, compartmentalized MAPK activity will mediate either cell survival or cell death (see Ref. 3 for a review). Thus, spatial signatures for the signal propagation of the MAPK/Erk signaling pathway have begun to be explored both at the experimental and theoretical levels in many different cellular models, from gametes to tissues.[158–160]

Scaffolding proteins and interactions of kinases with specific components of the cell architecture also determine the spatial and temporal integration of the cAMP/PKA signal.[161–163] In addition, the simple spatial organization of the cell also determines the spatiotemporal organization of the signal, leading to differential activation within the cell compartment. Indeed, in the case of the cAMP/PKA signaling cascade, modeling studies[164] and imaging experiments[165,146] have revealed the importance of cell morphology in the particular situation of neurons, where submembrane domains display faster and stronger responses than the bulk somatic cytosol.[166] It is therefore of great importance to address the question of intracellular signal integration, bearing in mind the multidimensional—space and time—parameters that determine the physiological outcome of an external signal.

Kinase activity reporter methodologies will definitely be able to generate enough data to be considered for modeling approaches. Still, while recording the dynamical pathway, which may exhibit variation from one cell to another, one needs to identify within these apparently variable responses reliable marks or features that are common between experiments and invariant to changes in cell morphologies or behaviors. These marks may be obtained through the acute stimulation/inhibition of a pathway,[167] but a response hierarchy may also be obtained from constitutive fluctuations in the activity of the biosensors.[168] The latter approach is performed in a nonperturbed pathway, under conditions closer to a physiological range than those in acute inhibition or activation. In concept, fluctuation analysis is quite scalable to a large number of components and requires further developed mathematical tools to ingrate this large set of component activities.

## 9. OUTLOOK AND PERSPECTIVE

### 9.1. What about multisensing?

In the prospect of detecting several parameters simultaneously in the same cell, the idea of using multiple probes is becoming more and more attractive. As the number of available tools is increasing, monitoring several modulators of the same pathway provides opportunities for the understanding of cell reaction dynamics. The principal challenge of these experiments is combining the well-chosen fluorophore pairs with the appropriate technique. Several parameters have to be considered and are well described and illustrated in the excellent review by Carlson and Campbell.[169]

Piljic and Schultz[170] have presented such an example of multisensing applied to the monitoring of several effectors involved in the regulation of calcium chloride conductance. It was achieved in an epithelial cell line using simultaneously four biosensors, including three FRET-based ones. These refined experiments were performed by ratiometric imaging, taking advantage of (i) the spectral shift between the two FRET pairs (namely CFP/YFP and mOrange/mCherry) and (ii) the subcellular targeting of some of these sensors to spatially discriminate signals from each reporter.

Taking into account the complexity, dynamics, and stress reactivity of the cellular environment, measurements of molecular dynamics could require the combination of fluorescence analysis techniques (such as fluorescence lifetime, spectrum or intensity fluctuations analysis) and a multimodal microscopy method suitable for the different complexity degrees of the living matter: cells, tissues, and whole organism. So far, biosensor recording has been achieved using mainly ratiometric techniques; however, FRET-FLIM could emerge as an interesting alternative.

### 9.2. Kinase activity measurements in living animals

Studying kinase activity profiles in living tissue or animals provides new insights on cell behavior in the intact biological context. A few publications describe the use of FRET-based biosensors in living animals. For example, ATP levels in a transgenic worm (*Caenorhabditis elegans* expressing ATeam[171]), cAMP levels in the fruit flies (transgenic *Drosophila melanogaster* expressing GFP-PKA[172,173]), as well as Rac GTPase activity in zebrafish (pRaichu RacFRET mRNA microinjected in *Danio rerio* embryos[174]) have already been performed.

However, these first experiments have highlighted some difficulties. Indeed, biosensor overexpression can lead to abnormal embryonic development or lethality, depending on promoter strength, simply because the cAMP biosensor is built on an active PKA enzyme bearing its intact catalytic activity.[172] It is therefore strongly suggested to use biosensors that have no biological effect, such as AKAR, and, indeed, the expression of these sensors have proved to be harmless and did not affect the memory processes in the living drosophila.[175] The depth at which measurements must be done, combined with the required low KAR expression levels, render measuring relevant variations more difficult.[174] The use of a two-photon excitation microscope, which increases the imaging depth and reduces the phototoxicity, must be preferred for opaque tissues or deep imaging for both ratio imaging and TD FLIM experiments, and indeed has provided new insights into cAMP/PKA signaling processes in living drosophila.[175]

Generation of transgenic mice expressing FRET-based biosensor remains a challenge,[176] for example, because of the two fluorophore's cDNAs recombination.[177] Although this has already been done, the low expression levels (probably due to gene silencing) complicate the imaging procedure.[178] Nevertheless, transgenic mice were generated for some KARs including EKAR and AKAR (with a high expression level), and successful kinase activity measurements were realized notably in the auricular skin and small intestine.[177] Another line has been developed with a ubiquitous cAMP sensor expression which allowed direct cAMP measurement on a variety of primary cell preparations.[179] As a perspective, transgenic animals can be designed for ubiquitous or tissue-specific KAR expression. Furthermore, the combination of animals expressing a KAR and mutated/KO/overexpressing gene involved in the regulation of the kinase of interest can increase our understanding of regulatory networks.

While somewhat challenging in a practical way, kinase activity measurements, and biosensors in general, have helped scientists in better their understanding of biological processes. These tools allow the generation of data that can be exploited by the mathematician for modeling approaches of regulatory networks. Finally, since biosensor experiments are performed within an intact physiological context, new light is being shed onto the spatiotemporal dynamic of molecular effectors involved in the regulation cellular functions.

## ACKNOWLEDGMENTS

This work was encouraged and supported by two CNRS national networks: the GDR2588 and the RTmFm, as well as the University Lille1. The opportunity for such a chapter arose thanks to the French biosensor workgroup. Financial contributions came from various sources for equipment, bench fees, and PhD student grants. This work was funded by the CNRS, the University Lille1, the GEFLUC, the Nord-Pas de Calais Région Council, the European Regional Developmental Funds, a Leica Microsystems partnership, and the French Research Agency ANR 07-PFTV-01101. We are grateful to the imaging facility BICeL (Bio Imaging Center of Lille) and the Interdisciplinary Research Institute (IRI CNRS USR3078).

## REFERENCES

1. Heim R, Tsien RY. Engineering green fluorescent protein for improved brightness, longer wavelengths and fluorescence resonance energy transfer. *Curr Biol* 1996;**6**(2):178–82.
2. Frommer WB, Davidson MW, Campbell RE. Genetically encoded biosensors based on engineered fluorescent proteins. *Chem Soc Rev* 2009;**38**(10):2833–41.
3. Riquet F, Vandame P, Sipieter F, Caillau-Maggio K, Spriet C, Héliot L, et al. Reporting kinase activities: paradigms, tools and perspectives. *BMB JBM* 2011;**1**(2):10–8.
4. Kunkel MT, Newton AC. Spatiotemporal dynamics of kinase signaling visualized by targeted reporters. *Curr Protoc Chem Biol* 2009;**1**(1):17–8.
5. Valeur B., *Molecular Fluorescence: Principles and Applications*. 2001: Molecular Fluorescence: Principles and Applications.
6. Lakowicz JR. *Principles of fluorescence spectroscopy*. Kluwer Academic/Plenium Publishers; 1999.
7. Periasamy A, Day RN. *Molecular imaging FRET microscopy and spectroscopy*. Oxford University Press, NY, pages 312; 2005.
8. Sun Y, Day RN, Periasamy A. Investigating protein-protein interactions in living cells using fluorescence lifetime imaging microscopy. *Nat Protoc* 2011;**6**(9):1324–40.
9. Camuzeaux B, Spriet C, Héliot L, Coll J, Duterque-Coquillaud M. Imaging Erg and Jun transcription factor interaction in living cells using fluorescence resonance energy transfer analyses. *Biochem Biophys Res Commun* 2005;**332**(4):1107–14.
10. Yockell-Lelièvre J, Spriet C, Cantin P, Malenfant P, Heliot L, de Launoit Y, et al. Functional cooperation between Stat-1 and ets-1 to optimize icam-1 gene transcription. *Biochem Cell Biol* 2009;**87**(6):905–18.
11. Trinel D, Leray A, Spriet C, Usson Y, Heliot L. Upgrading time domain FLIM using an adaptive Monte Carlo data inflation algorithm. *Cytometry A* 2011;**79**:528–37.
12. Pietraszewska-Bogiel A, Gadella TW. FRET microscopy: from principle to routine technology in cell biology. *J Microsc* 2011;**241**(2):111–8.
13. Forster T. Intermolecular energy migration and fluorescence. *Ann Phys* 1948;**2**:55–7.
14. Clegg RM, Murchie AI, Lilley DM. The 4-way DNA junction—a fluorescence resonance energy-transfer study. *Braz J Med Biol Res* 1993;**26**(4):405–16.
15. Clegg RM, Sener M, Govindjee M. From Foerster resonance energy transfer to coherent resonance energy transfer and back (Invited Paper). In: Alfano Robert R, editor. 2010. *Proceedings of SPIE* Vol. 7561. p. 7561–2 article CID Number 75610C, 21 pages.
16. Steinberg IZ. Long-range nonradiative transfer of electronic excitation energy in proteins and polypeptides. *Annu Rev Biochem* 1971;**40**:83–114.
17. Dale RE, Eisinger J, Blumberg WE. The orientation freedom of molecular probes. The orientation factor in intramolecular energy transfer. *Biophys J* 1979;**26**:161–94.
18. Dale RE, Eisinger J. Intramolecular energy transfer and molecular conformation. *Proc Natl Acad Sci USA* 1976;**73**(2):161–3.
19. Haas E, Wilchek M, Katchalski-Katzir E, Steinberg IZ. Distribution of end-to-end distances of oligopeptides in solution as estimated by energy transfer. *Proc Natl Acad Sci USA* 1975;**72**(5):1807–11.

20. Haas E, Katchalski-Katzir E, Steinberg IZ. Effect of the orientation of donor and acceptor on the probability of energy transfer involving electronic transitions of mixed polarizations. *Biochemistry* 1978;**17**:5064–70.

21. Pantano S. In silico description of fluorescent probes in vivo. *J Mol Graph Model* 2008;**27**(4):563–7.

22. Campbell RE. Fluorescent-protein-based biosensors: modulation of energy transfer as a design principle. *Anal Chem* 2009;**81**(15):5972–9.

23. Waharte F, Spriet C, Heliot L. Setup and characterization of a multiphoton FLIM instrument for protein-protein interaction measurements in living cells. *Cytometry A* 2006;**69**(4):299–306.

24. Peter M, Ameer-Beg SM, Hughes MK, Keppler MD, Prag S, Marsh M, et al. Multiphoton-FLIM quantification of the EGFP-mRFP1 FRET pair for localization of membrane receptor-kinase interactions. *Biophys J* 2005;**88**(2):1224–37.

25. Nagy P, Bene L, Balázs M, Hyun WC, Lockett SJ, Chiang NY, et al. EGF-induced redistribution of erbB2 on breast tumor cells: flow and image cytometric energy transfer measurements. *Cytometry* 1998;**32**(2):120–31.

26. Spriet C, Trinel D, Waharte F, Deslee D, Vandenbunder B, Barbillat J, et al. Correlated fluorescence lifetime and spectral measurements in living cells. *Microsc Res Tech* 2007;**70** (2):85–94.

27. Prasher DC, Eckenrode VK, Ward WW, Prendergast FG, Cormier MJ. Primary structure of the Aequorea victoria green-fluorescent protein. *Gene* 1992;**111**(2):229–33.

28. Shimomura O, Johnson FH, Saiga Y. Extraction, purification and properties of aequorin, a bioluminescent protein from the luminous hydromedusan, Aequorea. *J Cell Comp Physiol* 1962;**59**:223–39.

29. Tsien RY. The green fluorescent protein. *Annu Rev Biochem* 1998;**67**:509–44.

30. Shaner NC, Patterson GH, Davidson MW. Advances in fluorescent protein technology. *J Cell Sci* 2007;**120**(Pt 24):4247–60.

31. Wang Y, Shyy JY, Chien S. Fluorescence proteins, live-cell imaging, and mechanobiology: seeing is believing. *Annu Rev Biomed Eng* 2008;**10**:1–38.

32. Cubitt AB, Heim R, Adams SR, Boyd AE, Gross LA, Tsien RY. Understanding, improving and using green fluorescent proteins. *Trends Biochem Sci* 1995;**20**(11):448–55.

33. Zacharias DA, Violin JD, Newton AC, Tsien RY. Partitioning of lipid-modified monomeric GFPs into membrane microdomains of live cells. *Science* 2002;**296**(5569):913–6.

34. Rizzo MA, Springer GH, Granada B, Piston DW. An improved cyan fluorescent protein variant useful for FRET. *Nat Biotechnol* 2004;**22**(4):445–9.

35. Heim R, Cubitt AB, Tsien RY. Improved green fluorescence. *Nature* 1995;**373** (6516):663–4.

36. Miyawaki A, Griesbeck O, Heim R, Tsien RY. Dynamic and quantitative Ca2+ measurements using improved cameleons. *Proc Natl Acad Sci USA* 1999;**96**(5):2135–40.

37. Nagai T, Ibata K, Park ES, Kubota M, Mikoshiba K, Miyawaki A. A variant of yellow fluorescent protein with fast and efficient maturation for cell-biological applications. *Nat Biotechnol* 2002;**20**(1):87–90.

38. Matz MV, Fradkov AF, Labas YA, Savitsky AP, Zaraisky AG, Markelov ML, et al. Fluorescent proteins from nonbioluminescent Anthozoa species. *Nat Biotechnol* 1999;**17**(10):969–73.

39. Campbell RE, Tour O, Palmer AE, Steinbach PA, Baird GS, Zacharias DA, et al. A monomeric red fluorescent protein. *Proc Natl Acad Sci USA* 2002;**99**(12):7877–82.

40. Shaner NC, Campbell RE, Steinbach PA, Giepmans BN, Palmer AE, Tsien RY. Improved monomeric red, orange and yellow fluorescent proteins derived from Discosoma sp. red fluorescent protein. *Nat Biotechnol* 2004;**22**(12):1567–72.

41. Allen MD, Zhang J. Subcellular dynamics of protein kinase A activity visualized by FRET-based reporters. *Biochem Biophys Res Commun* 2006;**348**(2):716–21.

42. Harvey CD, Ehrhardt AG, Cellurale C, Zhong H, Yasuda R, Davis RJ, et al. A genetically encoded fluorescent sensor of ERK activity. *Proc Natl Acad Sci USA* 2008;**105**(49):19264–9.

43. Dunn TA, Wang CT, Colicos MA, Zaccolo M, DiPilato LM, Zhang J, et al. Imaging of cAMP levels and protein kinase A activity reveals that retinal waves drive oscillations in second-messenger cascades. *J Neurosci* 2006;**26**(49):12807–15.

44. Davidson MW, Campbell RE. Engineered fluorescent proteins: innovations and applications. *Nat Methods* 2009;**6**(10):713–7.

45. Shaner NC, Steinbach PA, Tsien RY. A guide to choosing fluorescent proteins. *Nat Methods* 2005;**2**(12):905–9.

46. Jares-Erijman EA, Jovin TM. FRET imaging. *Nat Biotechnol* 2003;**21**(11):1387–95.

47. Jares-Erijman EA, Jovin TM. Imaging molecular interactions in living cells by FRET microscopy. *Curr Opin Chem Biol* 2006;**10**(5):409–16.

48. Depry C, Zhang J. Visualization of kinase activity with FRET-based activity biosensors. *Curr Protoc Mol Biol* 2010;**Chapter 18**: Unit 18.15.

49. Jalink K, van Rheenen J. (2009) FilterFRET: quantitative imaging of sensitized emission. In: *Laboratory techniques in biochemistry and molecular biology* vol. 33: FRET and FLIM techniques. Ed. Gadella, T.W.J. Academic Press, Burlington, pp. 289–349.

50. Schechter MB, Burger G, Widefield application letter: FRET sensitized emission wizard widefield. reSOLUTION, vol. 4; 2009.

51. Borner S, Schwede F, Schlipp A, Berisha F, Calebiro D, Lohse MJ, et al. FRET measurements of intracellular cAMP concentrations and cAMP analog permeability in intact cells. *Nat Protoc* 2011;**6**(4):427–38.

52. Hempel CM, Vincent P, Adams SR, Tsien RY, Selverston AI. Spatio-temporal dynamics of cyclic AMP signals in an intact neural circuitm. *Nature* 1996;**384**(6605):166–9.

53. Garini Y, Young IT, McNamara G. Spectral imaging: principles and applications. *Cytometry A* 2006;**69**(8):735–47.

54. Booth MJ, Wilson T. Low-cost, frequency-domain, fluorescence lifetime confocal microscopy. *J Microsc (Oxford)* 2004;**214**:36–42.

55. Gadella TWJ, Jovin TM, Clegg RM. Fluorescence lifetime imaging microscopy (Flim)—spatial-resolution of microstructures on the nanosecond time-scale. *Biophys Chem* 1993;**48**(2):221–39.

56. Gratton E, Breusegem S, Sutin J, Ruan QQ. Fluorescence lifetime imaging for the two-photon microscope: time-domain and frequency-domain methods. *J Biomed Opt* 2003;**8**(3):381–90.

57. Herman P, Maliwal BP, Lin HJ, Lakowicz JR. Frequency-domain fluorescence microscopy with the LED as a light source. *J Microsc (Oxford)* 2001;**203**:176–81.

58. van Geest LK, Stoop KWJ. FLIM on a wide field fluorescence microscope. *Lett Pept Sci* 2003;**10**(5–6):501–10.

59. Leray A, Riquet FB, Richard E, Spriet C, Trinel D, Heliot L. Optimized protocol of a frequency domain fluorescence lifetime imaging microscope for FRET measurements. *Microsc Res Tech* 2009;**72**(5):371–9.

60. Schneider PC, Clegg RM. Rapid acquisition, analysis, and display of fluorescence lifetime-resolved images for real-time applications. *Rev Sci Instrum* 1997;**68**(11):4107–19.

61. Greger K, Neetz MJ, Reynaud EG, Stelzer EH. Three-dimensional fluorescence lifetime imaging with a single plane illumination microscope provides an improved signal to noise ratio. *Opt Express* 2011;**19**(21):20743–50.

62. Van Munster EB, Goedhart J, Kremers GJ, Manders EMM, Gadella TWJ. Combination of a spinning disc confocal unit with frequency-domain fluorescence lifetime imaging microscopy. *Cytometry A* 2007;**71A**(4):207–14.

63. Buranachai C, Kamiyama D, Chiba A, Williams BD, Clegg RM. Rapid frequency-domain FLIM spinning disk confocal microscope: lifetime resolution, image improvement and wavelet analysis. *J Fluoresc* 2008;**18**:929–42.

64. Gordon GW, Berry G, Liang XH, Levine B, Herman B. Quantitative fluorescence resonance energy transfer measurements using fluorescence microscopy. *Biophys J* 1998;**74**:2702–13.
65. Xia Z, Liu Y. Reliable and global measurement of fluorescence resonance energy transfer using fluorescence microscopes. *Biophys J* 2001;**81**:2395–402.
66. Hoppe A, Christensen K, Swanson JA. Fluorescence resonance energy transfer-based stoichiometry in living cells. *Biophys J* 2002;**83**:3652–64.
67. Jalink K, van Rheenen J. FilterFRET: quantitative imaging of sensitized emission. In: Gadella TWJ, editor. *FRET and FLIM techniques*. Durlington. Academic Press, 2009.
68. Hodgson L, Shen F, Hahn K. Biosensors for characterizing the dynamics of Rho family GTPases in living cells. *Curr Protoc Cell Biol* 2010;**46**:14.11.1–14.11.26.
69. Brito M, Guiot E, Vincent P. Imaging PKA activation inside neurons in brain slice preparations. *Neuromethods,* 2012;**68**:237–50.
70. Hodgson L, Nalbant P, Shen F, Hahn K. Imaging and photobleach correction of mero-CBD, sensor of endogenous Cdc42 activation. *Methods Enzymol* 2006;**406**:140–56.
71. Depry C, Zhang J. Visualization of kinase activity with FRET-based activity biosensors. *Curr Protoc Mol Biol* 2010;**18**:18.15.1–18.15.9.
72. Ting AY, Kain KH, Klemke RL, Tsien RY. Genetically encoded fluorescent reporters of protein tyrosine kinase activities in living cells. *PNAS* 2001;**98**(26):15003–8.
73. Klarenbeek JB, Goedhart J, Hink MA, Gadella TWJ, Jalink K. A mTurquoise-based cAMP sensor for both FLIM and ratiometric read-out has improved dynamic range. *PLoS One* 2011;**6**(4):e19170.
74. Grynkiewicz G, Poenie M, Tsien RY. A new generation of Ca2+ indicators with greatly improved fluorescence properties. *J Biol Chem* 1985;**260**(6):3440–50.
75. Hinde E, Digman MA, Welch C, Hahn KM, Gratton E. Biosensor Förster resonance energy transfer detection by the phasor approach to fluorescence lifetime imaging microscopy. *Microsc Res Tech* 2011;**75**(3):271–81.
76. Llères D, Swift S, Lamond AI. Detecting protein-protein interactions in vivo with FRET using multiphoton fluorescence lifetime imaging microscopy (FLIM). *Curr Protoc Cytom* 2007;**12**:12.10.1–12.10.19.
77. Elder AD, Frank JH, Swartling J, Dai X, Kaminski CF. Calibration of a wide-field frequency-domain fluorescence lifetime microscopy system using light emitting diodes as light sources. *J Microsc (Oxford)* 2006;**224**:166–80.
78. Squire A, Verveer PJ, Bastiaens PIH. Multiple frequency fluorescence lifetime imaging microscopy. *J Microsc* 2000;**197**:136–49.
79. Lakowicz JR, Laczko G, Cherek H, Gratton E, Limkeman M. Analysis of fluorescence decay kinetics from variable-frequency phase shift and modulation data. *Biophys J* 1984;**46**:463–77.
80. Verveer PJ, Bastiaens PIH. Evaluation of global analysis algorithms for single frequency fluorescence lifetime imaging microscopy data. *J Microsc* 2003;**209**(1):1–7.
81. Esposito A, Gerritsen HC, Wouters FS. Fluorescence lifetime heterogeneity resolution in the frequency domain by lifetime moments analysis. *Biophys J* 2005;**89**:4286–99.
82. Maus M, et al. An experimental comparison of the maximum likelihood estimation and nonlinear least-squares fluorescence lifetime analysis of single molecules. *Anal Chem* 2001;**73**(9):2078–86.
83. Laurence TA, Chromy BA. Efficient maximum likelihood estimator fitting of histograms. *Nat Methods* 2010;**7**(5):338–9.
84. Barber PR, Ameer-Beg SM, Pathmananthan S, Rowley M, Coolen ACC. A Bayesian method for single molecule, fluorescence burst analysis. *Biomed Opt Express* 2010;**1**(4):1148–58.
85. Köllner M, Wolfrum J. How many photons are necessary for fluorescence-lifetime measurements? *Chem Phys Lett* 1992;**200**:199–204.

86. Barber PR, Ameer-Beg SM, Gilbey J, Carlin LM, Keppler M, Ng TC, et al. Multiphoton time-domain fluorescence lifetime imaging microscopy: practical application to protein–protein interactions using global analysis. *J R Soc Interface* 2009;**6**:S93–S105.

87. Grecco HE, Roda-Navarro P, Verveer PJ. Global analysis of time correlated single photon counting FRET-FLIM data. *Opt Express* 2009;**17**(8):6493–508.

88. Harvey CD, Ehrhardt AG, Cellurale C, Zhong H, Yasuda R, Davis RJ, et al. A genetically encoded fluorescent sensor of ERK activity. *PNAS* 2008;**105**(49):19264–9.

89. Digman MA, Caiolfa VR, Zamai M, Gratton E. The phasor approach to fluorescence lifetime imaging analysis. *Biophys J* 2008;**94**(2):L14–L16.

90. Padilla-Parra S, Auduge N, Coppey-Moisan M, Tramier M. Quantitative FRET analysis by fast acquisition time domain FLIM at high spatial resolution in living cells. *Biophys J* 2008;**95**(6):2976–88.

91. Leray A, Spriet C, Trinel D, Heliot L. Three-dimensional polar representation for multispectral fluorescence lifetime imaging microscopy. *Cytometry A* 2009;**75**(12):1007–14.

92. Leray A, Spriet C, Trinel D, Blossey R, Usson Y, Heliot L. Quantitative comparison of polar approach versus fitting method in time domain FLIM image analysis. *Cytometry A* 2011;**79A**(2):149–58.

93. Elson DS, Munro I, Requejo-Isidro J, McGinty J, Dunsby C, Galletly N, et al. Real-time time-domain fluorescence lifetime imaging including single-shot acquisition with a segmented optical image intensifier. *New J Phys* 2004;**6**:180.

94. Ballew RM, Demas JN. An error analysis of the rapid lifetime determination method for the evaluation of single exponential decays. *Anal Chem* 1989;**61**:30–3.

95. Jameson DM, Gratton E, Hall RD. The measurement and analysis of heterogeneous emissions by multifrequency phase and modulation fluorometry. *Appl Spectrosc Rev* 1984;**20**(1):55–106.

96. Clayton AH, Hanley QS, Verveer PJ. Graphical representation and multicomponent analysis of single-frequency fluorescence lifetime imaging microscopy data. *J Microsc* 2004;**213**(Pt 1):1–5.

97. Redford GI, Clegg RM. Polar plot representation for frequency-domain analysis of fluorescence lifetimes. *J Fluoresc* 2005;**15**(5):805–15.

98. Fereidouni F, Esposito A, Blab GA, Gerritsen HC. A modified phasor approach for analyzing time-gated fluorescence lifetime images. *J Microsc* 2011;**244**(3):248–58.

99. Eichorst JP, Huang H, Clegg RM, Wang Y. Phase differential enhancement of FLIM to distinguish FRET components of a biosensor for monitoring molecular activity of membrane type 1 matrix metalloproteinase in live cells. *J Fluoresc* 2011;**21**:1763–77.

100. Kazi JU, Soh JW. Isoform-specific translocation of PKC isoforms in NIH3T3 cells by TPA. *Biochem Biophys Res Commun* 2007;**364**(2):231–7.

101. Way KJ, Chou E, King GL. Identification of PKC-isoform-specific biological actions using pharmacological approaches. *Trends Pharmacol Sci* 2000;**21**(5):181–7.

102. Gallegos LL, Kunkel MT, Newton AC. Targeting protein kinase C activity reporter to discrete intracellular regions reveals spatiotemporal differences in agonist-dependent signaling. *J Biol Chem* 2006;**281**(41):30947–56.

103. Violin JD, Zhang J, Tsien RY, Newton AC. A genetically encoded fluorescent reporter reveals oscillatory phosphorylation by protein kinase C. *J Cell Biol* 2003;**161**(5):899–909.

104. Kajimoto T, Sawamura S, Tohyama Y, Mori Y, Newton AC. Protein kinase C {delta}-specific activity reporter reveals agonist-evoked nuclear activity controlled by Src family of kinases. *J Biol Chem* 2010;**285**(53):41896–910.

105. Thiele A, Stangl GI, Schutkowski M. Deciphering enzyme function using peptide arrays. *Mol Biotechnol* 2011;**49**(3):283–305.

106. Luo K, Zhou P, Lodish HF. The specificity of the transforming growth factor beta receptor kinases determined by a spatially addressable peptide library. *Proc Natl Acad Sci USA* 1995;**92**(25):11761–5.

107. Tegge WJ, Frank R. Analysis of protein kinase substrate specificity by the use of peptide libraries on cellulose paper (SPOT-method). *Methods Mol Biol* 1998;**87**:99–106.
108. Dostmann WR, Tegge W, Frank R, Nickl CK, Taylor MS, Brayden JE. Exploring the mechanisms of vascular smooth muscle tone with highly specific, membrane-permeable inhibitors of cyclic GMP-dependent protein kinase Ialpha. *Pharmacol Ther* 2002;**93**(2–3):203–15.
109. Bardwell AJ, Abdollahi M, Bardwell L. Docking sites on mitogen-activated protein kinase (MAPK) kinases, MAPK phosphatases and the Elk-1 transcription factor compete for MAPK binding and are crucial for enzymic activity. *Biochem J* 2003;**370** (Pt 3):1077–85.
110. Fernandes N, Bailey DE, Vanvranken DL, Allbritton NL. Use of docking peptides to design modular substrates with high efficiency for mitogen-activated protein kinase extracellular signal-regulated kinase. *ACS Chem Biol* 2007;**2**(10):665–73.
111. Jacobs D, Glossip D, Xing H, Muslin AJ, Kornfeld K. Multiple docking sites on substrate proteins form a modular system that mediates recognition by ERK MAP kinase. *Genes Dev* 1999;**13**(2):163–75.
112. Sato M, Kawai Y, Umezawa Y. Genetically encoded fluorescent indicators to visualize protein phosphorylation by extracellular signal-regulated kinase in single living cells. *Anal Chem* 2007;**79**(6):2570–5.
113. Sharrocks AD, Yang SH, Galanis A. Docking domains and substrate-specificity determination for MAP kinases. *Trends Biochem Sci* 2000;**25**(9):448–53.
114. Smith JA, Poteet-Smith CE, Malarkey K, Sturgill TW. Identification of an extracellular signal-regulated kinase (ERK) docking site in ribosomal S6 kinase, a sequence critical for activation by ERK in vivo. *J Biol Chem* 1999;**274**(5):2893–8.
115. Holland PM, Cooper JA. Protein modification: docking sites for kinases. *Curr Biol* 1999;**9**(9):R329–R331.
116. Gavet O, Pines J. Activation of cyclin B1-Cdk1 synchronizes events in the nucleus and the cytoplasm at mitosis. *J Cell Biol* 2010;**189**(2):247–59.
117. Ibraheem A, Campbell RE. Designs and applications of fluorescent protein-based biosensors. *Curr Opin Chem Biol* 2010;**14**(1):30–6.
118. Zhang J, Ma Y, Taylor SS, Tsien RY. Genetically encoded reporters of protein kinase A activity reveal impact of substrate tethering. *Proc Natl Acad Sci USA* 2001;**98** (26):14997–5002.
119. Zhang J, Hupfeld CJ, Taylor SS, Olefsky JM, Tsien RY. Insulin disrupts beta-adrenergic signalling to protein kinase A in adipocytes. *Nature* 2005;**437** (7058):569–73.
120. Durocher D, Taylor IA, Sarbassova D, Haire LF, Westcott SL, Jackson SP, et al. The molecular basis of FHA domain:phosphopeptide binding specificity and implications for phospho-dependent signaling mechanisms. *Mol Cell* 2000;**6**(5):1169–82.
121. Lu PJ, Zhou XZ, Shen M, Lu KP. Function of WW domains as phosphoserine- or phosphothreonine-binding modules. *Science* 1999;**283**(5406):1325–8.
122. Nagai T, Miyawaki A. A high-throughput method for development of FRET-based indicators for proteolysis. *Biochem Biophys Res Commun* 2004;**319**(1):72–7.
123. VanEngelenburg SB, Palmer AE. Fluorescent biosensors of protein function. *Curr Opin Chem Biol* 2008;**12**(1):60–5.
124. Ibraheem A, Yap H, Ding Y, Campbell RE. A bacteria colony-based screen for optimal linker combinations in genetically encoded biosensors. *BMC Biotechnol* 2011;**11**:105.
125. Piljic A, de Diego I, Wilmanns M, Schultz C. Rapid development of genetically encoded FRET reporters. *ACS Chem Biol* 2011;**6**(7):685–91.
126. Kunkel MT, Ni Q, Tsien RY, Zhang J, Newton AC. Spatio-temporal dynamics of protein kinase B/Akt signaling revealed by a genetically encoded fluorescent reporter. *J Biol Chem* 2005;**280**(7):5581–7.

127. Komatsu N, Aoki K, Yamada M, Yukinaga H, Fujita Y, Kamioka Y, et al. Development of an optimized backbone of FRET biosensors for kinases and GTPases. *Mol Biol Cell* 2011;**22**(23):4647–56.

128. Levskaya A, Weiner OD, Lim WA, Voigt CA. Spatiotemporal control of cell signalling using a light-switchable protein interaction. *Nature* 2009;**461**(7266):997–1001.

129. Mitra RD, Silva CM, Youvan DC. Fluorescence resonance energy transfer between blue-emitting and red-shifted excitation derivatives of the green fluorescent protein. *Gene* 1996;**173**(1 Spec No):13–7.

130. Miyawaki A, Llopis J, Heim R, McCaffery JM, Adams JA, Ikura M, et al. Fluorescent indicators for Ca2+ based on green fluorescent proteins and calmodulin. *Nature* 1997;**388**(6645):882–7.

131. Goedhart J, von Stetten D, Noirclerc-Savoye M, Lelimousin M, Joosen L, Hink MA, et al. Structure-guided evolution of cyan fluorescent proteins towards a quantum yield of 93%. *Nat Commun* 2012;**3**:751.

132. Nguyen AW, Daugherty PS. Evolutionary optimization of fluorescent proteins for intracellular FRET. *Nat Biotechnol* 2005;**23**(3):355–60.

133. Ouyang M, Sun J, Chien S, Wang Y. Determination of hierarchical relationship of Src and Rac at subcellular locations with FRET biosensors. *Proc Natl Acad Sci USA* 2008;**105**(38):14353–8.

134. Kawai Y, Sato M, Umezawa Y. Single color fluorescent indicators of protein phosphorylation for multicolor imaging of intracellular signal flow dynamics. *Anal Chem* 2004;**76**(20):6144–9.

135. Nagai T, Yamada S, Tominaga T, Ichikawa M, Miyawaki A. Expanded dynamic range of fluorescent indicators for Ca(2+) by circularly permuted yellow fluorescent proteins. *Proc Natl Acad Sci USA* 2004;**101**(29):10554–9.

136. Ananthanarayanan B, Ni Q, Zhang J. Signal propagation from membrane messengers to nuclear effectors revealed by reporters of phosphoinositide dynamics and Akt activity. *Proc Natl Acad Sci USA* 2005;**102**(42):15081–6.

137. Sasaki K, Sato M, Umezawa Y. Fluorescent indicators for Akt/protein kinase B and dynamics of Akt activity visualized in living cells. *J Biol Chem* 2003;**278**(33):30945–51.

138. Palmer AE, et al. Bcl-2-mediated alterations in endoplasmic reticulum Ca2+ analyzed with an improved genetically encoded fluorescent sensor. *Proc Natl Acad Sci USA* 2004;**101**(50):17404–9.

139. Spriet C, Trinel D, Riquet F, Vandenbunder B, Usson Y, Heliot L. Enhanced FRET contrast in lifetime imaging. *Cytometry A* 2008;**73**(8):745–53.

140. Dunn KW, Mayor S, Myers JN, Maxfield FR. Applications of ratio fluorescence microscopy in the study of cell physiology. *FASEB J* 1994;**8**:573–83.

141. Tomazevic D, Likar B, Pernus F. Comparative evaluation of retrospective shading correction methods. *J Microsc* 2002;**208**(Pt 3):212–23.

142. Tsien RY, Harootunian AT. Practical design criteria for a dynamic ratio imaging system. *Cell Calcium* 1990;**11**(2–3):93–109.

143. Thevenaz P, Ruttimann UE, Unser M. A pyramid approach to subpixel registration based on intensity. *IEEE Trans Image Process* 1998;**7**(1):27–41.

144. Dross N, Spriet C, Zwerger M, Muller G, Waldeck W, Langowski J. Mapping eGFP oligomer mobility in living cell nuclei. *PLoS One* 2009;**4**(4):e5041.

145. Liu S, Zhang J, Xiang YK. FRET-based direct detection of dynamic protein kinase A activity on the sarcoplasmic reticulum in cardiomyocytes. *Biochem Biophys Res Commun* 2011;**404**(2):581–6.

146. Castro LR, Gervasi N, Guiot E, Cavellini L, Nikolaev VO, Paupardin-Tritsch D, et al. Type 4 phosphodiesterase plays different integrating roles in different cellular domains in pyramidal cortical neurons. *J Neurosci* 2010;**30**(17):6143–51.

147. Goedhart J, van Weeren L, Hink MA, Vischer NO, Jalink K, Gadella Jr TW. Bright cyan fluorescent protein variants identified by fluorescence lifetime screening. *Nat Methods* 2010;**7**(2):137–9.

148. Kremers GJ, Goedhart J, van Munster EB, Gadella Jr TW. Cyan and yellow super fluorescent proteins with improved brightness, protein folding, and FRET Forster radius. *Biochemistry* 2006;**45**(21):6570–80.

149. Kriegsheim A, Preisinger C, Kolch W. Mapping of signaling pathways by functional interaction proteomics. *Methods Mol Biol* 2008;**484**:177–92.

150. Santos SD, Verveer PJ, Bastiaens PI. Growth factor-induced MAPK network topology shapes Erk response determining PC-12 cell fate. *Nat Cell Biol* 2007;**9**(3):324–30.

151. Mebratu Y, Tesfaigzi Y. How ERK1/2 activation controls cell proliferation and cell death: is subcellular localization the answer? *Cell Cycle* 2009;**8**(8):1168–75.

152. Ebisuya M, Kondoh K, Nishida E. The duration, magnitude and compartmentalization of ERK MAP kinase activity: mechanisms for providing signaling specificity. *J Cell Sci* 2005;**118**(Pt 14):2997–3002.

153. Dong J, et al. EGFR-independent activation of p38 MAPK and EGFR-dependent activation of ERK1/2 are required for ROS-induced renal cell death. *Am J Physiol Renal Physiol* 2004;**287**(5):F1049–F1058.

154. Boutros T, Chevet E, Metrakos P. Mitogen-activated protein (MAP) kinase/MAP kinase phosphatase regulation: roles in cell growth, death, and cancer. *Pharmacol Rev* 2008;**60**(3):261–310.

155. Chen CH, Wang WJ, Kuo JC, Tsai HC, Lin JR, Chang ZF, et al. Bidirectional signals transduced by DAPK-ERK interaction promote the apoptotic effect of DAPK. *EMBO J* 2005;**24**(2):294–304.

156. Mebratu YA, Dickey BF, Evans C, Tesfaigzi Y. The BH3-only protein Bik/Blk/Nbk inhibits nuclear translocation of activated ERK1/2 to mediate IFNgamma-induced cell death. *J Cell Biol* 2008;**183**(3):429–39.

157. Trencia A, Perfetti A, Cassese A, Vigliotta G, Miele C, Oriente F, et al. Protein kinase B/Akt binds and phosphorylates PED/PEA-15, stabilizing its antiapoptotic action. *Mol Cell Biol* 2003;**23**(13):4511–21.

158. Maeder CI, Hink MA, Kinkhabwala A, Mayr R, Bastiaens PI, Knop M. Spatial regulation of Fus3 MAP kinase activity through a reaction-diffusion mechanism in yeast pheromone signalling. *Nat Cell Biol* 2007;**9**(11):1319–26.

159. Blossey R, Bodart JF, Devys A, Goudon T, Lafitte P. Signal propagation of the MAPK cascade in Xenopus oocytes: role of bistability and ultrasensitivity for a mixed problem. *J Math Biol* 2012;**64**(1–2):1–39.

160. Kholodenko BN, Birtwistle MR. Four-dimensional dynamics of MAPK information processing systems. *Wiley Interdiscip Rev Syst Biol Med* 2009;**1**(1):28–44.

161. Tasken K, Aandahl EM. Localized effects of cAMP mediated by distinct routes of protein kinase A. *Physiol Rev* 2004;**84**(1):137–67.

162. Willoughby D, Cooper DM. Organization and Ca2 + regulation of adenylyl cyclases in cAMP microdomains. *Physiol Rev* 2007;**87**(3):965–1010.

163. Wong W, Scott JD. AKAP signalling complexes: focal points in space and time. *Nat Rev Mol Cell Biol* 2004;**5**(12):959–70.

164. Neves SR, Tsokas P, Sarkar A, Grace EA, Rangamani P, Taubenfeld SM, et al. Cell shape and negative links in regulatory motifs together control spatial information flow in signaling networks. *Cell* 2008;**133**(4):666–80.

165. Gervasi N, Hepp R, Tricoire L, Zhang J, Lambolez B, Paupardin-Tritsch D, et al. Dynamics of protein kinase A signaling at the membrane, in the cytosol, and in the nucleus of neurons in mouse brain slices. *J Neurosci* 2007;**27**(11):2744–50.

166. Vincent P, Castro L, Gervasi N, Guiot E, Brito M, and Paupardin-Tritsch D. PDE4 control on cAMP/PKA compartmentation revealed by biosensor imaging in neurons. Hormone and Metabolic Research, 2012. in press.

167. Taylor RJ, Falconnet D, Niemisto A, Ramsey SA, Prinz S, Shmulevich I, Galitski T, et al. Dynamic analysis of MAPK signaling using a high-throughput microfluidic single-cell imaging platform. *Proc Natl Acad Sci USA* 2009;**106**(10):3758–63.

168. Machacek M, Hodgson L, Welch C, Elliott H, Pertz O, Nalbant P, et al. Coordination of Rho GTPase activities during cell protrusion. *Nature* 2009;**461**(7260):99–103.

169. Carlson HJ, Campbell RE. Genetically encoded FRET-based biosensors for multi-parameter fluorescence imaging. *Curr Opin Biotechnol* 2009;**20**(1):19–27.

170. Piljic A, Schultz C. Simultaneous recording of multiple cellular events by FRET. *ACS Chem Biol* 2008;**3**(3):156–60.

171. Kishikawa J, Fujikawa M, Imamura H, Yasuda K, Noji H, Ishii N, et al. MRT letter: expression of ATP sensor protein in Caenorhabditis elegans. *Microsc Res Tech* 2011;**75**(1):15–9.

172. Lissandron V, Rossetto MG, Erbguth K, Fiala A, Daga A, Zaccolo M. Transgenic fruit-flies expressing a FRET-based sensor for in vivo imaging of cAMP dynamics. *Cell Signal* 2007;**19**(11):2296–303.

173. Lissandron V, Terrin A, Collini M, D'Alfonso L, Chirico G, Pantano S, et al. Improvement of a FRET-based indicator for cAMP by linker design and stabilization of donor-acceptor interaction. *J Mol Biol* 2005;**354**(3):546–55.

174. Kardash E, Bandemer J, Raz E. Imaging protein activity in live embryos using fluorescence resonance energy transfer biosensors. *Nat Protoc* 2011;**6**(12):1835–46.

175. Gervasi N, Tchenio P, Preat T. PKA dynamics in a Drosophila learning center: coincidence detection by rutabaga adenylyl cyclase and spatial regulation by dunce phosphodiesterase. *Neuron* 2010;**65**(4):516–29.

176. Fan-Minogue H, Cao Z, Paulmurugan R, Chan CT, Massoud TF, Felsher DW, et al. Noninvasive molecular imaging of c-Myc activation in living mice. *Proc Natl Acad Sci USA* 2010;**107**(36):15892–7.

177. Kamioka Y, Sumiyama K, Mizuno R, Sakai Y, Hirata E, Kiyokawa E, et al. Live imaging of protein kinase activities in transgenic mice expressing FRET biosensors. *Cell Struct Funct* 2012;**37**:65–73.

178. Yamaguchi Y, Shinotsuka N, Nonomura K, Takemoto K, Kuida K, Yosida H, et al. Live imaging of apoptosis in a novel transgenic mouse highlights its role in neural tube closure. *J Cell Biol* 2011;**195**(6):1047–60.

179. Calebiro D, Nikolaev VO, Gagliani MC, de Filippis T, Dees C, Tacchetti C, et al. Persistent cAMP-signals triggered by internalized G-protein-coupled receptors. *PLoS Biol* 2009;**7**(8):e1000172.

CHAPTER SIX

# Fluorescent Sensors of Protein Kinases: From Basics to Biomedical Applications

## Thi Nhu Ngoc Van, May C. Morris
CRBM-CNRS-UMR 5237, Chemical Biology and Nanotechnology for Therapeutics, Montpellier, France

## Contents

217

## Abstract

Protein kinases constitute a major class of enzymes underlying essentially all biological processes. These enzymes present similar structural folds, yet their mechanism of action and of regulation vary largely, as well as their substrate specificity and their subcellular localization. Classical approaches to study the function/activity of protein kinases rely on radioactive endpoint assays, which do not allow for characterization of their dynamic activity in their native environment. The development of fluorescent biosensors has provided a whole new avenue for studying protein kinase behavior and regulation in living cells in real time with high spatial and temporal resolution. Two major classes of biosensors have been developed: genetically encoded single-chain fluorescence resonance energy transfer biosensors and peptide/protein biosensors coupled to small synthetic fluorophores which are sensitive to changes in their environment.

In this review, we discuss the developments in fluorescent biosensor technology related to protein kinase sensing and the different strategies employed to monitor protein kinase activity, conformation, or relative abundance, as well as kinase regulation and subcellular dynamics in living cells. Moreover, we discuss their application in biomedical settings, for diagnostics and therapeutics, to image disease progression and monitor response to therapeutics, in drug discovery programs, for high-throughput screening assays, for postscreen characterization of drug candidates, and for clinical evaluation of novel drugs.

## ABBREVIATIONS

**AFP** autofluorescent proteins
**AMPK** AMP-activated protein kinase
**ATM** Ataxia telangiectasia mutated kinase
**BiFC** bimolecular fluorescence complementation
**BTF** beta-turn-focused
**CaMKII** calcium/calmodulin-dependent protein kinase II
**CDK** cyclin-dependent kinase
**CFP** cyan fluorescent protein
**CHEF** chelation-enhanced fluorescence
**CML** chronic myelogenous leukemia
**CPP** cell-penetrating peptide
**EGFP** enhanced green fluorescent protein
**EGFR** epidermal growth factor receptor
**DPA** dipicolylamine
**ELISA** enzyme-linked immunosorbent assay
**ERK** extracellular signal-regulated kinase
**FAK** focal adhesion kinase
**FHA domain** forkhead-associated domain
**FLIM** fluorescence lifetime imaging
**FRET** fluorescence resonance energy transfer
**GFP** green fluorescent protein
**GLUT** Glutamate Transporter

**HCS**  high content screening
**HTS**  high-throughput screening
**KAR**  kinase activity reporter
**NBD**  nitrobenzofuran
**NES**  nuclear export sequence
**NIRF**  near-infrared fluorescence
**NLS**  nuclear localization sequence
**PAABD**  phosphoamino acid-binding domain
**PBD**  Polo Box Domain
**PKA**  cAMP-dependent protein kinase A
**PKB**  protein kinase B also known as Akt
**PKC**  protein kinase C
**PKD**  protein kinase D
**PTB**  phosphotyrosine-binding domain
**PTK**  protein tyrosine kinase
**RDF**  recognition-domain-focused
**RFP**  red fluorescent protein
**RGD**  Cyclic Arginine-Glycine-Aspartate
**SAPK**  stress-activated protein kinases
**SH2 domain**  Src-homology 2 domain
**SH3 domain**  Src-homology 3 domain
**Sox**  sulfonamide-oxine
**YFP**  yellow fluorescent protein

## 1. INTRODUCTION

### 1.1. Protein kinases—Structure, function, and mechanism of action

The first report of a phosphorylated protein was the identification of phosphorylated vitellin (phosvitin) in 1906, followed by the detection of phosphorylated casein in 1933.[1,2] Yet the enzymatic process of protein phosphorylation itself was described only in 1954, with the identification of an enzyme from rat liver mitochondria, namely, glycogen phosphorylase, that could catalyze the transfer of phosphate from adenosine triphosphate (ATP) to protein substrates, together with a complete characterization of the conditions underlying this process, and reference made to emphasize the high turnover rate of phosphorus and the reactivity of phosphoproteins in tumors.[3] Today, we know that protein phosphorylation is the consequence of a fine balance between the activities of protein kinases that add phosphate groups and protein phosphatases that remove phosphate groups from protein substrates, and that these activities are responsible for phosphorylation of up to 30% of all cellular proteins.[4–7]

Protein kinases catalyze the transfer of a phosphate group from ATP onto an amino acid of a substrate protein with a free hydroxyl group, namely, serine, threonine, and tyrosine (Fig. 6.1A). As such, protein kinases are classically divided into the families of serine/threonine kinases, tyrosine kinases, and dual-specificity kinases.[8] Protein kinases are further divided into either receptor or nonreceptor kinases, depending on whether they are localized at the surface of cells and play a role in integrating extracellular cues to be fed into the cell. The human genome is estimated to harbor 514 genes encoding protein kinases, known as the kinome, which corresponds to 2% of all eukaryotic genes.[9] Protein kinases are central to the regulation of most cellular processes, throughout development and cell differentiation, metabolism, gene transcription, cell growth and division, DNA replication, protein degradation, apoptosis, and immune response.[10,11] They play a major role in

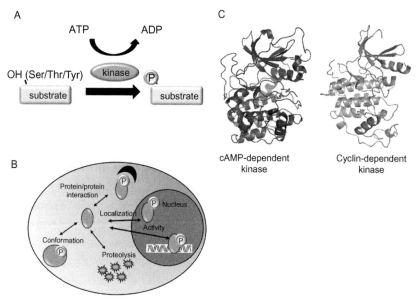

**Figure 6.1** *Protein kinase: structure, function, and mechanisms of action.* (A) A protein kinase is an enzyme that catalyzes the transfer of a phosphate group from ATP to a substrate protein, a process known as phosphorylation. This reaction consists in the removal of a phosphate group from ATP, followed by its covalent attachment to amino acids with a free hydroxyl group, namely, serine, threonine, and tyrosine. (B) Phosphorylation usually results in a functional change of the target protein (substrate) by changing enzyme activity, conformation, subcellular localization, or association with partners. (C) Structural fold of two protein kinases—cAMP-dependent kinase (PDB structure 2CPK) and cyclin-dependent kinase 2 (PDB structure 1B38). (For color version of this figure, the reader is referred to the online version of this chapter.)

signal transduction pathways in healthy cells, and their activity is normally tightly controlled and finely regulated. Phosphorylation affects substrate proteins in a variety of ways, by affecting their activity, subcellular localization, or their stability through routing to the ubiquitin–proteasome pathway for degradation, by promoting a conformational change in the structure of the substrate, its association with cofactors, partner proteins, inhibitors, or gene promoters[12] (Fig. 6.1B).

Protein kinases present a very similar structural fold, yet a high degree of structural plasticity.[13–16] Indeed, the structural conformation of active kinases differs from that of inactive kinases, yet the catalytic domains of active kinases adopt similar structures (Fig. 6.1C). Despite this structural homology, the primary sequences and modes of regulation vary widely from one kinase to another. Their specificity toward substrate recognition is equally very fine-tuned, and many kinases use binding sites that are remote from the catalytic site to recruit substrates with higher affinity, thereby leading to greater catalytic efficiency and specificity.[17,18]

Protein kinase activity is finely regulated through a variety of mechanisms, including activating phosphorylation (cyclin-dependent kinases, CDKs) or inhibitory phosphorylation (Src); through dimerization or multimerization (insulin receptor, EGF receptor); through allosteric regulation, upon binding of activator or inhibitor proteins, cofactors, or small molecules; through control of accessibility to their substrates; or through autoinhibition (PKA/PKI).[16] Protein kinases are "molecular switches" which may adopt at least two conformations corresponding to an "ON" state and an "OFF" state. Activation is associated with a net conformational change of a dynamic segment, termed "the activation loop", thereby providing an accessible binding site for the substrate.[16,19]

Protein kinases have been the focus of many studies not only because of their central role in major signaling pathways and biological processes but also because of their implication in disease. Indeed, a large number of genetic alterations result in deregulated protein kinase activity, thereby prompting or contributing to the development of pathological conditions, including cardiovascular diseases, neurodegenerative and endocrinological disorders, immune deficiency and viral infection, psoriasis, cancer, and diabetes.[20–23] As such, protein kinases constitute disease biomarkers and pharmacological targets in a variety of human diseases, and several small-molecule inhibitors have been developed to interfere with their function, some of which have become therapeutics administered in the clinic.[21–28]

## 1.2. Strategies for probing and studying protein kinases *in vitro* and *in cellulo*

Despite the central role of protein kinases in physiological pathways, and their implication as disease biomarkers in pathological disorders, there are very few means of studying their activity in real time in a noninvasive fashion and in their natural environment. Most kinase activity studies have been traditionally performed using biochemical assays based on purified enzymes produced as recombinant proteins from insect or mammalian cells in culture. Commonly implemented approaches to study protein kinase activity rely essentially on the incorporation of radioactive phosphate into artificial substrates, or on antigenic approaches, which further depend on highly specific antibodies to recognize the phosphorylated form of the kinase substrate(s). These assays have been widely used in the laboratory and further adapted to high-throughput drug screening, but they are irreversible endpoint assays, which do not allow for real-time analysis of kinase activity, and further lack the physiological cellular environment. Cell-based methods that monitor kinase activity have been developed that rely on the incorporation of $^{32}$P into cells; this approach requires cell lysis, substrate protein isolation, and purification to determine its relative degree of phosphorylation by measuring the amount of radioactivity incorporated. However, such cell-based assays are labor intensive, show poor sensitivity, and have the disadvantage of requiring high levels of radioactivity. Other cell-based assays for the study of kinase activity involve the use of radiolabeled phosphorylation-specific antibodies (i.e., antibodies that can discriminate between phosphorylated and nonphosphorylated proteins); the phosphorylated substrate can then be detected and quantified by immunoprecipitation, gel electrophoresis, or Western blotting after cell lysis. Although these assays generally require lower levels of radioactivity than $^{32}$P-based methods, they are equally labor intensive, time consuming, and complex to automate. Nonradioactive cell-based methods have emerged that use an ELISA (enzyme-linked immunosorbent assay) approach to measure the activation of specific kinase signaling pathways. These kinase assays, which employ phosphorylation-specific antibodies, have been demonstrated to be suitable for high-throughput drug screening.[29] More recently, a nonradioactive method for probing kinase activity has been developed, which is based on the use of a fluorogenic peptide substrate (rhodamine 110, bis peptide amide) that is cleaved in its unphosphorylated form, thereby releasing free rhodamine 110; phosphorylation prevents cleavage, and the compound remains as a nonfluorescent peptide conjugate.[30] This approach has been successfully applied to high-throughput screening (HTS) and does not require cell lysis. However, it does not allow for real-time studies, as the nonphosphorylated substrates are cleaved and degraded. Moreover, besides their disruptive character, these approaches merely provide a snapshot of a dynamic process and do not provide real insights into the dynamics and enzymatic kinetics of the target kinase in its native environment.

Faced with a real need for tools that enable qualitative and quantitative assessment of enzymatic activities in their native environment, biologists and chemists have designed a new generation of tools designated as "fluorescence-based reporters" or "fluorescent biosensors," which enable probing the biochemical function, activity, and dynamics of specific enzymes through sensitive changes in their fluorescent properties. Fluorescence detection is currently one of the most widely used approaches in biomolecular imaging because of its high intrinsic sensitivity and selectivity. Indeed, fluorescence lends itself to nondestructive imaging, thereby preserving the sample and the molecules of interest within. Moreover, fluorescence imaging provides a high degree of temporal and spatial resolution and allows for real-time tracking of a molecule in motion in a complex solution or environment.

As such, when fluorescent biosensors are designed to recognize a target with high specificity and to report on this event with high sensitivity, they constitute extremely useful tools for the detection of biomolecules, both *in vitro* and *in vivo*, and for monitoring dynamic molecular events, such as enzymatic activities, conformational changes, or protein/protein interactions.[31] High-resolution imaging and real-time measurements of fluorescent biosensors provide precious information on the spatial and temporal localization of a wide variety of intracellular targets and have proved a successful means of visualizing dynamic processes in living cells. Fluorescent biosensors that specifically probe protein kinases allow imaging the behavior of kinases in a dynamic and quantitative fashion, in space and in time, in cells, tissues, and whole organisms, thereby allowing tackling questions that could not be addressed previously. In addition, fluorescent reporters and biosensors can be exploited to develop high-throughput and high content screens for drug discovery purposes and for further assessing the efficacy and pharmacokinetics of lead compounds of potential therapeutic utility. Moreover, fluorescent biosensors offer promising perspectives for diagnostic strategies, for early detection of protein kinase dysregulation associated with the onset of disease, and for monitoring disease progression and response to therapeutics.

Two large classes of kinase biosensors have been developed: genetically encoded autofluorescent protein (AFP) biosensors and nongenetic fluorescent peptide biosensors. The discovery of the green fluorescent protein (GFP) led to the development of genetically encoded reporters, which have been successfully applied to the study of protein kinase behavior in living cells with high spatial and temporal resolution (for review, see Refs. 32–36). In parallel, combined efforts in fluorescence chemistry and in chemical biology have led to the design of an entirely different family of biosensors, based on peptide, polypeptide, or polymeric scaffolds, onto which small synthetic probes with particularly attractive photophysical properties can be incorporated. These nongenetic biosensors have been applied to monitor protein kinase activities *in vitro* and in more complex biological samples, including cell extracts and living cells, with an equally successful outcome (for review, see Refs. 35–38). While biosensors from each of these classes exhibit specific advantages, they equally suffer from characteristic drawbacks and shortcomings, which will be discussed further below and are summarized in Table 6.1. Genetically encoded biosensors are user-friendly, allowing for easy manipulation and transfection into cells. However, most of these biosensors are based on changes in fluorescence resonance energy transfer (FRET) between two AFPs, which are often fairly weak compared to the fluorescence changes associated with nongenetic, environmentally sensitive biosensors. In contrast, whereas peptide and protein biosensors are easy to engineer and handle *in vitro*, it is difficult to introduce these into living cells and *in vivo*. However, they offer complete control over the concentrations used and the timing of their availability *in cellulo*, whereas genetically encoded biosensors are heterogeneously expressed both in terms of levels and timing.

## 2. GENETICALLY ENCODED REPORTERS OF PROTEIN KINASES

The discovery of the GFP and its engineering into a wide variety of AFPs prompted the development of GFP-based reporters and the design of genetically encoded biosensors, which were rapidly applied to the field of protein kinases.[32–36,39–45] Several groups of genetically encoded biosensors have been developed (for review, see Refs. 34,44): (i) single-chain biosensors, which bear a pair of FRET AFPs within the same molecule that are brought together and undergo energy transfer due to intramolecular conformational changes in response to the recognition event; (ii) two-chain biosensors, in which two AFPs lie on two different molecules capable of interacting and undergoing intermolecular FRET

**Table 6.1** Pros and cons of genetically encoded versus fluorescent peptide biosensors

| Genetically encoded fluorescent biosensors | Fluorescent peptide/protein biosensors |
|---|---|
| Features | Features |
| • Genetically encoded constructs based on two AFPs<br>• Essentially FRET/FLIM | • Peptide or protein backbone onto which a synthetic fluorescent probe is introduced during or postsynthesis through chemical or enzymatic conjugation<br>• Environmentally sensitive, chelation-enhanced, self-reporting |
| Pros | Pros |
| • Easy to engineer through molecular biology approaches<br>• Easy to transfect or microinject into cultured cells | • Small size<br>• Versatility in design and engineering<br>• Easy and cheap to synthesize or express through recombinant protein engineering<br>• Allow control over concentration and time of addition to target or cells<br>• Allow technological improvements: introduction of quenchers, amino acid caging |
| Cons | Cons |
| • Large size—essentially due to the size of the autofluorescent proteins<br>• Require ectopic expression<br>• Lack of control over expression levels and timing<br>• Heterogeneous expression in a population of cells<br>• Difficult to obtain optimal expression for *in vivo* imaging | • Require means for efficient introduction into living cells and *in vivo*<br>• Require choice of NIRF probes for *in vivo* applications |

when the two chains are brought together for a protein/protein interaction; (iii) biosensors based on bimolecular fluorescence complementation (BiFC), in which two fragments of a split AFP are brought together as a consequence of the recognition event, thereby reconstituting the intact and fully fluorescent protein; (iv) single-chain biosensors which bear a single AFP whose spectral properties change in response to the recognition of a target by an encoded element other than the AFP; and (v) biosensors constituted by the AFP itself, whose spectral properties respond directly to the target or analyte, as pH–dependent or redox–dependent AFP biosensors (Fig. 6.2).

The most widely developed genetically encoded biosensors developed to probe protein kinase activities are single–chain FRET biosensors that report on kinase activity through phosphorylation–induced changes in FRET between

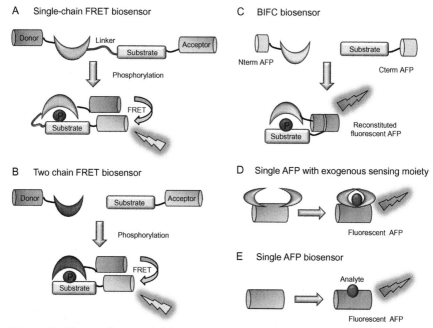

**Figure 6.2** Genetically encoded biosensors. Genetically encoded biosensors of kinase activity are derived from at least one genetically encoded autofluorescent protein. (A) Single-chain biosensors bear a pair of AFPs, within the same molecule, which are brought together and undergo changes in fluorescence resonance energy transfer due to intramolecular conformational changes in response to the phosphorylation event. (B) Two-chain biosensors, in which two AFPs lie on two different molecules susceptible for interacting and undergoing intermolecular FRET when the two chains are brought together for a protein/protein interaction. (C) Biosensors based on bimolecular fluorescence complementation (BiFC), in which two fragments of a split AFP are brought together as a consequence of target recognition, or in this case phosphorylation, thereby reconstituting the intact and fully fluorescent protein. (D) Single-chain biosensors composed of a single AFP whose spectral properties change in response to the recognition of a target by an exogenous sensing element (other than the AFP). (E) Single AFP biosensors, whose spectral properties respond directly to the target or analyte, such as pH-dependent or redox-dependent AFP biosensors. (For color version of this figure, the reader is referred to the online version of this chapter.)

two AFPs due to an intramolecular conformational change. A wide variety of these genetically encoded biosensors have been developed (see Table 6.2 and Figs. 6.3–6.6), but the basic structure of these kinase activity reporters (KARs) is essentially the same: a kinase-specific substrate sequence bearing a consensus phosphorylation site connected to a matching phosphoamino acid-binding domain (PAABD) by a flexible linker and flanked by a pair of AFPs that can transfer fluorescence resonance energy between each other. Following phosphorylation of the substrate sequence by the kinase, the PAABD binds the phosphorylated sequence, thereby promoting an intramolecular conformational change which alters the distance and orientation of the AFPs.

Several factors have to be considered when aiming to design an optimally responsive and selective genetically encoded FRET biosensor. Moreover, the specificity and selectivity of a kinase/biosensor couple should always be characterized *in vitro* and *in cellulo*.

First, needless to say, the size and sequence of the substrate have to be tailored to the most suitable sequence to gain the highest level of selectivity. The choice of the substrate sequence is essential for selectivity and may either be identified through an *in silico* approach through database mining from knowledge-based libraries, or simply through rational design of a sequence derived from a known substrate protein of a given kinase. The specificity of a kinase for a protein substrate is generally dependent on other regions of the protein that "dock" onto the kinase to offer higher affinity/recognition. Therefore, several designs include a docking domain distinct from the substrate sequence, so as to increase specificity for the target kinase while also increasing the efficiency of phosphorylation of the substrate by the kinase. This is well exemplified by MAPK protein kinases (ERK, p38, and JNK), which have very similar phosphorylation sites, yet distinct docking sites that can be made use of to improve substrate targeting and selectivity.[81,82]

Second, the choice of the PAABD is critical for an efficient and sensitive biosensor. The sequence of the PAABD should present high affinity and efficient recognition of the phosphorylated substrate, as opposed to poor affinity for the unphosphorylated substrate, and should display a fully reversible behavior between both forms. Several modular phosphobinding domains (PBDs) have been well characterized, including Src-homology 2/3 (SH2 SH3) and phosphotyrosine-binding (PTB) domains, 14.3.3 proteins, forkhead-associated (FHA) domains, WW domains, WD40 domains, and the PBD of Plk1 (for reviews, see Refs. 83–89).

**Table 6.2** Genetically encoded single-chain FRET biosensors of protein kinases

| Protein kinase | Biosensor name | AFP FRET pairs | Author and year | Reference |
|---|---|---|---|---|
| Abl/EGFR | CrkII–based reporter | CFP/YFP | Ting *et al.* (2001) | 46 |
| c–Abl | Picchu | CFP/YFP | Kurokawa *et al.* (2001) | 47 |
| Bcr–Abl | Pickles | ECFP/ Venus | Mizutani *et al.* (2010) | 48 |
| Aurora B | | CFP/YFP | Fuller *et al.* (2008) | 49 |
| AKT | AKTAR | Cerulean/ cpVenus | Gao and Zhang (2008) | 50 |
| AKT | AKTUS | CFP/YFP | Sasaki *et al.* (2003) | 51 |
| AKT | BKAR | ECFP/ citrine | Kunkel *et al.* (2005) | 52 |
| AMPK | AMPKAR | ECFP/ cpVenus | Tsou *et al.* (2011) | 53 |
| ATM | ATOMIC | CFP/YFP | Johnson *et al.* (2007) | 54 |
| CAMKII | Camui | CFP/YFP | Erickson *et al.* (2011) | 55 |
| CDK1/cyclin B activity | | mCerulean/ Ype | Gavet and Pines (2010) | 56 |
| ERK | EKAR | EGFP/ mRFP1 | Harvey *et al.* (2008) | 57 |
| ERK1 | Erkus | CFP/YFP | Sato *et al.* (2007) | 58 |
| IR | Phocus | CFP/YFP | Sato *et al.* (2002) | 59 |
| JNK kinase | JNKAR | EGFP/ citrine | Fosbrink *et al.* (2010) | 60 |
| FAK kinase | | ECFP/YPet | Seong *et al.* (2011) | 61 |
| Histone phosphorylation | | CFP/YFP | Lin and Ting (2004) | 62 |
| MLCK | MLCK–FIP ($Ca^{2+}$/ calmodulin) | CFP/YFP | Chew *et al.* (2002) | 63 |

**Table 6.2** Genetically encoded single-chain FRET biosensors of protein kinases—cont'd

| Protein kinase | Biosensor name | AFP FRET pairs | Author and year | Reference |
|---|---|---|---|---|
| PKA | ART | BGFP/ RGFP | Nagai *et al.* (2000) | 64 |
| PKA | AKAR1 | ECFP/YFP | Zhang *et al.* (2001) | 65 |
| PKA | AKAR2 | ECFP/ citrine | Zhang *et al.* (2005) | 66 |
| PKA | AKAR | EGFP/ cpVenus | Allen and Zhang (2006) | 67 |
| PKA (sarcoplasmic reticulum) | SR-AKAR3 | CFP/YFP | Liu *et al.* (2011) | 68 |
| PKC | CKAR | ECFP/ citrine | Violin *et al.* (2003) | 69 |
| PKC-delta | deltaCKAR | CFP/YFP | Kajimoto *et al.* (2010) | 70 |
| PKC | KPC-1 (pleckstrin based) | GFP/EYFP | Schleifenbaum *et al.* (2004) | 71 |
| PKA and PKC | KPAC-1 (pleckstrin based) | | Brumbaugh *et al.* (2006) | 72 |
| PKD | DKAR | CFP/YFP | Kunkel *et al.* (2007) | 73 |
| Plk1 | | CFP/YFP | Macůrek *et al.* (2008) | 74 |
| SAP3K | | Venus/ SECFP | Tomida *et al.* (2009) | 75 |
| Src | | CFP/YFP | Wang *et al.* (2005) | 76 |
| c-Src | Srcus | CFP/YFP | Hitosugi *et al.* (2007) | 77 |
| SYK | | ECFP/YPet | Xiang *et al.* (2011) | 78 |
| ZAP-70 | ROZA | CFP/YFP | Randriamampita *et al.* (2008) | 79 |

The linker between these domains is generally designed to be flexible (i.e., rich in glycine, alanine, and proline residues), and its length has proved to be a critical factor in affecting FRET efficiency.[90–92]

Last but not least, the choice of the AFP/FRET pair is essential. Ideally, AFPs that do not photobleach too rapidly, that have high quantum yield, and that are red-shifted should be chosen to maximize FRET efficiency and minimize phototoxicity.[93,94] Komatsu et al. recently proposed an optimized backbone for genetically encoded FRET biosensors bearing the cyan fluorescent protein (CFP) or Turquoise as donors, the yellow fluorescent protein (YFP) or YPet as an acceptor fluorophore, and a very long linker connecting the ligand to the sensor domain, which basically abolishes dimerization of the FRET pair, and showed that this backbone was ideal for both kinase and GTPase FRET biosensors reporting on ERK, PKA, S6K, RSK, Ras, and Rac1.[92]

In this section, we provide several examples to illustrate the different strategies that have been devised in designing genetically encoded FRET biosensors and describe how they have been employed to study the behavior of specific kinases in living cells, rather than attempt an exhaustive review of all the kinase biosensors that have been developed so far.

## 2.1. Phosphotyrosine kinase (PTK) biosensors

Ting et al. developed the first genetically encoded fluorescent reporters of tyrosine kinases to probe the activities of Src, Abl, and EGFR (epidermal growth factor receptor). These single-chain FRET biosensors consist of fusions of the CFP/YFP FRET pair, an SH2 phosphotyrosine-binding domain, and a substrate sequence, the phosphorylation of which on a critical tyrosine promotes binding to the SH2 domain, thereby inducing a conformational change that brings the two AFPs close enough to undergo FRET between the donor and the acceptor. While the EGFR and Src indicators were designed by combining an SH2 domain with an appropriate substrate, separated by a linker, the Abl reporter was directly engineered by introducing a domain derived from Crk protein, which includes an SH2 domain, two SH3 domains, and an intervening Abl phosphorylation site[46] (Fig. 6.3A and B). The Picchu sensor was constructed in a similar fashion to monitor specific phosphorylation by c-Abl, with a CFP/YFP FRET pair and an SH2–2 SH3 domain derived from the CrkII adaptor protein.[47] More recently, Mizutani et al. developed a sensitive and specific FRET biosensor, termed Pickles, for measuring the tyrosine kinase activity of Bcr–Abl, which

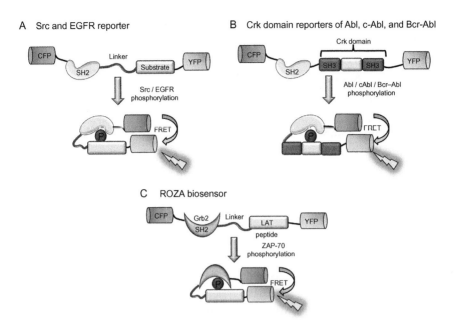

A   Src and EGFR reporter

B   Crk domain reporters of Abl, c-Abl, and Bcr-Abl

C   ROZA biosensor

**Figure 6.3** *Genetically encoded protein tyrosine kinase reporters.* Genetically encoded FRET biosensors of tyrosine kinases are composed of a substrate sequence and a matching phosphoaminoacid binding domain sandwiched between a FRET pair of GFP variants. (A) Fluorescent reporters of Src and EGFR (epidermal growth factor receptor), based on an SH2 phosphotyrosine-binding domain and a separate substrate sequence joined by a linker and flanked by the CFP/YFP couple.[46] (B) Structure of fluorescent reporters of Abl,[46] c-Abl (Picchu),[47] and Bcr–Abl (pickles),[48] derived from Crk adaptor proteins bearing an SH2 domain, two SH3 domains, and an intervening phosphorylation site. (C) ZAP-70 biosensor ROZA constituted of the SH2 domain of mouse Grb2 and a peptide substrate sequence derived from LAT sandwiched between the CFP–YFP pair.[79] (For color version of this figure, the reader is referred to the online version of this chapter.)

again incorporates one SH2 and two SH3 domains derived from CrkL, the most characteristic substrate of Bcr–Abl, sandwiched between Venus, a variant of YFP, and ECFP[48] (Fig. 6.3B). The high sensitivity of this biosensor allows the assessment of Bcr–Abl activity from patient cells to monitor the disease status and response to therapy, as well as the detection of the onset of drug-resistant cells within a heterogeneous population.[95,96]

Src is a nonreceptor tyrosine kinase that is activated by a variety of mechanisms and constitutes a major kinase in the signaling pathways involved in cancer and metastasis.[97] An Src-specifc biosensor based on an ECFP/YFP couple, an SH2 domain, and a substrate peptide derived from the c-Src substrate p130cas was developed to probe this kinase and visualize its dynamics of activation following mechanical stimuli.[76] More recently, a different variant, Srcus, was engineered including a nuclear export signal, which was employed to study Src activation by steroids in the cytosol and at the plasma membrane.[77] Another tyrosine kinase of particular relevance to cancer is the spleen tyrosine kinase Syk.[98] A FRET-based biosensor was developed to image Syk tyrosine kinase activity in living cells based on an ECFP/YPet FRET couple, a substrate derived from VAV2 and an SH2 PAABD.[78] This biosensor allowed following and quantifying real-time activation of Syk, associated with an increase in the FRET signal, upon immunoreceptor activation and following stimulation by the platelet-derived growth factor.

In order to examine the dynamics of the ZAP-70 tyrosine kinase activity at the immunological synapse, Randriamampita *et al.* developed the ROZA biosensor (Reporter Of ZAP-70 Activity). This single-chain FRET biosensor bears an SH2 domain from mouse Grb2 and a peptide substrate sequence from mouse LAT sandwiched between the CFP–YFP pair (Fig. 6.3C). ROZA was applied to image ZAP-70-dependent phosphorylation in T-cell lines and primary human lymphocytes with subcellular resolution during the formation of an immunological synapse.[79]

## 2.2. Biosensors for protein kinase PKA, PKB, PKC, and PKD

PKA (cAMP-dependent protein kinase) constitutes an essential enzyme involved in major biological processes including gene expression, metabolism, cell growth, and cell proliferation.[99] Genetically encoded single-chain FRET biosensors of PKA, combining the consensus kinase substrate sequence, with a matching PAABD sandwiched between a FRET pair of GFP variants, were amongst the first kinase reporters developed to characterize their endogenous activity in living cells in a dynamic fashion. The first genetically encoded biosensor of PKA activity, termed ART (cAMP-responsive tracer), was composed of BGFP/RGFP, the kinase-inducible domain (KID) of the CREB transcription factor (cAMP-responsive element-binding protein), which undergoes a conformational change upon phosphorylation[64] (Fig. 6.4A). As opposed to most other FRET biosensors, phosphorylation of KID by PKA disrupts the FRET between the flanking AFPs. An entirely different biosensor called AKAR (cAMP-dependent KAR) was later generated by Zhang *et al.*[65] (Fig. 6.4B). AKAR1 was based on a 14-3-3 phosphorecognition domain but was later optimized and replaced by an FHA domain in AKAR2,[66] so as to improve the reversibility of the biosensor and therefore its response to cellular phosphatases, for

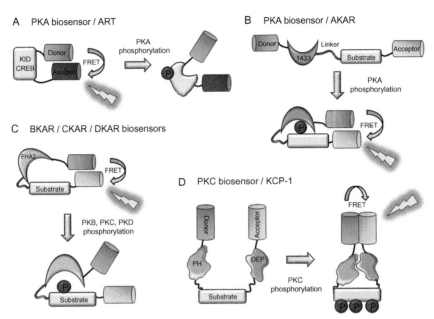

**Figure 6.4** *Genetically encoded biosensors of PKA, PKB, PKC, and PKD.* (A) ART: genetically encoded biosensor of PKA activity, composed of two AFPs joined by the kinase-inducible domain (KID) of the CREB transcription factor (cAMP-responsive element-binding protein).[64] Phosphorylation of KID by PKA promotes a conformational change which promotes FRET between the flanking AFPs. (B) PKA biosensor AKAR1 based on a 1433 PAABD and a substrate peptide of PKA sandwiched between a CFP/YFP pair.[65,66] (C) BKAR, CKAR, and DKAR biosensors of PKB, PKC, and PKD kinases, respectively.[52,69,73] These reporters are constituted of a consensus phosphorylation sequence and the FHA2 domain of Rad53 sandwiched between an mCFP/mYFP pair that undergoes FRET in the unphosphorylated state; phosphorylation of the substrate sequence results in a conformational change that alters the FRET ratio. (D) PKC biosensor KCP-1 based on a GFP/EYFP FRET pair, and a domain derived from pleckstrin constituted by a PH domain and a DEP domain flanking a loop that bears several phosphorylation sites.[71] Upon phosphorylation of this intervening loop by PKC, the pleckstrin moiety undergoes a conformational change, which prompts the PH and the DEP domains to interact. (For color version of this figure, the reader is referred to the online version of this chapter.)

measuring the cellular dynamics of PKA activity.[67,100] More recently, a variant SR-AKAR3 was developed to monitor PKA activity in the sarcoplasmic reticulum in cardiomyocytes.[68]

PKC kinases form a family of 15 isoforms that contribute to well-defined signaling pathways.[101] The first genetically encoded PKC biosensor, CKAR, was designed to incorporate a consensus phosphorylation sequence and the FHA2 domain of Rad53 sandwiched between the mCFP/mYFP pair, which undergo FRET in the unphosphorylated state, whereas phosphorylation of the substrate sequence results in a conformational change that alters the FRET ratio[69] (Fig. 6.4C). Targeting of this reporter to the plasma membrane where PKC is activated revealed an oscillatory phosphorylation in response to histamine, which is closely associated with a calcium oscillation. This original biosensor was not designed to discriminate between the different isoforms of PKC. However, more recent developments have led to reporters designed specifically to monitor PKCdelta activity.[70] A different strategy was employed by Schultz et al. To develop the KCP-1 biosensor, a PKC reporter that incorporates the GFP/EYFP FRET pair, together with a truncated form of pleckstrin that habors a PH domain (pleckstrin homology) and a DEP domain (Disheveled, Egl-10, pleckstrin) separated by a 14-amino acid loop kinase-specific substrate sequence bearing three phosphorylation sites was designed. Upon phosphorylation of this intervening loop by PKC, the pleckstrin moiety undergoes a conformational change, which prompts the PH and the DEP domains to interact. The major advantage of this reporter design is that it does not rely on interactions between the substrate and a distinct PAABD. This probe responds rapidly to PKC activation through phorbol ester stimulation or upon activation of physiologically relevant pathways[71] (Fig. 6.4D). The KPC-1 biosensor was further engineered to KCAP-1, which bears a PKA substrate sequence in the C-terminal loop just before the GFP moiety. This GFP/EYFP FRET biosensor consequently responds to PKC phosphorylation through an increase in FRET, as opposed to a decrease in FRET induced by PKA phosphorylation, thereby allowing monitoring PKA and PKC activities independently in living cells.[72]

PKB/Akt is a serine/threonine protein kinase that plays a key role in a wide variety of cellular processes, including glucose metabolism, apoptosis, cell proliferation, transcription, and cell migration, and integrates many oncogenic signals from the phosphatidylinositol 3-kinase pathway signaling pathway, thereby promoting cellular growth, survival, and proliferation together with decreased apoptosis. Akt is constitutively phosphorylated, activated, and recruited to the plasma membrane in a large variety of solid tumors and hematologic malignancies, and is known to be actively involved in tumor initiation and progression.[102,103] While several reporters based on bioluminescence have been developed, which will not be described here, several fluorescence-based reporters have also been engineered to monitor PKB/Akt signaling and study its dynamics in living cells.[50–52] The Aktus reporters are based on a CFP/YFP FRET pair, a protein kinase B (PKB) substrate sequence derived from Bad, and a PAABD domain derived from 14-3-3n, similar to the structure of AKAR. Aktus was further fused to the Golgi-targeting and mitochondrial-targeting domains of eNos and Bad, respectively, so as to generate variants that would preferentially colocalize with these endogenous substrates and thereby optimize their chances of phosphorylation by PKB compared to the nontargeted sensor.[51] Another PKB sensor, BKAR, was developed with the same structural design as the CKAR reporter[69] and employed to image phosphorylation catalyzed by PKB in real time and to investigate differences in the dynamics of this activity in the nucleus, cytosol, and when targeted to the plasma membrane.[52] This study revealed differences in biosensor response associated with different subcellular compartments, inferring differences in PKB signal propagation and termination. Another PKB reporter, AktAR, was developed to examine dynamic Akt activity in membrane microdomains and was based on the CFP/Venus FRET couple, an FHA1 domain, and a FOXO1 substrate.[50] This biosensor revealed that Akt activities are differentially regulated in different membrane microdomains and present overall higher activity within lipid rafts.

A reporter of protein kinase D, DKAR, based on the design of BKAR and CKAR,[52,69] was applied to investigate the dynamics of this kinase, revealing its dependence on calcium through positive feedback regulation of diacylglycerol production.[73]

## 2.3. Genetically encoded ERK/MAPK biosensors

ERK1/2 (extracellular signal-regulated kinases) play central roles in growth, cell proliferation, and differentiation. As their name suggests, these kinases are activated following different stimuli, including mitogenic factors, hormones, and cytokines, and they are themselves involved in regulating the activity of a wide variety of substrates including transcription factors and antiproliferative genes during G0 and G1 phases.[104] The first genetically encoded FRET biosensor of ERK, Erkus, was designed by Sato et al. in order to study the spatiotemporal dynamics of protein phosphorylation by activated ERK. Erkus is a single-chain biosensor encoding the CFP and YFP AFPs that flank an FHA2 domain from Rad53p, a short linker, and a substrate domain derived from the EGFR T669 peptide, as well as a short docking motif, the D-domain, derived from p90 ribosomal S6 kinase[58,80] (Fig. 6.5A). While this biosensor allowed the visualization of the dynamics of ERK kinase in real time, additional variants bearing either a nuclear localization sequence (NLS) or a nuclear export sequence (NES)

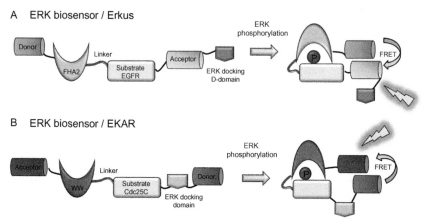

**Figure 6.5** *Genetically encoded biosensors of ERK kinases.* (A) The first genetically encoded FRET biosensor of ERK, Erkus, is a single-chain biosensor encoding the CFP/YFP FRET pair flanking an FHA2 domain from Rad53p, a short linker and a substrate domain derived from EGFR T669 peptide, together with a short docking motif, the D-domain, derived from p90 ribosomal S6 kinase.[58,80] (B) EKAR comprises a WW domain and a substrate derived from Cdc25C, immediately flanked by an ERK-specific docking domain, sandwiched between an mRFP/EGFP FRET pair.[57] (For color version of this figure, the reader is referred to the online version of this chapter.)

further allowed the comparison of the cytosolic and nuclear activities of ERK in living cells. More recently, Harvey *et al.* developed a different biosensor, termed EKAR, which comprises a WW domain and a substrate derived from Cdc25C, immediately flanked by an ERK-specific docking domain and separated by a 72-glycine linker, sandwiched between an mRFP and an enhanced green fluorescent protein (EGFP)[57] (Fig. 6.5B). Expression of EKAR biosensor was observed to be essentially nuclear; therefore, a cytoplasmic version of EKAR was developed, EKARcyto, by appending an NES to the C-terminus of the biosensor. Moreover, EKAR bears an ERK-specific docking site, which differs from that of Erkus, is adjacent to the substrate, and appears to contribute to the superior FRET signal of EKAR over Erkus. EKAR was successfully applied to measure the spatiotemporal signaling dynamics of this kinase in neurons from intact brain tissue by fluorescence lifetime imaging.

## 2.4. Cell cycle kinases: ATM, Plk, Aurora, CDK/cyclin biosensors

Several genetically encoded biosensors have been developed to study protein kinases involved in cell cycle regulation, although there have been fewer developments in this field than in the field of protein tyrosine kinases or of

PKA/PKC signaling. Aurora and Plk kinases are mitotic kinases whose central function and subcellular dynamics have recently been uncovered, thanks to fluorescent reporters of their activity. A biosensor of Aurora B kinase was developed to study the dynamics of protein phosphorylation by this mitotic kinase during anaphase.[49] This biosensor encodes a CFP/YFP FRET pair of AFPs, an Aurora B substrate peptide, and an FHA2 domain and was further targeted at chromosomes (histone H2B fusion) or centromeres (CENP-B fusion) (Fig. 6.6A). Fluorescence imaging of Aurora B, thanks to this FRET sensor in mitotic cells, revealed that Aurora kinase activity organizes the targeted microtubules to generate a structure-based feedback loop and a spatial phosphorylation gradient of multiple substrates during anaphase that predicts the cellular cleavage site. Along the same lines, a genetically encoded FRET biosensor of the mitotic Plk1 kinase was developed to probe the activity of this kinase in human cells in a physiological context and upon

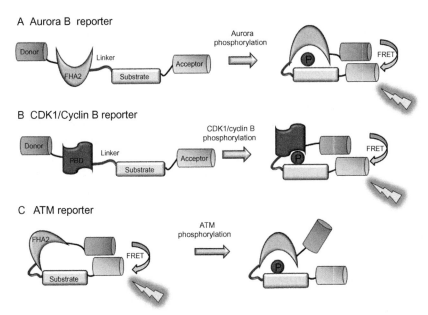

**Figure 6.6** *Genetically encoded biosensors cell cycle kinases.* (A) Aurora B kinase reporter encodes a CFP/YFP FRET pair, an Aurora B substrate peptide, and an FHA2 domain.[49] (B) CDK1/cyclin B kinase reporter comprises an autophosphorylation site from cyclin B1 fused to the PBD of Plk1 and flanked by mCerulean and YPet.[56] (C) ATM reporter encodes a CFP/YFP FRET pair, an FHA phosphobinding domain, and an ATM substrate sequence, the phosphorylation of which promotes a conformational change of the biosensor which disrupts FRET between the AFPs.[54] (For color version of this figure, the reader is referred to the online version of this chapter.)

checkpoint recovery.[74] This study revealed that Plk1 activation occurs several hours before entry into mitosis, thanks to Aurora A-mediated phosphorylation of Plk1, and that Bora/Aurora-A-dependent phosphorylation is a prerequisite for Plk1 to promote mitotic entry after a checkpoint-dependent arrest.

CDKs constitute the molecular engines that drive cell cycle progression. In particular, CDK1/cyclin B is essential for driving cells through mitosis. However, functional studies of this kinase have remained limited to classical antigenic approaches following cell fixation or extraction. More recently, Gavet and Pines developed a FRET sensor to probe CDK1/cyclin B kinase activity in living cells based on an autophosphorylation site from cyclin B1 fused to the Polo box domain (PBD) of Plk1 and flanked by mCerulean and YPet[56] (Fig. 6.6B). Thanks to this probe, imaging of CDK1/cyclin B activity was performed with high temporal precision, showing that this kinase is inactive in G2 until just before nuclear envelope breakdown, contributing to initiate prophase.

ATM (Ataxia telangiectasia mutated kinase) is an intracellular protein kinase that coordinates the DNA damage signaling pathway upon recognition of DNA double strand breaks. Activation of ATM leads to phosphorylation of transducer and effector proteins that coordinate a cellular response including DNA repair, cell cycle arrest, or apoptosis. A genetically encoded CFP–YFP FRET reporter was developed to probe ATM kinase activity in living cells, based on an FHA PBD and an ATM substrate sequence, the phosphorylation of which leads to changes in FRET efficiency between CFP and YFP associated with a conformational change of the biosensor[54] (Fig. 6.6C). This reporter allowed monitoring the ATM activity in a specific and quantitative fashion, providing a measurable output in response to double strand breaks, while also providing information on the spatiotemporal dynamics of ATM activity in living cells.

## 2.5. Other kinase biosensors

To visualize signal transduction based on protein phosphorylation in living cells, Sato et al. developed genetically encoded fluorescent indicators, named "phocuses" (fluorescent indicator for protein *pho*sphorylation that can be *cus*tom-made) that report on phosphorylation by the insulin receptor[59] (Fig. 6.7A). These reporters encode a CFP/YFP pair, a substrate sequence for the insulin receptor, a flexible linker sequence, and a phosphorylation recognition domain derived from SH2, PTB, or WW domains, or single-chain antibodies that recognize the phosphorylated substrate sequence.

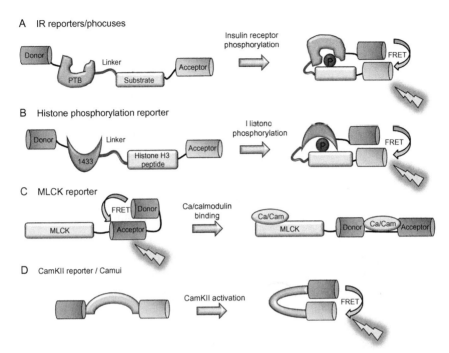

**Figure 6.7** *Genetically encoded biosensors of other kinases.* (A) Phocuses are fluorescent indicators that report on phosphorylation by the insulin receptor based on CFP/YFP FRET pair, a substrate peptide for the insulin receptor, a flexible linker sequence, and a phosphorylation recognition domain.[59] (B) Histone phosphorylation reporter by appending a substrate sequence from histone H3 bearing a phosphorylatable serine residue to a phosphoserine-binding domain, derived from 14-3-3, flanked by YFP and CFP, respectively.[62] (C) MLCK-FIP is a reporter of MLCK activity associated with $Ca^{2+}$/calmodulin binding. This reporter is based on a calmodulin-binding domain derived from MLCK and a BFP/GFP couple that undergoes FRET in the absence of $Ca^{2+}$ and calmodulin.[63] (D) Camui, a FRET biosensor of CaMK II, constituted of full-length CaMKII flanked by CFP and YFP, allows measurement of CaMKII activation associated with a conformational change.[55] (For color version of this figure, the reader is referred to the online version of this chapter.)

Upon phosphorylation, the intramolecular interaction between the substrate and the phosphorecognition domain promotes a conformational change which induces a consequent increase in FRET between CFP and YFP.

Lin and Ting engineered a histone phosphorylation reporter by appending a substrate sequence from histone H3 bearing a phosphorylatable serine residue to a phosphoserine-binding domain derived from 14-3-3 flanked by YFP and CFP, respectively[62] (Fig. 6.7B).

A biosensor of the $Ca^{2+}$/calmodulin-bound and active form of myosin light chain kinase (MLCK), MLCK-FIP, was developed by Chew et al., based on a calmodulin-binding domain derived from MLCK and a BFP/GFP couple that undergoes FRET in the absence of $Ca^{2+}$/calmodulin and displays a decrease in FRET upon activation of MLCK through binding of $Ca^{2+}$/calmodulin (Fig. 6.7C). This biosensor allowed imaging the spatial and temporal pattern of MLCK activation, revealing that it is enriched at the spindle equator during late metaphase and maximally activated just before cleavage furrow constriction.[63]

Calcium/calmodulin-dependent protein kinase II (CaMKII) is a key mediator of intracellular signaling in the heart. Erickson et al. developed Camui, a FRET biosensor of CaMKII, constituted of full-length CaMKII flanked at its N- and C-termini by CFP and YFP, respectively, which undergo robust FRET in the compactly folded autoinhibited state. Activation of the kinase results in a conformational change associated with a decrease in fluorescence transfer between CFP and YFP. This reporter was applied to study the activation of CamKII in living neurons and in cardiomyocytes[55] (Fig. 6.7D).

JNK kinase is involved in signal transduction of external stimuli such as stress and cytokines into functional responses that mediate apoptosis, proliferation, differentiation, and inflammation. Several genetically encoded fluorescent biosensors were engineered to detect endogenous JNK activity in living cells, based on an EGFP/citrine FRET pair; an FHA1, FHA2, or WW PAABD; a substrate domain derived from c-Jun or phosphoacceptor sites from JDP2 or ATF3; and docking domains from JDP2 or JIP1.[60] The JNKAR biosensor (for JNK activity reporter) specifically detects stress (ribotoxic and osmotic) and cytokine (TNF-α) induced JNK activity in living cells, in the cytoplasm, nucleus, mitochondria, and plasma membrane, associated with a phosphorylation-dependent increase in FRET between the two AFPs.

The stress-activated protein kinases (SAPKs) p38 and JNK are members of the mitogen-activated protein kinase family and are important determinants of cell fate when cells are exposed to environmental stresses such as UV and osmostress. SAPKs are activated by SAPK kinases (SAP2Ks), which are in turn activated by various SAP2K kinases (SAP3Ks). A genetically encoded FRET biosensor was developed to probe the dynamic behavior of SAP3K activity toward the MKK6 SAP2K, which was based on a Venus/SECFP pair, an FHA1 domain, and MKK6 substrate.[75] This biosensor allowed the study of the dynamic behavior of SAP3K activation in living cells, associated with stress stimuli such as epidermal growth factor and osmostress on the plasma membrane, anisomycin and UV in the cytoplasm, and etoposide in the nucleus.

AMP–activated protein kinase (AMPK) is activated when the AMP/ATP ratio in cells is elevated as a result of energy stress. A fluorescent biosensor of AMPK activity, AMPKAR, was developed to study the function of this kinase upon cellular stress.[53] This biosensor is based on an ECFP/cpVenus FRET pair, an FHA1 domain, and a substrate peptide. AMPKAR exhibits enhanced FRET in response to phosphorylation, allowing probing of the spatiotemporal dynamics of AMPK activity in living cells, thereby revealing that its activation takes place in the cytosol in response to energy stress but occurs in both the cytosol and the nucleus in response to calcium elevation.

Focal adhesion kinase (FAK) is crucial for many cellular processes. To visualize FAK activity and activation at different membrane microdomains, a genetically encoded FRET biosensor was developed, which is based on an ECFP/YPet FRET pair, an SH2 domain derived from c-Src, and a substrate peptide derived from FAK. This biosensor was either targeted into or outside of detergent-resistant membrane regions at the plasma membrane. This study revealed that FAK is activated at membrane microdomains but that its activation mechanisms vary in response to different physiological stimuli.[61]

## 3. FLUORESCENT PEPTIDE/PROTEIN BIOSENSORS

Concerted efforts of chemists and biologists in designing fluorescent probes for biological applications have led to the development of a very different class of biosensors that do not rely on genetically encoded AFPs. Fluorescent peptide, polypeptide, or protein biosensors constitute attractive alternatives to genetically encoded biosensors in that they offer a high degree of versatility, yet also a high degree of control. This class of biosensors is engineered by exploiting peptide substrate sequences or protein domains that bind a specific analyte or interface, which serves as platforms for site-selective coupling of one or more fluorescent probe(s). The fluorescent probe may be coupled by different means to the peptide backbone—most often chemically, enzymatically, or through replacement of a fluorescent amino acid analog (for review, see Ref. 105). The design allows for freedom in the choice of the fluorescent probe, amongst a wide variety of wavelengths and synthetic probes,[106] and for its incorporation at virtually any position within the peptide.

Peptides and polypeptides offer several major advantages inherent to their nature. They can be readily produced through synthetic chemistry or recombinant protein engineering; are easy to handle, store, and characterize; and can further be modified with a wide variety of unnatural substituents, such as modified amino acids, to improve specificity and selectivity, as well as small synthetic fluorescent probes. Peptides are very small yet can serve as substrates, docking sequences, or complementary biomolecular recognition interfaces, thereby offering a wide array of possibilities and strong potential for development of fluorescent biosensors. Peptides also constitute scaffolds for a wide variety of technological improvements, including introduction of quenchers or caging of specific amino acids. Moreover, compared to antibodies, they are cheaper to produce, display low antigenicity, and are rapidly eliminated by the organism, thereby generating relatively low cytotoxicity. Peptide biosensors are readily applicable *in vitro* and in cell lysates. Their applicability in living cells and *in vivo*, however, remains challenging, requiring suitable methods to facilitate their intracellular delivery. Notwithstanding, major advances in the field of protein and peptide delivery over the past 20 years have provided efficient means of introducing this class of biomolecules into living cells and *in vivo* based on protein transduction domains and cell-penetrating peptides (CPPs).[107,108] Once the issue of delivery is solved, peptide biosensors offer a major advantage over genetically encoded biosensors, in that they allow for immediate and controlled use, compared to genetically encoded biosensors, which require long periods of time for their expression and/or maturation (Table 6.1).

Several strategies have been employed to design and engineer peptide biosensors of kinase activity, which are quite different from the strategies developed to generate genetically encoded KARs. In all cases, phosphorylation induces changes in the spectral properties of the fluorophore(s) incorporated into the peptide scaffold, which may involve a shift in its emission wavelength and/or an important change in its quantum yield. Fluorescent peptide biosensors can be broadly divided into three different groups, depending on the mechanism through which fluorescence reports on phosphorylation: environmentally sensitive biosensors, chelation-enhanced fluorescence through metal-ion binding, and biosensors involving quenching/unquenching strategies. We will describe the different categories of fluorescent peptide biosensors that have been developed to probe kinase activities, together with some representative examples to illustrate their mechanism of action and their applications (summarized in Table 6.3). Additional details may be found in the original papers or in comprehensive reviews on this class of sensors.[35–38,134–136]

## 3.1. Environmentally sensitive kinase biosensors

Environment-sensitive fluorophores respond to changes in the polarity of their environment.[105] A good probe will be poorly fluorescent in a fairly polar aqueous solution but will become highly fluorescent in nonpolar solvents or upon docking into a hydrophobic protein pocket or membrane. Several solvatochromic dyes have been described in detail, and their features

**Table 6.3** Fluorescent peptide and protein biosensors of protein kinases

| Protein kinase | Biosensor features | Author and year | Reference |
|---|---|---|---|
| | *Environmentally sensitive peptide biosensors* | | |
| PKC | NBD probe proximal to phosphorylation site | Yeh *et al.* (2002) | 109 |
| PKC | NBD probe proximal to phosphorylation site, light activatable | Veldhuyzen *et al.* (2003) | 110 |
| Src | Phosphorylation-driven protein–protein interaction, based on an SH2 domain | Wang and Lawrence (2005) | 111 |
| Src | Merobody: fibronection monobody conjugated to merocyanine dye | Gulyani *et al.* (2011) | 112 |
| CDKSENS | Reports on CDK/cyclin abundance/cyanine probe | Kurzawa *et al.* (2011) | 113 |
| CDKACT | Reports on CDK/cyclin activity/WW phosphobinding domain/cyanine probe | Van *et al.* (in preparation) | 114 |
| | *Metal ion-induced and chelation-enhanced fluorescence* | | |
| PKCalpha | $Ca^{2+}$ dependent | Chen *et al.* (2002) | 115 |
| PKA, PKC, Abl | $Mg^{2+}$ dependent, Sox based | Shults *et al.* (2003), Shults and Imperiali (2003) | 116,117 |

*Continued*

**Table 6.3**  Fluorescent peptide and protein biosensors of protein kinases—cont'd

| Protein kinase | Biosensor features | Author and year | Reference |
|---|---|---|---|
| Akt, PKA, MK2 | $Mg^{2+}$ dependent, Sox based, cell lysates | Shults et al. (2006) | 118 |
| PKC, PKA, Akt1, MK2, CDK2, PIM2 | $Mg^{2+}$ dependent, Sox based, multiplexed assay in cell lysates | Shults et al. (2005) | 119 |
| PKC, Pim2, Akt1, MK2, PKA, Abl, Src, IRK | $Mg^{2+}$ dependent, Sox based, recognition-domain focused chemosensors | Lukovic et al. (2008), Gonzalez-Vera et al. (2010) | 120,121 |
| ERK1, 2 | $Mg^{2+}$ dependent, Sox based, protein-based docking domain (Sox-PNT) | Lukovic et al. (2009) | 122 |
| p38 alpha | $Mg^{2+}$ dependent, Sox based, protein-based docking domain (Sox-MEF2A) | Stains et al. (2011) | 123 |
| CDK9/cyclin T | Hybrid biosensor: WW domain of Pin1 and a $Zn^{2+}$–DPA complex | Anai et al. (2007) | 124 |
| Src and Abl | Lanthanide biosensors: carbostyril group sensitizer, $Tb^{3+}$ or $Eu^{3+}$ chelator | Tremblay et al. (2008) | 125 |
| | *Quenching/unquenching strategies* | | |
| Src kinase peptide biosensor | Self-reporting biosensor: tyrosine quencher/pyrene group | Wang et al. (2006) | 126 |
| Src kinase peptide biosensor | Self-reporting biosensor: tyrosine quencher/Cascade Yellow, Cascade Blue, or Oregon Green | Wang et al. (2006) | 127 |
| Bcr–Abl and Lyn | Self-reporting biosensor: tyrosine quencher/Cascade Yellow and Oxazine | Wang et al. (2010) | 128 |
| PKA | Deep quench: pyrene/Rose Bengal/14-3-3 phosphoserine binding domain | Sharma et al. (2007) | 129 |
| PKA | Deep quench: pyrene/Coumarin probe and Acid | Agnes et al. (2010) | 130 |

**Table 6.3** Fluorescent peptide and protein biosensors of protein kinases—cont'd

| Protein kinase | Biosensor features | Author and year | Reference |
|---|---|---|---|
| | Green/14-3-3 phosphoserine binding domain | | |
| PKA, PKC, p38, MAPKAP K2, Akt, Erk1, Src | Micelle/auto-assembly | Sun *et al.* (2005) | 131 |
| | *Ratio metric strategies* | | |
| PKA | Single probe/wavelength shift/ phosphorylation-sensitive Cd (II)-cylcen aminocoumarin | Kikuchi *et al.* 2009 | 132 |
| Src | Dual probe/merocyanine and mCerulean | Gulyani *et al.* (2011) | 112 |
| | *Photoactivatable biosensors* | | |
| PKC | NBD probe proximal to phosphorylation site, caged serine | Veldhuyzen *et al.* (2003) | 110 |
| PKC beta | NBD probe proximal to phosphorylation site, caged serine | Dai *et al.* (2007) | 133 |
| Src | Self-reporting biosensor, caged tyrosine | Wang *et al.* (2006) | 127 |

when coupled to peptide biosensors for biological applications have been readily characterized by several groups of chemists (see Ref. 105 and references therein). Although most of these probes possess excellent photophysical properties for *in vitro* applications, their excitation and emission wavelengths tend to be incompatible with *in vivo* applications. Notwithstanding, several groups of chemists are currently devoting their effort to synthesize environmentally sensitive red-shifted dyes with suitable properties for *in vivo* sensing applications.

Environmentally sensitive kinase biosensors constitute a unique class of sensors in that they rely on changes in the environment of the fluorescent probe to report on phosphorylation of the kinase biosensor. Different means have been employed to achieve this goal: phosphorylation itself, phosphorylation-induced conformational changes, and binding of the phosphorylation site by a phosphorecognition-binding domain (for review, see Refs. 134–136). Environmentally sensitive kinase biosensors have been generated through incorporation of fluorescent probes on the phosphorylatable residue itself, immediately adjacent to the residue or proximal to it (i.e., 2–5 residues away). One of the first set of fluorescent peptide probes based on the propinquity effect was generated by labeling a PKC peptide substrate at its N-terminal serine hydroxyl moiety, thereby yielding a library of 417 compounds through incorporation of different fluorescent labels. Although many of the dyes used did not result in a significant increase in fluorescence of the peptide probe upon phosphorylation by PKC *in vitro*, an NBD (nitrobenzofuran)-labeled variant displayed a robust increase in fluorescence intensity (2.5-fold) (Fig. 6.8A). Moreover, this peptide probe was successfully applied to monitor PKC activity in cell lysates and further microinjected into living cells to visualize and monitor the spatiotemporal dynamics of PKC activity.[109] To further improve this biosensor, Lawrence and coworkers synthesized a light-activatable variant of the NBD–PKC probe through incorporation of a single photolytically sensitive cage on the phosphorylatable serine group, thereby offering a precise means of controlling biosensor availability in living cells through photoactivation.[110]

Environmentally sensitive dyes will also respond to changes in their local environment associated with protein–protein interactions. In the case of kinase sensors, this has been exploited by using phosphopeptide recognition domains, including 14-3-3, SH2, or WW domains,[83–89] that bind the phosphorylated peptide but not the unphosphorylated species. The fluorescent probe is consequently embedded in a hydrophobic environment, which modifies the local polarity and consequently affects its spectral properties. This strategy tends to enhance the fluorescence of environmentally sensitive probes more significantly than simply proximal phosphorylation, and further increases the selectivity of the biosensor, because the PAABD is more likely to bind the phosphorylated substrate than any other nonspecific protein found in cell extracts. This strategy was employed to generate a very sensitive probe for Src tyrosine kinase based on a fluorescent peptide labeled with dapoxyl, NBD, or Cascade Yellow two residues away from the phosphorylation site, thereby leading to a fivefold increase in fluorescence upon phosphorylation, thanks to interaction with an SH2 domain[111] (Fig. 6.8B). More recently, Gulyani et al. generated a new Src biosensor based on a fibronectin monobody scaffold, which was employed to quantify the dynamics of Src activity at the edge of living cells, in correlation with protrusion and retraction activities.[112] This merobody

A Fluorescent dye proximal to phosphorylation site

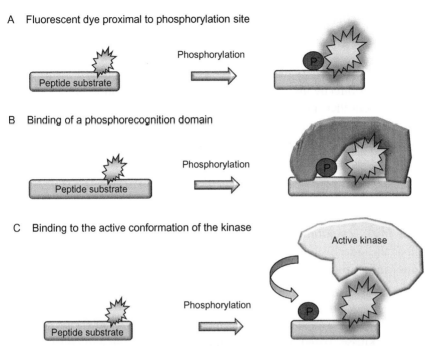

B Binding of a phosphorecognition domain

C Binding to the active conformation of the kinase

**Figure 6.8** *Environmentally sensitive peptide-based kinase biosensors.* Environmentally sensitive peptide biosensors bear a fluorophore whose spectral properties are directly modified/affected by proximal phosphorylation (at ±1–5 residues). Environmentally sensitive peptide biosensors of kinase activity have been developed with a fluorescent dye coupled directly at or proximal to the phosphorylation site whose fluorescence is enhanced (A) by phosphorylation itself,[109] (B) upon binding of a phosphopeptide recognition domain such as SH2 for phosphotyrosine,[111] and (C) upon binding of the kinase itself through an interface that does not affect its catalytic activity.[112] (For color version of this figure, the reader is referred to the online version of this chapter.)

biosensor was engineered by a HTS approach, by derivatizing a monobody that specifically recognizes the active conformation of Src kinases with an environmentally sensitive merocyanine dye, the fluorescence of which is enhanced upon target binding, yet without interfering with target binding or phosphorylation (Fig. 6.8C).

## 3.2. Metal-ion-induced, chelation-enhanced fluorescence

Metal-ion-induced, chelation-enhanced fluorescence (CHEF) relies on fluorophores that can coordinate metal ions ($Ca^{2+}$, $Mg^{2+}$, $Zn^{2+}$, or lanthanides) and whose ability to chelate these ions is further aided by proximal phosphorylation of a peptide substrate. Upon chelation, the electronic structure of the fluorophore is modified, and consequently its spectral properties.[137]

### 3.2.1 $Ca^{2+}$-chelating biosensors

Lawrence and coworkers developed a set of PKC peptide probes through incorporation of a $Ca^{2+}$-sensitive fluorophore with a tetracarboxylate-chelating site previously developed by Tsien and collaborators.[138] The fluorophore was directly incorporated onto the N-terminal phosphorylation site of a consensus PKC peptide substrate, lying within only a couple of angstroms from the phosphorylation site. The unphosphorylated probe has poor affinity for $Ca^{2+}$, but phosphorylation by PKC creates a divalent metal-ion-binding site, and coordination of $Ca^{2+}$ by the fluorophore induces a twist that affects its spectral properties, promoting a 3.6-fold increase in fluorescence[115] (Fig. 6.9A).

### 3.2.2 $Mg^{2+}$-chelating biosensors—Sox biosensors

Imperiali and collaborators developed a solvatochromic probe, the Sox (sulfonamide-oxine) dye, derived from 8-hydroxyquinoline, that chelates $Mg^{2+}$ and undergoes fluorescent enhancement upon coordination of a phosphate group on a phosphopeptide.[116] The Sox dye is resistant to photobleaching, is small, and causes minimal perturbation of substrate affinities with the kinase. Based on the Sox amino acid, Imperiali and collaborators developed a "versatile kinase kinase activity scaffold" that incorporates a consensus phosphorylation site coupled to a β-turn sequence that preorganizes the binding site for $Mg^{2+}$ and a Sox dye, together with a kinase recognition motif[117] (Fig. 6.9B). This β-turn-focused (BTF) design allows the phosphorylation site to be either N- or C-terminal of the kinase recognition motif. $Mg^{2+}$ binding is weak in absence of phosphorylation (Kd 100–300 mM) and considerably increased upon phosphorylation (Kd 4–20 mM). Sox-based BTF biosensors were successfully developed for PKC, PKA, and Abl substrate peptides with a phosphorylatable serine, threonine, or tyrosine residue with consequent increases in fluorescence intensity between three- and fivefold.[117] Further, this class of biosensors was applied to develop a multiplex fluorescence-based assay of Akt, PKA, and MK2 kinase activity in cell lysates.[118,119] To further increase the specificity of recognition, recognition-domain-focused (RDF) chemosensors were engineered with extended binding sequences to maximize recognition of a variety of well-characterized Ser/Thr and Tyr kinases, including PKC, Pim2, Akt1, MK2, PKA, Abl, Src, and IRK[120] (Fig. 6.9C). These biosensors reported on phosphorylation with robust fluorescence enhancement and high sensitivity, and were further used in 96-well plate formats, indicating that they were amenable to HTS of kinase inhibitors.

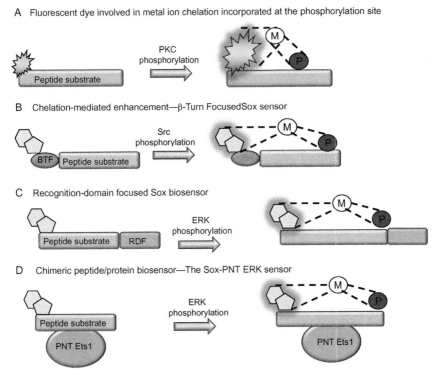

**Figure 6.9** *Metal-ion-induced, chelation-enhanced fluorescence.* (A) Fluorescent dye involved in metal-ion chelation incorporated at the phosphorylation site, such as the $Ca^{2+}$-sensitive PKC probe.[115] (B) Chelation-mediated enhancement—β-turn-focused (BTF) Sox sensor.[117] (C) Recognition-domain-focused (RDF) Sox biosensor.[120] (D) Chimeric peptide/protein biosensor—the Sox-PNT ERK sensor.[122,123] (For color version of this figure, the reader is referred to the online version of this chapter.)

Moreover, a high-throughput approach was developed to generate Sox-based chemosensors with the highest possible selectivity using combinatorial libraries.[121] More recently, the RDF strategy has been extended to engineer Sox protein biosensors that incorporate a kinase-specific docking domain distal to the phosphorylation site, which enhances the specificity of recognition significantly more than an extended peptide substrate motif. For instance, a Sox-based ERK1/2 sensor (Sox-PNT) engineered through incorporation of the PNT domain of the Ets1 transcription factor into the Sox-ERK1/2 peptide substrate sequence was developed and readily applied to probe ERK1/2 kinase in cell lysates, displaying high selectivity against other kinases from the JNK, p38, and CDK families.[122] Likewise, a Sox-based p38α-selective sensor was developed to probe p38α kinase in cell lysates in a selective fashion, by incorporating an MEF2A docking peptide and p38α phosphorylation site bearing a CSox amino acid two residues before the phosphorylation site into the same scaffold[123] (Fig. 6.9D).

Although this strategy is extremely powerful, it should be kept in mind that the spatial constraint and proximity required between the phosphorylation site and the position of the chelating fluorophore may be a limitation for recognition by certain kinases.

### 3.2.3 $Zn^{2+}$-chelating biosensors

$Zn^{2+}$ chelators have been developed based on binuclear Zn(II)–dipicolylamine (DPA) complexes spanned by a fluorescent module (stilbene, anthracene, phenyl, or biphenyl) that is sensitive to the recognition of phosphate by the $Zn^{2+}$ moiety[139] (Fig. 6.10A). Based on these $Zn^{2+}$ chelators, a hybrid biosensor was designed for enhanced phosphopeptide recognition based on the phosphopeptide-binding WW domain of Pin1 coupled with the $Zn^{2+}$ chemosensor[124] (Fig. 6.10B). This combinatorial strategy was successfully employed to monitor phosphorylation of a pSer-CTD peptide by CDK9/cyclin T1.

### 3.2.4 Lanthanide-chelating biosensors

Yet another class of environmentally sensitive probes worth mentioning report on lanthanide ions ($Eu^{3+}$, $Tb^{3+}$), and constitute extremely attractive tools to develop biosensors, given their photophysical features.[140] Lanthanide biosensors display poor affinity for lanthanides (such as $Eu^{3+}$ or $Tb^{3+}$) in their unphosphorylated state and are essentially nonfluorescent, compared to their phosphorylated counterparts whose affinity for $Eu^{3+}$ or $Tb^{3+}$ increases significantly owing to the presence of the phosphate which coordinates with the metal[140] (Fig. 6.10C). Several kinase biosensors have been developed, including a kinase recognition domain for Src and Abl, a carbostyril group that acts as a luminescence sensitizer, and an iminodiacetate-chelating moiety that displayed up to 10-fold increase in signal upon phosphorylation[125] (Fig. 6.10D). The major limitations of this approach are that it requires the presence of free $Tb^{3+}$ or $Eu^{3+}$ as well as photosensitizing groups coupled to the peptide biosensor and that it cannot be applied *in vivo* because of the very short excitation wavelength (262 nm).

## 3.3. Biosensors involving quenching–unquenching strategies

### 3.3.1 Self-reporting biosensors

Aromatic amino acids, such as tyrosine and tryptophan, constitute good quenchers of organic dyes through a process that involves π/π stacking between the aromatic groups. In particular, tryptophan has been used in a countless number of assays to quench the fluorescence of probes.[141] A similar

A  Zn$^{2+}$ chelators–Zn$^{2+}$-dependent enhancement

B  Zn$^{2+}$ DPA receptors for detection of phosphopeptides—hybrid biosensor

C  Lanthanide sensors

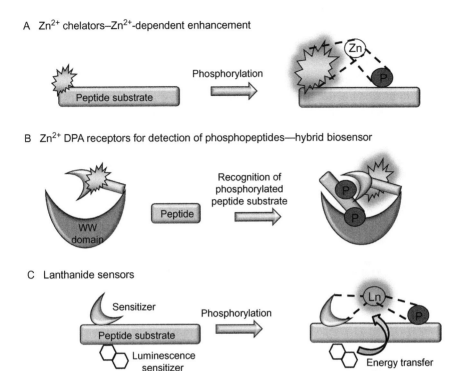

**Figure 6.10** *Metal-ion-induced fluorescence: Zn$^{2+}$ and lanthanides.* (A) Zn$^{2+}$ chelators— Zn$^{2+}$-dependent enhancement of fluorescence.[139] (B) Zn$^{2+}$–DPA receptors for detection of phosphopeptides—hybrid biosensor between a phosphopeptide-binding domain (WW) and a fluorescent Zn$^{2+}$–DPA chemosensor that binds phosphate through cooperative recognition.[124] (C) Lanthanide sensors.[140] (D) Kinase biosensor including a recognition domain for Src and Abl, a carbostyril group that acts as a luminescence sensitizer, and an iminodiacetate-chelating moiety.[125] (For color version of this figure, the reader is referred to the online version of this chapter.)

strategy was employed to generate a fluorescent "self-reporting" peptide biosensor of protein tyrosine kinase Src, in which the phosphorylatable tyrosine directly quenches the fluorescence of a proximal dye (pyrene) until phosphorylation of tyrosine disrupts the stacking interaction, thereby releasing the fluorophore and promoting full enhancement of fluorescence[126] (Fig. 6.11A). Further variants of this biosensor were then successfully developed for applications *in cellulo*, through incorporation of Cascade Yellow, Cascade Blue, or Oregon Green.[127] More recently, two orthogonal biosensors of Lyn and Abl kinases bearing Cascade Yellow and Oxazine were developed and successfully applied to visualize these kinases in chronic myelogenous leukemia (CML) drug-resistant cell lines.[128]

### 3.3.2 The deep quench strategy

The strategy described above is not applicable to residues that are not aromatic. Therefore, for Ser/Thr kinase biosensors, Lawrence and collaborators developed the "deep quench" strategy, which involves shielding of the fluorophore by a quencher in solution, which will be displaced upon phosphorylation of the biosensor because of binding of a phosphorecognition domain to the phosphorylated serine or threonine, which will in turn promote

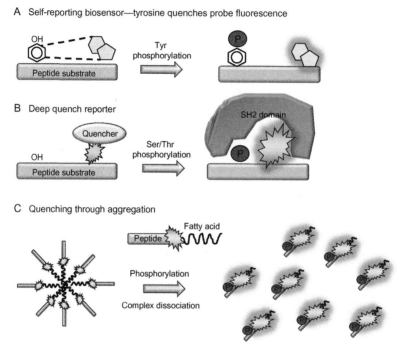

**Figure 6.11** *Biosensors involving quenching–unquenching strategies.* (A) Self-reporting biosensor—the phosphorylatable tyrosine residue serves as a quencher that silences the fluorescence of a proximal fluorophore; phosphorylation disrupts this interaction, thereby leading to complete enhancement of fluorescence.[126–128] (B) Deep-quench reporter developed for Ser/Thr peptides—fluorescence of a probe is quenched by a compound in solution; phosphorylation promotes recruitment of a phosphoSer/Thr-binding domain (e.g., 14-3-3) which promotes enhancement of probe fluorescence.[129,130] (C) Quenching through aggregation—an amphiphilic peptide biosensor composed of a substrate domain with a phosphorylatable residue, labeled with fluorescein, and an N-terminal carbon tail that promotes auto-assembly or aggregation of the peptides. Peptide aggregation leads to fluorescence quenching within the micelle; phosphorylation disrupts the micelle structure, thereby releasing the individual probes whose fluorescence is enhanced.[131] (For color version of this figure, the reader is referred to the online version of this chapter.)

significant enhancement of the fluorescence of the environmentally sensitive probe (Fig. 6.11B). This strategy has proved extremely successful in generating a PKA biosensor bearing a pyrene probe, using Rose Bengal as a quencher and a 14-3-3pi as a phosphoserine-binding domain, which displays 64-fold increase in fluorescence,[129] and more recently, another PKA biosensor bearing a Coumarin probe and Acid Green as a quencher, which displays up to 150-fold increase in fluorescence, was used to monitor PKA activity in mitochondria.[130] Despite its apparent potency, the complexity of this approach makes it difficult to apply in living cells and *in vivo*.

### 3.3.3 Quenching through aggregation

A variant of these strategies was devised by Sun *et al.* based on an "auto-assembly" or "micelle" approach to quench fluorescence of peptide kinase biosensors for PKA, PKC, p38, MAPKAP K2, Akt, Erk1, and Src.[131] This involves the design of an amphiphilic peptide biosensor labeled with fluorescein, composed of a substrate domain with a phosphorylatable residue and of an N-terminal carbon tail that promotes auto-assembly or aggregation of the peptides, thereby quenching their fluorescence. Phosphorylation of the peptide biosensor disrupts the micelle structure, thereby leading to fluorescence (Fig. 6.11C).

## 3.4. Ratiometric strategies

### 3.4.1 Intramolecular ratiometric strategy—Single-probe biosensor

Two different ratiometric strategies have been developed. The first consists of a single probe that is sensitive to environmental changes that will affect its spectral properties sufficiently to allow for ratiometric quantification. A metal-ion, anion-sensitive ratiometric peptide kinase biosensor has been developed to probe PKA activity, consisting of a Cd(II)-cylcen appended aminocoumarin coupled to a substrate peptide for PKA. Upon phosphorylation by PKA, the metal-complex moiety binds to a phosphorylated residue, which in turn displaces the coumarin fluorophore, resulting in ratiometric change of the probe's excitation spectrum[132] (Fig. 6.12A). A different strategy is based on the incorporation of two fluorophores within the same biosensor, the first displaying sensitivity to target recognition and the second being inert. This concept was employed to generate the Src merobody biosensor, through incorporation of a merocyanine dye onto the fibronectin biosensor scaffold, as well as the mCerulean autofluorescent protein for ratio imaging *in vivo*[112] (Fig. 6.12B).

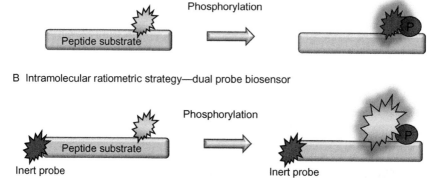

A  Intramolecular ratiometric strategy—single probe biosensor-wavelength shift

B  Intramolecular ratiometric strategy—dual probe biosensor

**Figure 6.12** *Ratiometric strategies.* (A) Intramolecular ratiometric strategy—single-probe biosensor based on a coumarin/$Cd^{2+}$-cyclen probe that undergoes a wavelength shift upon phosphorylation (from 360 to 410 nm) due to chelation of the phosphate group by the $Cd^{2+}$ center—anion sensor-based ratiometric probe.[132] (B) Intramolecular ratiometric strategy—based on incorporation of two different fluorescent probes on a single biosensor scaffold—one involved in the sensing process, the other serving as an internal control, such as the Src merobody biosensor based on a fibronectin monobody that is conjugated to a merocyanine dye and fused with a mCerulian which serves as an internal control.[112] (For color version of this figure, the reader is referred to the online version of this chapter.)

## 3.5. Optimizing peptide biosensors through quenching and caging strategies

### 3.5.1 Intramolecular quenching

One of the means of optimizing the signal-to-noise ratio in applications involving the use of peptide or protein kinase biosensors consists in silencing the basal fluorescence of the fluorescent probe, thanks to a quencher moiety conjugated within the same biosensor. This strategy implies that the quencher is in spatial proximity with the fluorescent probe and that the quenching mechanism will be reversed upon phosphorylation of the biosensor by the kinase. Different quencher strategies have been employed (Fig. 6.13). The simplest strategy involves incorporation of a synthetic quencher into the peptide backbone so that it quenches the fluorescent probe until phosphorylation of the biosensor allows the disruption of the interaction between the quencher and the fluorophore, much in the same way as tryptophan or tyrosine residues quench proximal fluorophores[126,141] (Fig. 6.13A). An alternative approach, which consists in intramolecular homoquenching between two fluorophores (through homo- or heteroquenching) conjugated to the peptide biosensor, can be employed to reduce basal fluorescence of the biosensor, although this does not necessarily ensure that fluorescence will be completely quenched[142] (Fig. 6.13B).

### 3.5.2 Light activation

Another means of optimizing the signal-to-noise ratio and increasing the sensitivity of the response involves generating light-activatable probes that can be selectively activated in response to UV irradiation. Light-activatable probes are generated through molecular caging, by masking the main function with a photolabile group, which will be selectively released upon illumination. This provides spatial and temporal control over the activation of the probe, thereby allowing dampening or silencing any nonspecific fluorescence and reducing nonspecific noise attributable to basal fluorescence prior to activation[143] (Fig. 6.13C). The first caged fluorescent reporter of intracellular enzymatic activity was a light-activatable variant of the NBD–PKC probe, engineered through incorporation of a single photolytically sensitive cage on the phosphorylatable serine group of the peptide, which precluded

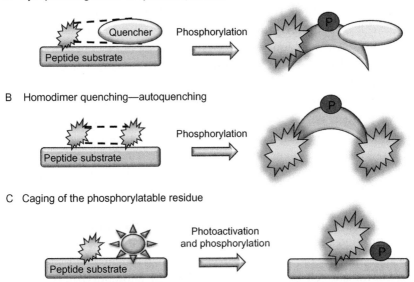

A  Dye quenching thanks to synthetic quencher

B  Homodimer quenching—autoquenching

C  Caging of the phosphorylatable residue

**Figure 6.13** *Quenching and photoactivation strategies for peptide biosensors.* (A) Incorporation of a synthetic quencher into the peptide backbone. (B) Intramolecular homoquenching between two fluorophores conjugated to the peptide biosensor. (C) Selective photoactivation of caged compounds by UV irradiation.[110,127,133] (For color version of this figure, the reader is referred to the online version of this chapter.)

its phosphorylation by PKC.[110] This variant offers the advantage of controlling both the timing and the relative concentration of biosensor released upon illumination and was readily microinjected into HeLa cells to monitor intracellular PKC activity. Further, a caged biosensor of PKC was microinjected into PTK2 cells to follow its kinase activity at mitosis, with high temporal precision, thereby revealing that kinase activity is required for the progression from prophase into metaphase and that robust kinase activity of the PKCb isoform occurs just prior to nuclear envelope breakdown. This biosensor provides "visual snapshots of intracellular kinase activity" with high spatial and temporal control.[133] Similarly, a caged Src reporter was generated by modifying the phosphorylatable tyrosine of a self-reporting peptide biosensor with a photolabile nitrobenzyl moiety and microinjected into A549 cells expressing high levels of Src, allowing the control of the timing of kinase sensing.[127]

## 3.6. Peptide biosensors for cell cycle targets: CDKSENS and CDKACT

Enzymes involved in regulation of cell cycle progression are central for processes such as cell growth, DNA replication, and cell division, and therefore constitute pillars of cellular physiology. Many kinases involved in coordination of cell cycle progression have been well documented to undergo mutations that alter their function, aberrant expression, or hyperactivation in human disorders, and thereby constitute attractive pharmacological targets and biomarkers of disease.[144,145,27] However, despite the pharmacological importance of these kinases, there are barely any means for probing these kinases, which limits the development of diagnostic approaches to monitor alterations in their levels or activities associated with human cancer. The development of cell cycle reporters and biosensors to probe some of these kinases is now paving the way for developing strategies to detect anomalies in these kinases associated with cancer.[146]

### 3.6.1 CDKSENS and CDKACT—Fluorescent peptide biosensors for probing the relative abundance and activity of CDKs

Cyclin-dependent kinases (CDK/cyclins) are heterodimeric protein kinases that coordinate cell cycle progression through phosphorylation of well-defined enzymatic and structural targets.[147–149] CDK/cyclins are known to contribute to sustain aberrant cell proliferation in human cancers and constitute attractive pharmacological targets for the development of anticancer drugs. Unfortunately, there are no direct means of studying the levels and activities of these kinases in their natural environment in living cells. From a fundamental perspective, this limits our understanding of the physiology of these enzymes in their natural environment. From a diagnostic perspective, reporters that provide information on alterations in CDK/cyclin levels and activity would provide the means of identifying cancer cells or tumors in which these kinases constitute relevant targets for therapeutic intervention.

We have developed two new families of peptide biosensors that report on the relative abundance and activity of CDK/cyclins, respectively, through changes in fluorescence of an environmentally sensitive probe.[113,114] CDKSENS is a biligand peptide composed of a CDK-binding pseudosubstrate sequence and a cyclin-binding sequence that allows docking of the biosensor onto the CDK/cyclin complex with high affinity and specificity. This peptide bears a fluorescent probe (cyanine dye) which undergoes significant enhancement upon recognition and docking onto CDK/cyclin complexes.[113] CDKACT is a modular peptide biosensor constituted of a peptide substrate, onto which a fluorescent probe is coupled proximal to the phosphorylation site, and a PBD that recognizes the phosphorylated peptide sequence, separated by a short proline/glycine-rich linker. This biosensor undergoes changes in fluorescence intensity upon phosphorylation by active CDK/cyclin complexes and is fully reversible, thereby allowing monitoring kinase activity in real time through changes in fluorescence intensity of an environmentally sensitive probe.[114] Both CDKSENS and CDKACT were validated using recombinant CDK/cyclin complexes as well as cell extracts derived from healthy fibroblasts and cancer cell lines expressing endogenous CDK/cyclins. The design of CDKSENS and CDKACT is depicted in Fig. 6.14.

CDKSENS and CDKACT constitute sensitive tools for reporting on CDK/cyclin levels and activity *in vitro* and in cell extracts. Furthermore, these peptide biosensors can be introduced into living cells, thanks to CPPs,[107,108] thereby allowing monitoring of the CDK/cyclin levels and activities in their natural environment through live-cell fluorescence imaging and ratiometric quantification. Indeed, CDKSENS and CDKACT allow the detection of subtle differences when tampering with a single CDK or cyclin (through siRNA downregulation or through ectopic expression) or following treatment with CDK/cyclin drugs. They further allow the detection of the differences between different healthy and cancer cell lines, thereby enabling identification of cells that express

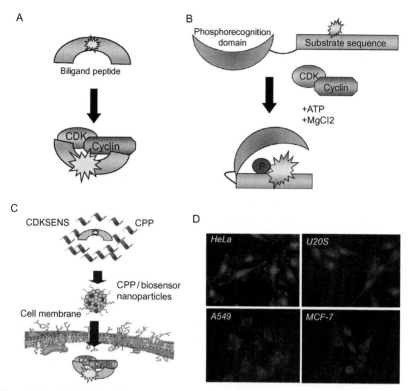

**Figure 6.14** *CDKSENS and CDKACT biosensors.* (A) CDKSENS and (B) CDKACT biosensors are fluorescent peptide biosensors that allow probing the relative abundance and the kinase activity of cyclin-dependent kinases, respectively, *in vitro* and (C) in living cells following delivery mediated by cell-penetrating peptides. (D) A representative example of CDKSENS labeled with Cy3 internalized into different cancer cell types. (See Color Insert.)

high levels or activities of CDK/cyclin kinases from cells that present decreased or defective assemblies.

CDKSENS provides information on CDK/cyclin complexes that is not conveyed by antibodies that recognize individual CDK or cyclin subunits. Likewise, CDKACT allows circumventing the requirement for antigenic or radioactive assays to report on CDK/cyclin kinase activity. As such, these technologies offer promising perspectives for the development of fluorescence-based cancer diagnostics, present strong potential for monitoring response to therapeutic strategies targeting CDKs or cyclins, and constitute attractive tools for drug discovery programs.

## 4. APPLICATIONS

Protein kinases are central to most biological signaling pathways and constitute disease biomarkers in many pathological conditions. As such, they have been the focus of intensive studies using different technologies, mostly based on antigenic and radioactive assays. The more recent development of fluorescent reporters and biosensors has provided the means of imaging the activity and dynamic behavior of protein kinases in real time in their natural environment, with high spatial and temporal resolution, without perturbing their native structure or function. Fluorescent biosensors provide a wealth of opportunities over other approaches used for fundamental studies of protein kinases and allow for countless practical applications in analytical chemistry and biotechnology, as well as in biomedical and drug discovery programs (Fig. 6.15).

### 4.1. Fundamental studies of protein kinase localization, function, and regulation

From a fundamental perspective, fluorescent biosensors have proved useful for studying the dynamics of subcellular localization, function, and regulation of individual kinases in real time. While these tools may be employed to study the kinetics of protein kinase activation *in vitro*, they may equally well be used to probe associated activities in cell extracts and further be applied to characterize the spatiotemporal dynamics and activities of these enzymes in living cells with high spatial and temporal resolution. Biosensor technology allows addressing those aspects of protein kinase biology that had not been characterized previously because of the lack of appropriate technologies. Indeed, fluorescent biosensors allow imaging the behavior of protein kinases within distinct subcellular compartments and, in response to stimuli, involving different cofactors, agonists, or coregulators, thereby contributing to unravel complex signaling cascades that lead to kinase activation in a physiological context and providing a comprehensive understanding of their overall function in a cellular context. Moreover, fluorescent biosensors are very useful for comparative studies of protein kinases in healthy and pathological conditions, highlighting molecular and cellular alterations associated with disease in a qualitative and quantitative fashion (for review, see Refs. 150–152).

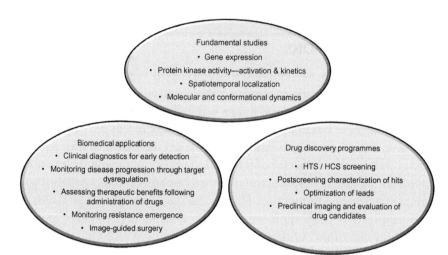

**Figure 6.15** *Applications of fluorescent kinase biosensors.* Fluorescent biosensors can be applied to probe the function, regulation, and spatiotemporal dynamics of protein kinases in fundamental studies. Aside from their utility in fundamental research, they constitute potent tools for biomedical and drug discovery approaches. Biomedical applications include development of clinical diagnostics, through imaging disease biomarkers, monitoring disease progression, and response to therapeutics. Drug discovery approaches include high-throughput screening and postscreening evaluation of lead compounds and drug candidates.

## 4.2. Monitoring protein kinase biomarkers in clinical diagnostics

From a biomedical perspective, fluorescent biosensors constitute very attractive tools for probing differential or aberrant activation of protein kinases in a pathological context and have a number of practical applications in several areas of health and disease, including clinical diagnostics and monitoring response to therapeutics. For instance, imaging protein kinases that constitute disease biomarkers, thanks to biosensors whose fluorescence is affected by their activity, can provide a wealth of information on the functional status of these enzymes. In this respect, fluorescent biosensors provide a basis for the development of diagnostic strategies involving detection of protein kinase biomarkers whose activity is dysregulated in pathological disorders or involved in disease pathogenesis. Although there are still very few fluorescent biosensors that have actually been applied to detect protein kinases in a diagnostic perspective, a couple of noteworthy examples are currently employed for clinical diagnostics of dysregulated protein kinases in cancer.

The genetically encoded FRET biosensor developed by Mizutani *et al.* for detection of Bcr–Abl kinase activity allows the detection of cancerous and drug-resistant cells and can further be applied for direct evaluation of kinase inhibitor efficacy.[48] This highly sensitive biosensor was therefore applied to assess Bcr–Abl activity from patient cells and further employed to establish a correlation with the disease status: a clinical diagnosis of Bcr–Abl kinase activity in CML. Moreover, this probe has been used to monitor response to therapy and to detect the onset of drug-resistant cells, thereby allowing foreseeing when and how to employ second-generation inhibitors or novel compounds to treat drug-resistant mutants.[95,96] Wang et al. developed fluorescent peptide biosensors that are enzymatically and photophysically distinct, allowing the simultaneous monitoring of Bcr–Abl and Lyn tyrosine kinase activity through multicolor imaging. In particular, this combination of orthogonal probes revealed significant differences in Lyn kinase activity, but not in Abl kinase activity, between CML cell lines that are sensitive to imatinib or which develop resistance to this drug.[128]

It can be expected that the potential of fluorescent biosensors will be harnessed for clinical diagnostics in the coming years, not only for early stage detection but also for monitoring kinase activities throughout disease progression and for assessing the benefits of therapeutic intervention, drug efficiency, and resistance, in particular, but not exclusively in the field of cancer.[152]

## 4.3. Drug discovery strategies—HTS/HCS assays and postscreening evaluation

Protein kinases constitute one of the major classes of therapeutic targets for drug discovery programs developed by academia and the pharmaceutical industry. In this respect, fluorescent biosensors constitute sensitive and selective tools that provide a means of monitoring protein kinase activity in a continuous fashion in vitro or in cell-based assays, in high-throughput and high content screening assays. Indeed, fluorescent biosensors are particularly well suited to screen large, complex libraries for inhibitors that affect enzymatic activity or function in high-throughput formats because of the high sensitivity of response to compounds that affect kinase activity. Moreover, as fluorescent reporters allow visualization of the activity of target enzymes in a continuous fashion, they allow establishing screens that follow the action of compounds on kinase activity over time, rather than at a static point. Fluorescent biosensors are further well suited to the qualitative and quantitative evaluation of kinase inhibitor efficacy in living cells at a postscreening stage, to gain insight into the pharmacokinetics and pharmacodynamics of hits and to determine the efficacy of novel leads and optimize derivative compounds for therapeutic applications. Finally, biosensor technology can be employed to assess the potential of candidate drugs in preclinical evaluation trials and to monitor response over time as well as the emergence of resistance.

To this aim, several strategies have been developed successfully: FRET-based approaches and positional biosensors whose changes in subcellular localization reflect the activity or the inhibition of the target enzymes. The gold standard for high-throughput formats is an assay that can be easily miniaturized without losing signal intensity or signal-to-noise ratio and that does not suffer from off-target effects.[153]

Several HTS assays have been developed for screening small-molecule compounds from complex libraries, based on genetically encoded FRET biosensors.[154] Likewise, nongenetic, environmentally sensitive peptide biosensors are perfectly well suited to HTS assays and Sox biosensors have been readily applied to probe kinase activities in HTS formats using cell lysates.[118,119,122,123] As such, fluorescent biosensors constitute attractive tools for drug discovery and, in particular, for assays based on cellular and molecular imaging.[155–159] Fluorescence imaging technologies based on biosensors facilitate drug discovery in a complex physiological environment of a cell in a quantitative fashion. Fluorescence near-infrared and multispectral imaging modalities are also becoming increasingly popular in preclinical studies in animal models to image disease and drug targets, drug distribution, and therapeutic response in a noninvasive fashion, thereby contributing to accelerated preclinical drug development *in vivo*.

## 5. CONCLUSIONS AND PERSPECTIVES

The opportunities offered by fluorescent biosensors for fundamental research and biomedical applications are countless. Although a large number of challenges have been overcome, there are still many technological developments required to move toward further applications. Future directions include the development of biosensors for multiplexed detection of protein kinases; strategies for improving delivery, targeting, and activation of biosensors; and *in vivo* imaging and biomedical applications, fluorescence-based diagnostics, theragnostic approaches, and image-guided surgery.

The future of these nanotools looks very bright, as their versatility and potency allows for their application in a myriad of different applications. Fluorescent biosensors can be expected to contribute to early detection of disease, in particular, cancer, thereby paving the way for personalized diagnostics and theranostic applications, multiplexed sensing technologies, and *in vivo* applications.

## 5.1. Multicolor imaging—Multisensing, multiplex, or multiparameter imaging

Multicolor imaging approaches have already been well described for studying protein/protein interactions. However, multiplex analysis of kinase activities through multicolor imaging constitutes a challenging objective, as it implies using different sets of fluorescent probes, which are sufficiently distinct with respect to their spectral properties and which allow imaging the complex dynamic patterns of kinases from different signaling pathways. Multisensing, multiplex, or multiparameter imaging consists in devising means of probing several targets using different fluorescent probes or biosensors simultaneously by measuring different excitation and emission wavelengths. This strategy is extremely useful for fundamental research, as it potentially allows the following of several signaling pathways in parallel and in real time, to gain insight into their hierarchy of activation and study the dynamics of their interconnections or their independent behavior. Obviously, from a diagnostics perspective, the multisensing approach allows a more detailed readout of the biomarker status. Indeed, most pathologies are characterized by the concomitant activation of multiple kinases from different signaling pathways. Moreover, in drug discovery strategies, multisensing allows gaining a larger set of information concerning the specificity of pharmacological inhibitors to characterize the underlying differences in potency, off-target effects, and genetic backgrounds.

The main challenge, however, consists in combining different tools into one single experimental design and finding suitable means to acquire, discriminate, and quantify their signals. This proves extremely challenging from a technological perspective when using FRET-based biosensors, as it requires the use of spectrally distinct FRET pairs for multiparameter live-cell fluorescence imaging.[160,161] Notwithstanding, this strategy has been successfully developed for imaging of dual FRET-based caspase-3 biosensors[162] for simultaneous imaging of cAMP and cGMP in single cells, thanks to the sapphire/RFP (red fluorescent protein) and CFP/YFP set of FRET biosensors, respectively,[163] and for monitoring three calcium-dependent signaling events, thanks to a cytosolic CamKIIα-CFP/YFP biosensor, a membrane-bound PKC-CFP/YFP sensor, and a translocating annexin A4-mOrange/mCherry FRET biosensor.[161] Likewise, multiplex strategies have been developed using nongenetic biosensors coupled to synthetic probes with different spectral properties. For example, Lawrence and collaborators developed a set of orthogonal fluorescent peptide biosensors whose combination allows for simultaneous multicolor monitoring of

Bcr–Abl and Lyn tyrosine kinase activity in cell extracts.[128] The combination of probes that provide a readout of different enzymatic activities offers a means of assessing the differential behavior of distinct targets in a given signaling pathway. Moreover, from a therapeutic perspective, it can provide information on the efficacy of drug cocktails and dosing regimens prior to therapy and during therapy, depending on disease progression and emergence of resistance.

## 5.2. *In vivo* imaging of protein kinase activities

Despite the complexity of enzymatic reactions in living systems, major advances have been made in probe design and application for *in vivo* imaging.[164,165] Efforts have been made in developing fluorescent proteins for imaging in living cells and tissues,[94] and several synthetic near-infrared fluorescence (NIRF) probes have been developed by chemists for *in vivo* application.[166] Moreover, imaging technologies have improved significantly[167] and strategies have been designed for *in vivo* imaging through engineering of activatable or targeted probes.[168–173] Nevertheless, *in vivo* imaging of fluorescent biosensors remains extremely challenging for several reasons. With respect to genetically encoded biosensors, the challenge resides essentially in the heterogeneity and lack of control associated with ectopic expression. This said, recent developments are contributing to establish stable expression of FRET biosensors in cell lines and expression in transgenic mouse models.[174,175] With respect to nongenetic biosensors, the challenge lies in the choice of synthetic probes that are compatible with *in vivo* imaging and in delivering peptide or polypeptide scaffolds *in vivo* in an efficient fashion. As mentioned above, the development of NIRF probes which are appropriate for *in vivo* applications[166] solves part of the problem. Moreover, the development of nanoparticle strategies for delivery of peptides and proteins into cells[107,108] will be of considerable help in promoting the successful application of nongenetic biosensors *in vivo*.

## 5.3. Delivery, targeting, and activation strategies

A major bottleneck for *in vivo* applications concerns the delivery of fluorescent biosensors into cells, tissues, and organs. In practice, this involves engineering biosensor formulations with nanocarriers which are stable in bodily fluids and which can be delivered to their target passively, through prolonged circulation and enhanced permeability retention in the tumor, for instance, or actively, thanks to specific targeting sequences. One of the future priorities for biosensor application *in vivo* therefore concerns the refinement of noninvasive strategies for the delivery of fluorescent biosensors into living organisms.

With respect to targeting strategies, there are two different levels that may have a major impact on the signal reported by the biosensor. The first consists in targeting a specific intracellular compartment, thanks to subcellular targeting sequences. Indeed, several kinase biosensors have been specifically targeted to the nucleus, the Golgi, the endoplasmic reticulum, the mitochondria, or the plasma membrane, revealing differences in kinase activities and dynamics, as well as differences attributable to differential localization of discrete subpopulations or specific kinase isoforms. Hence, subcellular targeting allows monitoring the differential activity associated with subcellular localization, whereas it may not be as clear-cut if the biosensor is simply distributed randomly within the cellular cytoplasm. The second level of targeting consists in targeting specific cell types, which involves a more complex approach, requiring conjugation or complexation of the biosensors with formulations that direct the biosensor to cell-surface receptors.

Aside from targeting strategies, "smart" strategies based on specific activation of probes are being developed, notably for imaging cancer and metastasis, and some of these have been successfully applied to image-guided surgery.[168,176] These activatable probes make use of properties inherent to cancer cells for specific penetration and/or activation within tumor vicinity. For instance, metalloproteases secreted in the vicinity of tumors cleave target sequences which normally prevent the probe from penetrating cells in a nonspecific fashion, thereby releasing a cell-penetrating sequence that allows for endosomal uptake of the probe.[169] Other systems are activated by the acidic pH that characterizes the tumor environment, such as pH-activatable fluorescent moieties coupled to cancer-targeting antibodies[172] and pH-activatable CPPs conjugated to a fluorescent probes (pHLIP technology).[173] Yet another strategy consists in silencing a fluorescent probe through molecular quenching until specific enzymes release the molecular cage. This strategy has been applied to generate a fluorescent probe activated by tumor-specific γ-glutamyl transferases (GGT),[177] which can be applied topically to probe fluorescence at local tumor sites and which allows for rapid visualization of surface lesions upon intraoperative detection of tumors and metastases.

Other examples include targeting of NIRF probes to specific cell-surface antigens and receptors, such as integrins and Glutamate Transporter receptors (GLUT), through conjugation with Cyclic Arginine-Glycine-Aspartate (RGD) and 2-deoxyglucose, respectively, to image tumors *in vivo*.[178,179]

## 5.4. Looking toward the future—Biomedical imaging for diagnostics

Visualizing the activities of protein kinases in disease such as cancer and metastasis can provide insights into its pathogenesis while also providing a means for early detection of target dysregulation. Fluorescent biosensors unquestionably constitute one of the most promising classes of tools for detection of biomarkers *in situ*, as they allow circumventing the biopsy and extraction required for traditional *ex vivo* detection of biomarkers using antigenic approaches.

While the perspective of developing diagnostics through molecular imaging may still seem far-fetched, advances in fluorescence imaging technologies, associated with ongoing developments in medical devices for whole-animal and patient imaging, are paving the way toward this possibility. Moreover, together with the development of novel synthetic probes and smart chemical biology approaches, new generations of molecular tracers and fluorescent biosensors are being proposed to report on disease biomarkers. The concomitant development of nanocarriers and nanoparticles that are suited to introduce these fluorescent probes *in vivo* is now providing new opportunities to "illuminate disease" through molecular imaging. Fluorescent imaging technology can easily be combined with endoscopy or tomographic optical imaging methods, thereby raising expectations for the application of fluorescent biosensors to detect target dysfunction or hyperactivation *in vivo* in diseased tissues, tumors, or metastatic cancer cells. Finally, this approach can potentially be applied to monitor response to therapeutics and thereby develop a theragnostic approach allowing the determination of the status of a given biomarker over time while administrating a drug and assessing the benefits of treatment.

## ACKNOWLEDGMENTS

Research in M.C. Morris group is supported by the CNRS (Centre National de la Recherche Scientifique) and grants from the Association de Recherche contre le Cancer (ARC), the Region Languedoc-Roussillon (Subvention "Chercheuse d'Avenir"), and the Institut National du Cancer (INCA). N. V. T. N. is supported by a fellowship from USTH.

## REFERENCES

1. Levene PA, Alsberg CL. The cleavage products of vitellin. *J Biol Chem* 1906;**2**:127–33.
2. Lipmann FA, Levene PA. Serinephosphoric acid obtained on hydrolysis of vitellinic acid. *J Biol Chem* 1932;**98**:109–14.
3. Burnett G, Kennedy EP. The enzymatic phosphorylation of proteins. *J Biol Chem* 1954;**211**:969–80.

4. Rubin CS, Rosen OM. Protein phosphorylation. *Annu Rev Biochem* 1975;**44**:831–87.
5. Krebs EG, Beavo JA. Phosphorylation-dephosphorylation of enzymes. *Annu Rev Biochem* 1979;**48**:923–59.
6. Hunter T. Protein kinases and phosphatases: the yin and yang of protein phosphorylation and signaling. *Cell* 1995;**80**:225–36.
7. Cohen P. The origins of protein phosphorylation. *Nat Cell Biol* 2002;**4**:E127–E130.
8. Hunter T. Protein kinase classification. *Methods Enzymol* 1991;**200**:3–37.
9. Manning G, Whyte DB, Martinez R, Hunter T, Sudarsanam S. The protein kinase complement of the human genome. *Science* 2002;**298**:1912–34.
10. Hunter T. A thousand and one protein kinases. *Cell* 1987;**50**:823–9.
11. Maller JL. On the importance of protein phosphorylation in cell cycle control. *Mol Cell Biochem* 1993;**127–128**:267–81.
12. Johnson LN, Barford D. The effects of phosphorylation on the structure and function of proteins. *Annu Rev Biophys Biomol Struct* 1993;**22**:199–232.
13. Taylor SS, Radzio-Andzelm E. Three protein kinase structures define a common motif. *Structure* 1994;**2**:345–55.
14. Véron M, Radzio-Andzelm E, Tsigelny I, Taylor SS. Protein kinases share a common structural motif outside the conserved catalytic domain. *Cell Mol Biol* 1994;**40**:587–96.
15. Johnson LN, Noble ME, Owen DJ. Active and inactive protein kinases: structural basis for regulation. *Cell* 1996;**85**:149–58.
16. Huse M, Kuriyan J. The conformational plasticity of protein kinases. *Cell* 2002;**109**:275–82.
17. Ubersax JA, Ferrell Jr. JE. Mechanisms of specificity in protein phosphorylation. *Nat Rev Mol Cell Biol* 2007;**8**:530–41.
18. Turk BE. Understanding and exploiting substrate recognition by protein kinases. *Curr Opin Chem Biol* 2008;**12**:4–10.
19. Nolen B, Taylor S, Ghosh G. Regulation of protein kinases: controlling activity through activation segment conformation. *Mol Cell* 2004;**15**:661–75.
20. Brognard J, Hunter T. Protein kinase signaling networks in cancer. *Curr Opin Genet Dev* 2011;**21**:4–11.
21. Kumar R, Singh VP, Baker KM. Kinase inhibitors for cardiovascular disease. *J Mol Cell Cardiol* 2007;**42**:1–11.
22. Blease K. Targeting kinases in asthma. *Expert Opin Investig Drugs* 2005;**14**:1213–20.
23. Ben-Bassat H. Biological activity of tyrosine kinase inhibitors: novel agents for psoriasis therapy. *Curr Opin Investig Drugs* 2001;**2**:1539–45.
24. Cohen P. Protein kinases—the major drug targets of the twenty-first century? *Nat Rev Drug Discov* 2002;**1**:309–15.
25. Manning BD. Challenges and opportunities in defining the essential cancer kinome. *Sci Signal* 2009;**2**:pe15.
26. Johnson L. Protein kinases and their therapeutic exploitation. *Biochem Soc Trans* 2007;**35**:7–11.
27. Lapenna S, Giordano A. Cell cycle kinases as therapeutic targets for cancer. *Nat Rev Drug Discov* 2009;**8**:547–66.
28. Druker BJ, Tamura S, Buchdunger E, Ohno S, Segal GM, Fanning S, et al. Effects of a selective inhibitor of the Abl tyrosine kinase on the growth of Bcr–Abl positive cells. *Nat Med* 1996;**2**:561–6.
29. Versteeg HH, Nijhuis E, van den Brink GR, Evertzen M, Pynaert GN, van Deventer SJ, et al. A new phosphospecific cell-based ELISA for p42/p44 mitogen-activated protein kinase (MAPK), p38 MAPK, protein kinase B and cAMP-response-element-binding protein. *Biochem J* 2000;**350**(Pt 3):717–22.
30. Kupcho K, Somberg R, Bulleit B, Goueli SA. A homogeneous, nonradioactive high-throughput fluorogenic protein kinase assay. *Anal Biochem* 2003;**317**:210–7.

31. Lemke EA, Schultz C. Principles for designing fluorescent sensors and reporters. *Nat Chem Biol* 2011;**7**:480–3.
32. Tsien RY. The green fluorescent protein. *Annu Rev Biochem* 1998;**67**:509–44.
33. Zhang J, Campbell RE, Ting AY, Tsien RY. Creating new fluorescent probes for cell biology. *Nat Rev Mol Cell Biol* 2002;**3**:906–18.
34. Ibraheem A, Campbell RE. Designs and applications of fluorescent protein-based biosensors. *Curr Opin Chem Biol* 2010;**14**:30–6.
35. Wang H, Nakata E, Hamachi I. Recent progress in strategies for the creation of protein-based fluorescent biosensors. *Chembiochem* 2009;**10**:2560–77.
36. Morris MC. Fluorescent biosensors of intracellular targets from genetically encoded reporters to modular polypeptide probes. *Cell Biochem Biophys* 2010;**56**:19–37.
37. González-Vera JA. Probing the kinome in real time with fluorescent peptides. *Chem Soc Rev* 2012;**41**:1652–64.
38. Pazos E, Vázquez O, Mascareñas JL, Vázquez ME. Peptide-based fluorescent biosensors. *Chem Soc Rev* 2009;**38**:3348.
39. Lippincott-Schwartz J, Patterson GH. Development and use of fluorescent protein markers in living cells. *Science* 2003;**300**:87–91.
40. Shaner NC, Steinbach PA, Tsien RY. A guide to choosing fluorescent proteins. *Nat Methods* 2005;**2**:905–9.
41. Tsien RY. Breeding and building molecules to spy on cells and tumors. *FEBS Lett* 2005;**579**:927–32.
42. Giepmans BN, Adams SR, Ellisman MH, Tsien RY. The fluorescent toolbox for assessing protein location and function. *Science* 2006;**312**:217–24.
43. Wu B, Piatkevich KD, Lionnet T, Singer RH, Verkhusha VV. Modern fluorescent proteins and imaging technologies to study gene expression, nuclear localization, and dynamics. *Curr Opin Cell Biol* 2011;**23**:310–7.
44. Gaits F, Hahn K. Shedding light on cell signaling: interpretation of FRET biosensors. *Sci STKE* 2003;**165**:PE3.
45. Aye-Han N-N, Qiang N, Zhang J. Fluorescent biosensors for real-time tracking of post-translational modification dynamics. *Curr Opin Chem Biol* 2009;**13**:392–7.
46. Ting AY, Kain KH, Klemke RL, Tsien RY. Genetically encoded fluorescent reporters of protein tyrosine kinase activities in living cells. *Proc Natl Acad Sci USA* 2001;**98**:15003–8.
47. Kurokawa K, Mochizuki N, Ohba Y, Mizuno H, Miyawakiand A, Matsuda M. A pair of fluorescent resonance energy transfer-based probes for tyrosine phosphorylation of the CrkII adaptor protein in vivo. *J Biol Chem* 2001;**276**:31305–10.
48. Mizutani T, Kondo T, Darmanin S, Tsuda M, Tanaka S, Tobiume M, et al. A novel FRET-based biosensor for the measurement of BCR-ABL activity and its response to drugs in living cells. *Clin Cancer Res* 2010;**16**:3964–75.
49. Fuller BG, Lampson MA, Foley EA, Rosasco-Nitcher S, Le KV, Tobelmann P, et al. Midzone activation of aurora B in anaphase produces an intracellular phosphorylation gradient. *Nature* 2008;**453**:1132–6.
50. Gao X, Zhang J. Spatiotemporal analysis of differential Akt regulation in plasma membrane microdomains. *Mol Biol Cell* 2008;**19**:4366–73.
51. Sasaki K, Sato M, Umezawa Y. Fluorescent indicators for Akt/protein kinase B and dynamics of Akt activity visualized in living cells. *J Biol Chem* 2003;**278**:30945–51.
52. Kunkel MT, Ni Q, Tsien RY, Zhang J, Newton AC. Spatio-temporal dynamics of protein kinase B/Akt signaling revealed by a genetically encoded fluorescent reporter. *J Biol Chem* 2005;**280**:5581–7.
53. Tsou P, Zheng B, Hsu C-H, Sasaki AT, Cantley LC. A fluorescent reporter of AMPK activity and cellular energy stress. *Cell Metab* 2011;**13**:476–86.

54. Johnson SA, Zhongsheng Y, Hunter T. Monitoring ATM kinase activity in living cells. *DNA Repair (Amst)* 2007;**6**:1277–84.
55. Erickson JR, Patel R, Ferguson A, Bossuyt J, Bers DM. Fluorescence resonance energy transfer-based sensor Camui provides new insight into mechanisms of calcium/calmodulin-dependent protein kinase II activation in intact cardiomyocytes. *Circ Res* 2011;**109**:729–38.
56. Gavet O, Pines J. Progressive activation of CyclinB1-Cdk1 coordinates entry to mitosis. *Dev Cell* 2010;**18**:533–43.
57. Harvey CD, Ehrhardt AG, Cellurale C, Zhong H, Yasuda R, Davis RJ, et al. A genetically encoded fluorescent sensor of ERK activity. *Proc Natl Acad Sci USA* 2008;**105**:19264–9.
58. Sato M, Kawai Y, Umezawa Y. Genetically encoded fluorescent indicators to visualize protein phosphorylation by extracellular signal-regulated kinase in single living cells. *Anal Chem* 2007;**79**:2570–5.
59. Sato M, Ozawa T, Inukai K, Asano T, Umezawa Y. Fluorescent indicators for imaging protein phosphorylation in single living cells. *Nat Biotechnol* 2002;**20**:287–94.
60. Fosbrink M, Aye-Han NN, Cheong R, Levchenko A, Zhang J. Visualization of JNK activity dynamics with a genetically-encoded fluorescent biosensor. *Proc Natl Acad Sci USA* 2010;**107**:5459–64.
61. Seong J, Ouyang M, Kim T, Sun J, Wen PC, Lu S, et al. Detection of focal adhesion kinase activation at membrane microdomains by fluorescence resonance energy transfer. *Nat Commun* 2011;**2**:406.
62. Lin CW, Ting AY. A genetically encoded fluorescent reporter of histone phosphorylation in living cells. *Angew Chem Int Ed Engl* 2004;**43**:2940–3.
63. Chew TL, Wolf WA, Gallagher PJ, Matsumura F, Chisholm RL. A fluorescent resonant energy transfer-based biosensor reveals transient and regional myosin light chain kinase activation in lamella and cleavage furrows. *J Cell Biol* 2002;**156**:543–53.
64. Nagai Y, Miyazaki M, Aoki R, Zama T, Inouye S, Hirose K, et al. A fluorescent indicator for visualizing cAMP-induced phosphorylation in vivo. *Nat Biotechnol* 2000;**18**:313–6.
65. Zhang J, Ma Y, Taylor SS, Tsien RY. Genetically encoded reporters of protein kinase A activity reveal impact of substrate tethering. *Proc Natl Acad Sci USA* 2001;**98**:14997–5002.
66. Zhang J, et al. Insulin disrupts beta-adrenergic signalling to protein kinase A in adipocytes. *Nature* 2005;**437**:569–73.
67. Allen MD, Zhang J. Subcellular dynamics of protein kinase A activity visualized by FRET-based reporters. *Biochem Biophys Res Commun* 2006;**348**:716–21.
68. Liu S, Zhang J, Xiang YK. FRET-based direct detection of dynamic protein kinase A activity on the sarcoplasmic reticulum in cardiomyocytes. *Biochem Biophys Res Commun* 2011;**404**:581–6.
69. Violin JD, Zhang J, Tsien RY, Newton AC. A genetically encoded fluorescent reporter reveals oscillatory phosphorylation by protein kinase C. *J Cell Biol* 2003;**161**:899–909.
70. Kajimoto T, Sawamura S, Tohyama Y, Mori Y, Newton AC. Protein kinase C (delta)-specific activity reporter reveals agonist-evoked nuclear activity controlled by Src family of kinases. *J Biol Chem* 2010;**285**:41896–910.
71. Schleifenbaum A, Stier G, Gasch A, Sattler M, Schultz C. Genetically encoded FRET probe for PKC activity based on pleckstrin. *J Am Chem Soc* 2004;**126**:11786–7.
72. Brumbaugh J, Schleifenbaum A, Gasch A, Sattler M, Schultz C. A dual parameter FRET probe for measuring PKC and PKA activity in living cells. *J Am Chem Soc* 2006;**128**:24–5.

73. Kunkel MT, Toker A, Tsien RY, Newton AC. Calcium-dependent regulation of protein kinase D revealed by a genetically encoded kinase activity reporter. *J Biol Chem* 2007;**282**:6733–42.

74. Macůrek L, Lindqvist A, Lim D, Lampson MA, Klompmaker R, Freire R, et al. Polo-like kinase-1 is activated by aurora A to promote checkpoint recovery. *Nature* 2008;**455**:119–24.

75. Tomida T, Takekawa M, O'Grady P, Saito H. Stimulus-specific distinctions in spatial and temporal dynamics of stress-activated protein kinase kinase kinases revealed by a fluorescence resonance energy transfer biosensor. *Mol Cell Biol* 2009;**29**:6117–27.

76. Wang Y, Botvinick EL, Zhao Y, Berns MW, Usami S, Tsien RY, et al. Visualizing the mechanical activation of Src. *Nature* 2005;**434**:1040–5.

77. Hitosugi T, Sasaki K, Sato M, Suzuki Y, Umezawa Y. Epidermal growth factor directs sex-specific steroid signaling through Src activation. *J Biol Chem* 2007;**282**:10697–706.

78. Xiang X, Sun J, Wu J, He HT, Wang Y, Zhu C. A FRET-based biosensor for imaging Syk activities in living cells. *Cell Mol Bioeng* 2011;**4**:670–7.

79. Randriamampita C, Mouchacca P, Malissen B, Marguet D, Trautmann A, Lellouch AC. Dependent FRET based biosensor reveals kinase activity at both the immunological synapse and the antisynapse. *PLoS One* 2008;**3**:e1521.

80. Smith JA, et al. Identification of an extracellular signal-regulated kinase (ERK) docking site in ribosomal S6 kinase, a sequence critical for activation by ERK in vivo. *J Biol Chem* 1999;**274**:2893–8.

81. Holland PM, Cooper JA. Protein modification: docking sites for kinases. *Curr Biol* 1999;**9**:R329–R331.

82. Fernandes N, et al. Use of docking peptides to design modular substrates with high efficiency for mitogen-activated protein kinase extracellular signal-regulated kinase. *ACS Chem Biol* 2007;**2**:665–73.

83. Yaffe MB, Elia AE. Phosphoserine/threonine-binding domains. *Curr Opin Cell Biol* 2001;**13**:131–8.

84. Yaffe MB. Phosphotyrosine-binding domains in signal transduction. *Nat Rev Mol Cell Biol* 2002;**3**:177–86.

85. Seet BT, Dikic I, Zhou MM, Pawson T. Reading protein modifications with interaction domains. *Nat Rev Mol Cell Biol* 2006;**7**:473–83.

86. Pawson T, Kofler M. Kinome signaling through regulated protein-protein interactions in normal and cancer cells. *Curr Opin Cell Biol* 2009;**21**:147–53.

87. Durocher D, et al. The molecular basis of FHA domain:phosphopeptide binding specificity and implications for phospho-dependent signaling mechanisms. *Mol Cell* 2000;**6**:1169–82.

88. Lu PJ, et al. Function of WW domains as phosphoserine- or phosphothreonine-binding modules. *Science* 1999;**283**:1325–8.

89. Lowery DM, Mohammad DH, Elia AE, Yaffe MB. The Polo-box domain: a molecular integrator of mitotic kinase cascades and Polo-like kinase function. *Cell Cycle* 2004;**3**:128–31.

90. Ibraheem A, et al. A bacteria colony-based screen for optimal linker combinations in genetically encoded biosensors. *BMC Biotechnol* 2011;**11**:105.

91. Piljic A, et al. Rapid development of genetically encoded FRET reporters. *ACS Chem Biol* 2011;**6**:685–91.

92. Komatsu N, Aoki K, Yamada M, Yukinaga H, Fujita Y, Kamioka Y, et al. Development of an optimized backbone of FRET biosensors for kinases and GTPases. *Mol Biol Cell* 2011;**22**:4647–56.

93. Nguyen AW, Daugherty PS. Evolutionary optimization of fluorescent proteins for intracellular FRET. *Nat Biotechnol* 2005;**23**:355–60.

94. Chudakov DM, Matz MV, Lukyanov S, Lukyanov KA. Fluorescent proteins and their applications in imaging living cells and tissues. *Physiol Rev* 2010;**90**:1103–63.
95. Lu S, Wang Y. Fluorescence resonance energy transfer biosensors for cancer detection and evaluation of drug efficacy. *Clin Cancer Res* 2010;**16**:3822–4.
96. Tunceroglu A, Matsuda M, Birge RB. Real-time fluorescent resonance energy transfer analysis to monitor drug resistance in chronic myelogenous leukemia. *Mol Cancer Ther* 2010;**9**:3065–73.
97. Martin GS. The road to Src. *Oncogene* 2004;**23**:7910–7.
98. Coopman PJ, Mueller SC. The Syk tyrosine kinase: a new negative regulator in tumor growth and progression. *Cancer Lett* 2006;**241**:159–73.
99. Taylor SS, Yang J, Wu J, Haste NM, Radzio-Andzelm E, Anand G. PKA: a portrait of protein kinase dynamics. *Biochim Biophys Acta* 2004;**1697**:259–69.
100. Depry C, Zhang J. Using FRET-based reporters to visualize subcellular dynamics of protein kinase A activity. *Methods Mol Biol* 2011;**756**:285–94.
101. Way KJ, Chou E, King GL. Identification of PKC-isoform specific biological actions using pharmacological approaches. *Trends Pharmacol Sci* 2000;**21**:181–7.
102. Manning BD, Cantley LC. AKT/PKB signalling: navigating downstream. *Cell* 2007;**129**:1261–74.
103. Yoeli-Lerner M, Toker A. Akt/PKB signalling in cancer: a function in cell motility and invasion. *Cell Cycle* 2006;**5**:603–5.
104. Roux PP, Blenis J. ERK and p38 MAPK-activated protein kinases: a family of protein kinases with diverse biological functions. *Microbiol Mol Biol Rev* 2004;**68**:320–44.
105. Loving GS, Sainlos M, Imperiali B. Monitoring protein interactions and dynamics with solvatochromic fluorophores. *Trends Biotechnol* 2010;**28**:73–83.
106. Lavis LD, Raines RT. Bright ideas for chemical biology. *ACS Chem Biol* 2008;**3**: 142–55.
107. Morris MC, Deshayes S, Heitz F, Divita G. Cell-penetrating peptides: from molecular mechanisms to therapeutics. *Biol Cell* 2008;**100**:201–17.
108. Heitz F, Morris MC, Divita G. Twenty years of cell-penetrating peptides: from molecular mechanisms to therapeutics. *Br J Pharmacol* 2009;**157**:195–206.
109. Yeh RH, Yan X, Cammer M, Bresnick AR, Lawrence DS. Real time visualization of protein kinase activity in living cells. *J Biol Chem* 2002;**277**:11527–32.
110. Veldhuyzen WF, Nguyen Q, McMaster G, Lawrence DS. A light-activated probe of intracellular protein kinase activity. *J Am Chem Soc* 2003;**125**:13358–9.
111. Wang Q, Lawrence DS. Phosphorylation-driven protein-protein interactions: a protein kinase sensing system. *J Am Chem Soc* 2005;**127**:7684–5.
112. Gulyani A, Vitriol E, Allen R, Wu J, Gremyachinskiy D, Lewis S, et al. A biosensor generated via high-throughput screening quantifies cell edge Src dynamics. *Nat Chem Biol* 2011;**7**:437–44.
113. Kurzawa L, Pellerano M, Coppolani JB, Morris MC. Fluorescent peptide biosensor for probing the relative abundance of cyclin-dependent kinases in living cells. *PLoS One* 2011;**6**:e26555.
114. Van Thi Nhu N, Lykaso S, Pellerano M, Morris MC. CDKACT, a fluorescent biosensor for real-time measurement of CDK/cyclin activity; 2012 [in preparation].
115. Chen CA, Yeh RH, Lawrence DS. Design and synthesis of a fluorescent reporter of protein kinase activity. *J Am Chem Soc* 2002;**124**:3840–1.
116. Shults MD, Pearce DA, Imperiali B. Modular and tunable chemosensor scaffold for divalent zinc. *J Am Chem Soc* 2003;**125**:10591–7.
117. Shults MD, Imperiali B. Versatile fluorescence probes of protein kinase activity. *J Am Chem Soc* 2003;**125**:14248–9.
118. Shults MD, Carrico-Moniz D, Imperiali B. Optimal Sox-based fluorescent chemosensor design for serine/threonine protein kinases. *Anal Biochem* 2006;**352**:198–207.

119. Shults MD, Janes KA, Lauffenburger DA, Imperiali B. A multiplexed homogeneous fluorescence-based assay for protein kinase activity in cell lysates. *Nat Methods* 2005;**2**:277–83.

120. Luković E, González-Vera JA, Imperiali B. Recognition-domain focused chemosensors: versatile and efficient reporters of protein kinase activity. *J Am Chem Soc* 2008;**130**:12821–7.

121. González-Vera JA, Luković E, Imperiali B. A rapid method for generation of selective sox-based chemosensors of Ser/Thr kinases using combinatorial peptide libraries. *Bioorg Med Chem Lett* 2010;**19**:1258–60.

122. Lukovic E, Vogel Taylor E, Imperiali B. Monitoring protein kinases in cellular media with highly selective chimeric reporters. *Angew Chem Int Ed Engl* 2009;**48**:6828–31.

123. Stains CI, Luković E, Imperiali B. A p38α-selective chemosensor for use in unfractionated cell lysates. *ACS Chem Biol* 2011;**6**:101–5.

124. Anai T, Nakata E, Koshi Y, Ojida A, Hamachi I. Design of a hybrid biosensor for enhanced phosphopeptide recognition based on a phosphoprotein binding domain coupled with a fluorescent chemosensor. *J Am Chem Soc* 2007;**129**:6232–9.

125. Tremblay MS, Lee M, Sames D. A luminescent sensor for tyrosine phosphorylation. *Org Lett* 2008;**10**:5–8.

126. Wang Q, Cahill SM, Blumenstein M, Lawrence DS. Self-reporting fluorescent substrates of protein tyrosine kinases. *J Am Chem Soc* 2006;**128**:1808–9.

127. Wang Q, Dai Z, Cahill SM, Blumenstein M, Lawrence DS. Light-regulated sampling of protein tyrosine kinase activity. *J Am Chem Soc* 2006;**128**:14016–7.

128. Wang Q, Zimmerman EI, Toutchkine A, Martin TD, Graves LM, Lawrence DS. Multicolor monitoring of dysregulated protein kinases in chronic myelogenous leukemia. *ACS Chem Biol* 2010;**5**:887–95.

129. Sharma V, Agnes RS, Lawrence DS. Deep quench: an expanded dynamic range for protein kinase sensors. *J Am Chem Soc* 2007;**129**:2742–3.

130. Agnes RS, Jernigan F, Shell JR, Sharma V, Lawrence DS. Suborganelle sensing of mitochondrial cAMP-dependent protein kinase activity. *J Am Chem Soc* 2010;**132**:6075–80.

131. Sun H, Low KE, Woo S, Noble RL, Graham RJ, Connaughton SS, et al. Real-time protein kinase assay. *Anal Chem* 2005;**77**:2043–9.

132. Kikuchi K, Hashimoto S, Mizukami S, Nagano T. Anion sensor-based ratiometric peptide probe for protein kinase activity. *Org Lett* 2009;**11**:2732–5.

133. Dai Z, Dulyaninova NG, Kumar S, Bresnick AR, Lawrence DS. Visual snapshots of intracellular kinase activity at the onset of mitosis. *Chem Biol* 2007;**14**:1254–60.

134. Chen CA, Yeh RH, Yan X, Lawrence DS. Biosensors of protein kinase action: from *in vitro* assays to living cells. *Biochim Biophys Acta* 2004;**1697**:39–51.

135. Lawrence DS, Wang Q. Seeing is believing: peptide-based fluorescent sensors of protein tyrosine kinase activity. *Chembiochem* 2007;**8**:373–8.

136. Sharma V, Wang Q, Lawrence DS. Peptide-based fluorescent sensors of protein kinase activity: design and applications. *Biochim Biophys Acta* 2008;**1784**:94–9.

137. Haas KL, Franz KJ. Application of metal coordination chemistry to explore and manipulate cell biology. *Chem Rev* 2009;**109**:4921–60.

138. Minta A, Kao JPY, Tsien RY. Fluorescent indicators for cytosolic calcium based on rhodamine and fluorescein chromophores. *J Biol Chem* 1989;**264**:8171–8.

139. Sakamoto T, Ojida A, Hamachi I. Molecular recognition, fluorescence sensing, and biological assay of phosphate anion derivatives using artificial Zn(II)-Dpa complexes. *Chem Commun* 2009;**2**:141–52.

140. Allen KN, Imperiali B. Lanthanide-tagged proteins—an illuminating partnership. *Curr Opin Chem Biol* 2010;**14**:247–54.

141. Marme N, Knemeyer JP, Sauer M, Wolfrum J. Inter- and intramolecular fluorescence quenching of organic dyes by tryptophan. *Bioconjug Chem* 2003;**14**:1133–9.

142. Ogawa M, Kosaka N, Choyke PL, Kobayashi H. H-type dimer formation of fluorophores: a mechanism for activatable, in vivo optical molecular imaging. *ACS Chem Biol* 2009;**4**:535–46.

143. Lee HM, Larson DR, Lawrence DS. Illuminating the chemistry of life: design, synthesis, and applications of "caged" and related photoresponsive compounds. *ACS Chem Biol* 2009;**4**:409–27.

144. Malumbres M. Physiological relevance of cell cycle kinases. *Physiol Rev* 2011;**91**: 973–1007.

145. Malumbres M, Barbacid M. Cell cycle kinases in cancer. *Curr Opin Genet Dev* 2007;**17**:60–5.

146. Kurzawa L, Morris M. Cell-cycle markers and biosensors. *Chembiochem* 2010;**11**: 1037–47.

147. Morgan DO. Cyclin-dependent kinases: engines, clocks, and microprocessors. *Annu Rev Cell Dev Biol* 1997;**13**:261–91.

148. Obaya AJ, Sedivy JM. Regulation of cyclin-Cdk activity in mammalian cells. *Cell Mol Life Sci* 2002;**59**:126–42.

149. Satyanarayana A, Kaldis P. Mammalian cell-cycle regulation: several Cdks, numerous cyclins and diverse compensatory mechanisms. *Oncogene* 2009;**28**:2925–39.

150. Ni Q, Titov DV, Zhang J. Analyzing protein kinase dynamics in living cells with FRET reporters. *Methods* 2006;**40**:279–86.

151. Zhang J, Allen MD. FRET-based biosensors for protein kinases: illuminating the kinome. *Mol Biosyst* 2007;**3**:759–65.

152. Morris MC. *Fluorescent biosensors for cancer cell imaging and diagnostics*. Biosensor and Cancer. Ed. Victor Preedy: Science Publishers; 2012.

153. Von Ahsen O, Bomer U. High throughput screening for kinase inhibitors. *Chembiochem* 2005;**6**:481–90.

154. Allen MD, DiPilato LM, Rahdar M, Ren YR, Chong C, Liu JO, et al. Reading dynamic kinase activity in living cells for high-throughput screening. *ACS Chem Biol* 2006;**1**:371–6.

155. Giuliano KA, Taylor DL. Fluorescent-protein biosensors: new tools for drug discovery. *Trends Biotechnol* 1998;**16**:135–40.

156. Wolff M, Wiedenmann J, Nienhaus GU, Valler M, Heilker R. Novel fluorescent proteins for high-content screening. *Drug Discov Today* 2006;**11**:1054–60.

157. Lang P, Yeow K, Nichols A, Scheer A. Cellular imaging in drug discovery. *Nat Rev Drug Discov.* 2006;**5**:343–56.

158. El-Deiry WS, Sigman CC, Kelloff GJ. Imaging and oncologic drug development. *J Clin Oncol* 2006;**24**:3261–73.

159. Willmann JK, van Bruggen N, Dinkelborg LM, Gambhir SS. Molecular imaging in drug development. *Nat Rev Drug Discov* 2008;**7**:591–607.

160. Carlson HJ, Campbell RE. Genetically encoded FRET-based biosensors for multiparameter fluorescence imaging. *Curr Opin Biotechnol* 2009;**20**:19–27.

161. Piljic A, Schultz C. Simultaneous recording of multiple cellular events by FRET. *ACS Chem Biol* 2008;**3**:156–60.

162. Ai HW, Hazelwood KL, Davidson MW, Campbell RE. Fluorescent protein FRET pairs for ratiometric imaging of dual biosensors. *Nat Methods* 2008;**5**:401–3.

163. Niino Y, Hotta K, Oka K. Simultaneous live cell imaging using dual FRET sensors with a single excitation light. *PLoS One* 2009;**4**:e6036.

164. Rao J, Dragulescu-Andrasi A, Yao H. Fluorescence imaging in vivo: recent advances. *Curr Opin Biotechnol* 2007;**18**:17–25.

165. Razgulin A, Ma N, Rao J. Strategies for in vivo imaging of enzyme activity: an overview and recent advances. *Chem Soc Rev* 2011;**40**:4186–216.
166. Ho N, Weissleder R, Tung CH. Development of water-soluble far-red fluorogenic dyes for enzyme sensing. *Tetrahedron* 2006;**62**:578–85.
167. Ntziachristos V. Fluorescence molecular imaging. *Annu Rev Biomed Eng* 2006;**8**:1–33.
168. Weissleder R, Pittet MJ. Imaging in the era of molecular oncology. *Nature* 2008;**452**:580–9.
169. Jiang T, Olson ES, Nguyen QT, Roy M, Jennings PA, Tsien RY. Tumor imaging by means of proteolytic activation of cell-penetrating peptides. *Proc Natl Acad Sci USA* 2004;**101**:17867–72.
170. Bloch S, Lesage F, McIntosh L, Gandjbakhche A, Liang K, Achilefu S. Whole-body fluorescence lifetime imaging of a tumor-targeted near-infrared molecular probe in mice. *J Biomed Opt* 2005;**10**:054003.
171. Kunkel MT, Newton AC. Spatiotemporal dynamics of kinase signaling visualized by targeted reporters. *Curr Protoc Chem Biol* 2009;**1**:17–8.
172. Urano Y, Asanuma D, Hama Y, Koyama Y, Barrett T, Kamiya M, et al. Selective molecular imaging of viable cancer cells with pH-activatable fluorescence probes. *Nat Med* 2009;**15**:104–9.
173. Reshetnyak YK, Yao L, Zheng S, Kuznetsov S, Engelman DM, Andreev OA. Measuring tumor aggressiveness and targeting metastatic lesions with fluorescent pHLIP. *Mol Imaging Biol* 2011;**13**:1146–56.
174. Aoki K, Komatsu N, Hirata E, Kamioka Y, Matsuda M. Stable expression of FRET biosensors: a new light in cancer research. *Cancer Sci* 2012;**103**:614–9.
175. Kamioka Y, Sumiyama K, Mizuno R, Sakai Y, Hirata E, Kiyokawa E, et al. Live imaging of protein kinase activities in transgenic mice expressing FRET biosensors. *Cell Struct Funct* 2012;**37**:65–73.
176. Keereweer S, Kerrebijn JD, van Driel PB, Xie B, Kaijzel EL, Snoeks TJ, et al. Optical image-guided surgery—where do we stand? *Mol Imaging Biol* 2011;**13**:199–207.
177. Urano Y, Sakabe M, Kosaka N, Ogawa M, Mitsunaga M, Asanuma D, et al. Rapid cancer detection by topically spraying a γ-glutamyltranspeptidase-activated fluorescent probe. *Sci Transl Med* 2011;**3**:110ra119.
178. Chen K, Xie J, Chen X. RGD-human serum albumin conjugates as efficient tumor targeting probes. *Mol Imaging* 2009;**8**:65–73.
179. Kovar JL, Volcheck W, Sevick-Muraca E, Simpson MA, Olive DM. Characterization and performance of a near-infrared 2-deoxyglucose optical imaging agent for mouse cancer models. *Anal Biochem* 2009;**384**:254–62.

# CHAPTER SEVEN

# Time-Resolved Förster Resonance Energy Transfer-Based Technologies to Investigate G Protein-Coupled Receptor Machinery: High-Throughput Screening Assays and Future Development

**Pauline Scholler**[*,†,‡,§], **Jurriaan M. Zwier**[§], **Eric Trinquet**[§],
**Philippe Rondard**[*,†,‡], **Jean-Philippe Pin**[*,†,‡], **Laurent Prézeau**[*,†,‡],
**Julie Kniazeff**[*,†,‡]

*CNRS, UMR-5203, Institut de Génomique Fonctionelle, Montpellier, F-34094, France
†INSERM, U661, Montpellier, F-34094, France
‡Universités de Montpellier 1&2, UMR-5203, Montpellier, F-34094, France
§Cisbio Bioassays, Parc Marcel Boiteux, BP 84175, Codolet, France

## Contents

*Progress in Molecular Biology and Translational Science*, Volume 113
ISSN 1877-1173
http://dx.doi.org/10.1016/B978-0-12-386932-6.00007-7

## Abstract

High-throughput screening requires easy-to-monitor, rapid, robust, reliable, and miniaturized methods to test thousands of compounds on a target in a short period, in order
to find active drugs. Only a few methods have been proved to fulfill all these requirements. New screening approaches based on fluorescence and especially on the principle of resonance energy transfer are being developed to study one of the main
targets in the pharmaceutical industry, namely, the G protein-coupled receptors
(GPCRs). Two types of approaches are clearly defined: generic approaches that are immediately applicable to a lot of targets such as second messenger kits or kinase kits;
target-specific approaches that sense the receptor itself such as fluorescent ligands
or fluorescent partners. This chapter focuses on sensors and approaches using the
time-resolved Förster resonance energy transfer and homogeneous time-resolved
fluorescence principle, their use, and their prospective applications for screening drugs
acting on GPCRs.

# 1. INTRODUCTION

Drug discovery remains a challenge in the pharmaceutical industry. To
be validated as a future medicine, a compound must fulfill certain criteria to
ensure its specificity for the target, its mode of action, and limited side effects.
To assess whether it fulfills all these criteria, the molecules are tested in various
assays starting from hit identification by screening large libraries of compounds
using rapid and accurate tests, and followed by a precise analysis of the properties of a reduced number of molecules using animal models and clinical tests.
While the duration of the later part of the process is rather incompress-
ible, efforts have been made to reduce the experimental time spent on the
identification of lead compounds and also to reach good pharmacological
characterization before qualifying a compound for further tests. Accordingly, several different assays have been developed to rapidly test a large
number of molecules during high-throughput screening (HTS) campaigns. Over the past decades, the assays used for HTS have constantly
evolved, together with the new findings and concepts arising from basic
research on the target (identification of new mechanisms, new signaling

pathways, new partners, etc.) and also on the development of new technologies.

Because of the increasing number of available compounds to be tested, the format of HTS assays had to adapt to miniaturized formats (384-, 1536-, and even 3456-well microplates) and automated platforms. This miniaturization must come with increased accuracy of the assays to limit false positives. Also, the will to use environment-friendly tests had required scientists to find alternatives to the radioactive tests that generate wastes that have to be handled at high cost. A good alternative was found in fluorescent probes and fluorescent assays. The first HTS assays using these probes appeared rather accurate and almost as good as the previous assays and are now being widely used by pharmaceutical companies in screening campaigns. However, among the screened molecules, some are colored and interfere with the detection of the fluorescent probe.[1] This can lead to false positives, and these compounds are often excluded from further fluorescent analysis and thus from further characterization. To circumvent this, some solutions are available. Indeed, in time-resolved Förster resonance energy transfer (TR-FRET), which is a specific type of fluorescent method, the peculiar timeframe of signal recordings permits getting rid of most of fluorescent contamination arising from molecules and other components of the assay.[2,3] TR-FRET has additional advantages over the other techniques, such as a higher signal-to-noise ratio, for example. This technology has been successfully applied to offer kits to measure the activation of different cellular processes and solutions to measure the binding of molecules to a specific target. This chapter presents how TR-FRET and the ratiometric "homogeneous time-resolved fluorescence" (HTRF®) technology are an interesting alternative in HTS assays and provides examples using a model of the targets used in drug discovery, namely, G protein-coupled receptors (GPCRs).

GPCRs are cell-surface proteins that recognize a large variety of stimuli and that are key signaling molecules participating in most cellular processes.[4] Today, up to 30% of the therapeutic molecules available on the market and many drugs of abuse act on GPCRs,[5] and this family of receptors still represents a promising target for the development of new therapeutic solutions for the treatment of several pathologies. The GPCR family contains more than 850 members in humans classified on the basis of their sequence homology into several classes, mainly class-A (rhodopsin-like), class-B (secretin receptor-like), and class-C (metabotropic glutamate receptor (mGluR)-like) receptors[6] A structural feature common to all GPCRs is a bundle of seven membrane-spanning α-helices with the amino terminus at the extracellular side and the carboxy terminus in the cytosol. The length and the structure as well as the sequence of amino and carboxy termini largely differ between receptors.[4] Binding of the agonist at the level of the transmembrane domain and/or of the amino terminus promotes conformational changes that allow the activation of a heterotrimeric G protein, that is, the exchange from guanosine diphosphate (GDP) to guanosine-5'-triphosphate (GTP), which will in turn modulate several intracellular signaling pathways. Several subtypes of heterotrimeric G proteins exist, and each promotes different signaling events. While primarily thought to activate one specific subtype of G proteins and downstream signaling cascades, most GPCRs are now described as complex signaling proteins. Indeed, most of the receptors were reported to activate more than one type of G proteins but also to be capable of modulating signaling pathways independently of the G proteins.[7,8] Moreover, in addition to the classical agonists and antagonists, synthetic molecules acting as allosteric modulators of GPCRs have been identified.[9] Furthermore, the formation of receptor oligomers, that is, the physical interaction between several GPCRs identical or not, and the participation of GPCRs in signaling complexes, involving intracellular proteins and other types of membrane proteins, were proposed to alter the different properties of a given receptor.[10] These concepts have had a real impact on the understanding of GPCR pharmacology and on drug discovery aiming at identifying new compounds acting on these receptors.

In the prospect of drug discovery, the complete characterization of a lead compound requires the analysis of its pharmacological properties as well as its effect on the target and on the signaling. In addition, off-target effects have to be assessed to exclude molecules that could have drastic adverse effects. In the case of GPCRs, a full characterization of molecules requires two types of HTS-compatible assays: (i) functional assays to determine the effect of compounds in terms of signaling and (ii) mechanistic assays to detect the direct effects of compounds on their target, for example, binding properties or oligomerization. TR-FRET was shown to be a suitable method to explore both aspects in an HTS-compatible format.

## 2. OVERVIEW OF THE TR-FRET PRINCIPLE AND ITS ADVANTAGES

Fluorescent techniques represent nowadays the majority of detection technologies used in HTS, and, among them, FRET, particularly TR-FRET, has many advantages to offer and thus plays an important role in the HTS campaigns. The principle of FRET and TR-FRET will be only briefly described here. For more comprehensive explanations, readers may refer to specific reviews.[3,11,12]

## 2.1. Principle of FRET

FRET is a nonradiative transfer of energy between two fluorophores, a donor and an acceptor, and depends on three conditions: First, the donor emission spectrum must partially overlap the acceptor excitation spectrum to allow a transfer of energy; second, the two fluorophores must be in close proximity, usually at a distance of less than 100 Å; and, third, the orientations of the fluorophore dipole moments must not be perpendicular to each other (otherwise no energy transfer can occur). The efficacy of energy transfer ($E$) that is measured is inversely proportional to the sixth power of the distance between the two chromophores ($R$):

$$E = \frac{1}{1 + \frac{R^6}{R_0^6}}$$

where $R_0$ is the Förster radius for a given fluorophore couple. This is a constant depending on the spectral properties of the dyes as well as their relative orientation and is defined as the distance at which $E = 50\%$, such that it can be calculated or determined experimentally. Because of the distance dependence, FRET measurement is an ideal technique to follow movements occurring at the level of proteins, in the nanometer range, and it is the reason why Stryer and colleagues defined it as a "spectroscopic ruler."[13] Depending on the position of the dyes on the protein, it can be used to analyze conformational changes within a single protein (e.g., one GPCR) or to monitor protein–protein interactions (e.g., association between two GPCRs). Historically, genetically encoded fluorescent proteins such as CFP and YFP were fused to the protein of interest, but nowadays, because of their size, genetically encoded fluorescent proteins are often replaced by small organic dyes that can be introduced at more precise locations in the protein. An overview of some techniques available to label proteins with FRET-compatible organic fluorophores will be presented in Chapter 4.

## 2.2. Principle of TR-FRET

Just like classic FRET, TR-FRET consists of a nonradiative energy transfer from a donor molecule to an acceptor fluorophore, but it is based on the use of particular donors: lanthanide ions (also called rare earths). The peculiarity of lanthanides is their long-lived fluorescence, usually in the millisecond range, a lifetime at least 100,000-fold longer as compared to that of classical fluorophores with typical lifetimes of upto 10 ns. Upon excitation, TR-FRET donors can induce a delayed emission from the classic acceptor fluorophores. This peculiarity enables the detection of a fluorescent signal in a time-resolved manner, allowing the excitation and the detection processes to be separated temporally (see Fig. 7.1). Usually, a 50–100 μs delay is applied between excitation and recording of the emission signal, which rejects most of the background short-lived emission arising from biological media, instrumentation, and microtiter plates, or by the chemical compounds to be tested in HTS. This time-resolved detection is a first major improvement over the actual fluorescence measurements in which the fluorescence of some molecules often impairs their analysis in screening assays.

## 2.3. Advantages of TR-FRET over classic FRET

Besides the main advantage of allowing time-resolved detection, TR-FRET lanthanide donors also present peculiar emission spectra, allowing really high signal-to-noise ratios. In fact, as for classic FRET, the energetic compatibility between the two fluorophores of the FRET pair is needed in TR-FRET, which will be the case if the absorption spectrum of the acceptor overlaps the emission spectrum of the donor. Lanthanide TR-FRET donors, especially europium and terbium, present very large Stokes shifts, that is, the difference in wavelength between positions of the band maxima of the absorption and emission spectra of the fluorophore, and emit at several wavelengths with sharp emission peaks (see Fig. 7.1). These specificities confer several advantages. First is better spectral selectivity, as absorption and emission spectra of the donor and acceptor fluorophores are well separated as compared to the nonnegligible overlap of classic FRET or BRET pairs which results, for classic FRET, in the direct excitation of the acceptor by the excitation light and thus further increases the background signal. Second, a variety of acceptors can be used, thanks to the multiple emission peaks of the donors. In the case of europium, the acceptors are preferably near-infrared fluorophores, while terbium complexes offer more flexibility, as multiple acceptors evenly disposed in the emission spectrum can be used (e.g., fluorescein-, rhodamine-, or indocyanine-derived acceptors). Third, just like classic FRET, TR-FRET is dependent on the distance between the fluorophore pair $(R)$ relative to their $R_0$. The variety of donor–acceptor couples that can be used in TR-FRET

A    Temporal selectivity

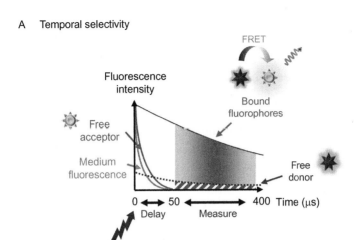

B    Spectral selectivity

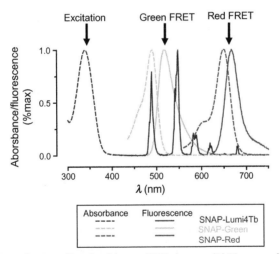

**Figure 7.1** Specificities of lanthanides as FRET donors. (A) Time-resolved detection of fluorescent/FRET signals. Thanks to long fluorescence lifetime of the lanthanides used in TR-FRET, a delay is applied between excitation by a laser pulse and recordings of the fluorescence/FRET signals, allowing suppression of all interfering short-lifetime emissions, a phenomenon known as "temporal selectivity." (B) Absorption and emission spectra of SNAP-Lumi4Tb, a donor used in TR-FRET (purple), and SNAP-Green (green) and SNAP-Red (Red), two acceptors compatible with the donor. Characteristics of SNAP-Lumi4Tb spectra allow to transfer on several acceptor fluorophores and induce a high spectral selectivity between the donor fluorescence emission at 620 nm and the acceptor fluorescence emissions at 520 nm (SNAP-Green) and 665 nm (SNAP-Red). (See Color Insert.)

gives access to pairs with different intrinsic $R_0$ while keeping the high signal-to-noise ratio, thanks to the lanthanides donor. From a biological point of view, it makes it possible to detect and follow molecular events or conformational changes occurring in cells at different amplitudes, with a better accuracy as compared to classical FRET, in the range from 1 to 10 nm. A considerable advantage used in the HTRF technology is the ratiometric approach where the acceptor signal is divided by the donor signal to correct for compound interference.[14] A final advantage is that lanthanide emission is unpolarized because of its long excited-state lifetime. This phenomenon suppresses the possibility that the acceptor and the lanthanide-cryptate dipole moments become perpendicular to each other and thus prevents the eventuality of no transfer of energy between them. This allows a more accurate interpretation of the TR-FRET signals measured as compared to studies with classic FRET signals, where this parameter is rarely taken into consideration.

## 2.4. TR-FRET in HTS campaigns

All these characteristics give TR-FRET strong advantages over other HTS screening technologies and over other fluorescent techniques. First of all, its combination of high sensitivity and robustness makes miniaturization possible, even up to a 1536-well plate format. It is a safe method as compared to radioactive assays, and waste handling is not a major issue anymore. A reduced quantity of lanthanide-labeled entities is needed when compared to classical fluorescent probes, thanks to the high signal-to-noise ratio, which can yield important savings in an HTS campaign. And, nowadays, the variety of assays and ligands available is continuously increasing, offering TR-FRET solutions to assess almost all events occurring during activation and signaling of a GPCR. A few drawbacks still remain with TR-FRET. It is still less sensitive than radioactive assays, and it can be cumbersome to develop TR-FRET-compatible fluorescent ligands that retain a good affinity for the target. On the same aspect, the labeling of the target with TR-FRET fluorophores is not always possible or easy, and in the case of GPCRs, it is often necessary to modify the receptor, which could affect its properties (this will be developed in part 4.1 and 4.2). Finally, the range of linearity for some quantitative assays is limited, and assays are often more expensive than those based on genetically encoded fluorescent proteins or classical fluorophores. However, when compared to the limitations of other techniques, it is no surprise that TR-FRET assays are more and more widely used.

In the next sections, generic and target-specific TR-FRET methods to analyze GPCR signaling machinery will be presented, along with what we consider as good recent examples to illustrate the possibilities and specificities of TR-FRET-based technologies.

# 3. SCREENING WITH GENERIC METHODS

GPCR signaling consists of a series of spatial and temporal events, including, of course, the cascade known as the G protein cycle and downstream events but also other pathways that are G protein independent. Most of these signaling cascades are common to many GPCRs, which makes it worth developing generic methods to measure activation or production of one of the metabolites in these cascades to screen for GPCR ligands in HTS. Similar to all screening methods for GPCRs, TR-FRET-based generic methods must be simple, homogeneous with minimal reagent additions, nonradioactive, robust, and adaptable to 384- or 1536-well plate format with robotic automation. But the question is whether to look for proximal or distal events. Measuring proximal events can reduce false positives, but most techniques are not sensitive enough to avoid moving down the signal transduction cascade to amplify the response. In this part, we will briefly present the TR-FRET-based generic techniques currently available, and which could offer a really good compromise between sensitivity, detection of an early event in the GPCR activation cascade, pharmacological information about the type of ligands or the signaling pathway involved, and in some cases, more biologically relevant assays.

## 3.1. Sandwich or competitive assay format

Almost all functional generic assays based on TR-FRET and used for GPCR HTS are based on two formats: a sandwich format or a competition format, illustrated in Fig. 7.2 (see also Ref. 12).

In a sandwich assay, the TR-FRET signal comes from the proximity of two antibodies, one labeled with a lanthanide cryptate or chelate as donor of energy and the second one labeled with a compatible acceptor. The two antibodies recognize the same target but at different regions, one recognizing a specific modification of the target such as phosphorylated site, for example, and the second recognizing a part of the protein that remains unmodified. Thus, the larger the amount of modified target produced, the more the antibodies bound on the two sites of the target, and thus the more the TR-FRET signal measured by the proximity of the two fluorophores.

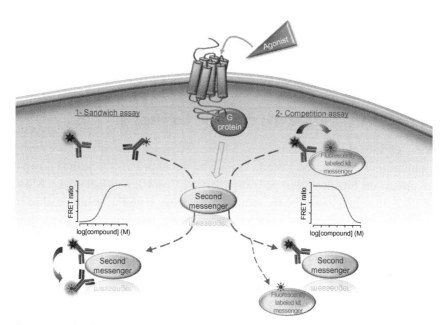

**Figure 7.2** Sandwich and competitive assay format. In a sandwich assay, normalized FRET signals are generated through energy transfer between two antibodies, one labeled with a donor and the other with an acceptor, and recognizing the target to assay at different regions. The larger the amount of target produced, the higher the normalized FRET signal. In a competitive assay, the target generated through G protein activation competes with an acceptor-labeled version of itself for binding to a donor-labeled antibody, decreasing the possible energy transfer. Thus, the normalized FRET signal decreases as more of the target is produced. (See Color Insert.)

The TR-FRET signal measured ratiometrically is then proportional to the quantity of target produced, and it can sometimes be translated into a proper concentration of target produced, as the quantity of each labeled antibodies introduced is controlled.

In a competitive assay, the target (e.g., a metabolite such as cAMP) produced through a biological event competes with an acceptor-labeled version of it (usually provided in the kit used) for binding to a donor-labeled antibody. Thus, the TR-FRET signal measured decreases as more endogenous target is produced because it displaces the labeled target, separating the two fluorophores which can no longer transfer energy to each other. Again, for some assays, the TR-FRET signal measured ratiometrically is proportional to the quantity of target produced and can indicate a proper concentration of target in the medium.

These two formats have been deployed to assay many metabolites and targets in the GPCR activation cascade and will be listed below.

## 3.2. G protein activation and secondary messengers

Upon ligand binding, the receptor undergoes conformational changes leading to a stabilized, active form, responsible for initiating the signaling cascade, which occurs mainly by the activation of the heterotrimeric G protein.

### 3.2.1 Protein activation and exchange of GDP for GTP

When the G protein cycle is initiated, it results in the exchange of GDP with
GTP in the Gα subunit, but unfortunately, no TR-FRET-based assay to
measure GTP binding is yet available. However, a method was developed
by Perkin Elmer based on the time-resolved measurement of the fluorescence
of a europium chelate (no acceptor fluorophore is present in this case).[15] Very
similar to what is classically done with radiolabeled [[35]S]GTPγS, upon agonist
binding to the GPCR expressed in the membranes, the G proteins are activated and incorporate the nonhydrolyzable europium-labeled GTP analog.
After washing or filtration steps to remove unbound labeled GTP, the fluorescence is measured. Because it is measured in a time-resolved manner, the
signal is higher than with classical fluorophores, avoiding background fluorescence from the membranes, for example, and it represents a powerful alternative to radioisotopic assays. However, this assay is not widely used in
HTS yet.

### 3.2.2 Heterotrimeric G protein subfamilies

Following G protein activation and dissociation (or rearrangement), the
Gα-GTP on one side and the Gβγ subunits on the other modulate several effectors. G proteins have been classified into four subfamilies, namely, Gs, Gi/o,
Gq/11, and G12/13, based on the structural similarity of their α subunits and
on the type of modulatory response they induce.[16] Each GPCR preferentially
couples to one subfamily of G protein, thus stimulating preferentially one signaling cascade. TR-FRET-based assays are being developed to investigate
each of G protein signaling pathway, offering a catalog of assays for screening
activation of each GPCR.

### 3.2.3 Screening for Gs and Gi/o signaling pathways

Both Gs and Gi/o families modulate adenylate cyclase activity, leading, respectively, to an increase and a decrease in the intracellular concentration of
$3',5'$-cyclic adenosine monophosphate (cAMP). Two main TR-FRET-
based assays are currently available to detect cAMP level variations in cells,
both based on a competition format: the HTRF-based cAMP detection kit
from Cisbio Bioassays[17] and the LANCE® cAMP detection kit from Perkin
Elmer[18] (see Table 7.1). Both assays rely on the use of acceptor-labeled

**Table 7.1** Summary of TR-FRET-compatible assays for GPCR

| GPCR signaling cascade and downstream events | Biological measurement | Available TR-FRET assay (company) | Basis | Illustrative paper |
|---|---|---|---|---|
| Ligand binding | Displacement of a reference ligand | Tag-lite (Cisbio Bioassays) | Ligand binding and/or displacement of a reference fluorescent ligand that generates TR-FRET signal with a Lumi4–Tb labeled SNAP-tagged GPCR | Zwier et al.[19] |
| | | DELFIA receptor–ligand-binding assays (Perkin Elmer) | An europium-labeled ligand is competed off with a tested unlabeled ligand. The amount of specifically bound labeled ligand is revealed by TR-fluorescence after addition of an enhancement solution and wash steps to remove unbound ligand | Handl et al.[20] |
| G protein–dependent events | Guanine nucleotide exchange | DELFIA Eu-GTP binding assay (Perkin Elmer) | Binding of a nonhydrolyzable europium-labeled GTP analog to receptor-activated G proteins | Frang et al.[15] |
| Gq pathway | Inositol monophosphate (IP1) accumulation | IP-One (Cisbio Bioassays) | Competition assay between endogenously produced IP1 and acceptor-labeled IP1 for binding on donor-labeled IP1 antibody (TR-FRET signal decreases as endogenous IP1 is produced) | Trinquet et al.[21] Trinquet et al.[22] |

| Gs/Gi pathway | Intracellular cAMP level variation | LANCE cAMP (Perkin Elmer) | Competition assay between endogenously produced cAMP and acceptor-labeled cAMP for binding on donor-labeled cAMP antibody (TR–FRET signal decreases as endogenous cAMP is produced) | van Borren et al.[18] |
|---|---|---|---|---|
| | | HTRF cAMP (Cisbio Bioassays) | Competition assay between endogenously produced cAMP and acceptor-labeled cAMP for binding on donor-labeled cAMP antibody (TR–FRET signal decreases as endogenous cAMP is produced) | Thomsen et al.[17] Comparison: Gabriel et al.[23] |
| | | QRET-based cAMP assay (Perkin Elmer) | Competition assay between labeled and endogenously produced cAMP: soluble quenchers reduce the signal of unbound europium-labeled cAMP in solution, whereas the antibody-bound fraction is fluorescent | Martikkala et al.[24] |
| G protein–dependent/independent pathways | ERK1/2 phosphorylation | Cellul'erk (Cisbio Bioassays) | Sandwich immunoassay: detection of phosphorylated ERK by the production of TR–FRET signal between a donor-labeled anti-ERK antibody and an acceptor-labeled anti-phospho-ERK antibody | Thomsen et al.[17] |
| | Ras activation | Not commercially available yet | Assay based on homogeneous quenching resonance energy transfer technique (QRET): terbium-labeled GTP binding to small GTPase protein H-Ras protects the signal of the label from quenching, whereas the signal of the nonbound fraction of Tb–GTP is quenched by a soluble quencher | Martikkala et al.[25] |

*Continued*

**Table 7.1** Summary of TR-FRET-compatible assays for GPCR—cont'd

| GPCR signaling cascade and downstream events | Biological measurement | Available TR-FRET assay (company) | Basis | Illustrative paper |
|---|---|---|---|---|
| G protein–independent pathways | Akt phosphorylation | Phospho–Akt (Ser473) (Cisbio Bioassays) Phospho–Akt (Thr308) (Cisbio Bioassays) | Sandwich immunoassay: detection of phosphorylated Akt by production of TR–FRET signal between an acceptor–labeled anti-Akt antibody and a donor–labeled anti–phospho–Akt antibody | Degorce et al.[12] |
| | Specific metabolites | Examples: Insulin, cortisol, cGMP, ADP, etc. (Cisbio Bioassays) | Competition assay | Degorce et al.[12] |
| Dimerization | | GPCR Tag-lite (Cisbio Bioassay) | TR–FRET signal detection between a donor–labeled SNAP-tagged GPCR A and an acceptor–labeled SNAP-, CLIP-, or HALO-tagged GPCR B | Maurel et al.[26] Doumazane et al.[27] |

cAMP and donor-labeled anti-cAMP antibody, with the difference that, in the HTRF® assay, the donor is a cryptate of lanthanide, whereas in the LANCE® assay, it is a lanthanide chelate. The first one has, for example, been validated both for Gs-coupled receptors such as the β2-adrenergic receptor or the histamine H2 receptor and for Gi/o-coupled receptors such as the histamine H3 receptor.[28,29] Both assays have been reported as more sensitive technologies compared to reporter gene techniques, for example, with concentrations below 10 fmol cAMP/well being quoted.[30] Moreover, a study by Jean and colleagues recently showed that the sensitivity of these methodologies could enable the detection of cAMP in the native cerebral tissues of rats.[31] Knowing how strongly cell phenotypes can influence many aspects of GPCR pharmacology, studying unmodified receptors in their native environment would be the optimal assay to screen for, and methodologies that enable this have to be developed. Another approach, developed by Martikkala and colleagues and also based on a competition assay, uses the quenching resonance energy transfer technique (QRET) instead of TR-FRET to measure changes in cAMP levels in HEK293 cells overexpressing either β2-adrenergic or δ-opioid receptors (DOP).[24] In this assay, soluble quenchers reduce the signal of unbound europium-labeled cAMP in solution, whereas the antibody-bound fraction is fluorescent. It is an HTS-compatible method and has the advantage of using only a single-label probe.

It has to be emphasized that screening Gs-coupled receptors is generally straightforward, whereas screening Gi/o-coupled receptors in cAMP assays can be considerably more difficult. To maximize the inhibition signal (reflecting the inhibition of cAMP production by adenylyl cyclase), it is often necessary to stimulate adenylyl cyclase with forskolin (a direct activator of adenylyl cyclase) or a Gs-coupled GPCR expressed in the same cells, and this has to be carefully titrated during optimization of the assay so as not to miss out on nonpotent and partial agonists, for example.

### 3.2.4 Screening for Gq/11 signaling pathway

Gq/11 family of G proteins modulates phospholipase C (PLC), which promotes the production of inositol 1,4,5-trisphosphate (IP3) that interacts with the inositol-3-phosphate or ryanodine receptors (ligand-gated $Ca^{2+}$ channels) located in the endoplasmic and/or sarcoplasmic reticulum, resulting in the transient release of calcium ions into the cytosol. One of the most widely used methods to screen for GPCR ligand is to detect this calcium ion release, usually with fluorescent probes changing their emission properties when bound to $Ca^{2+}$ (FLIPR calcium assay from Molecular Device is probably the most widely used).[32] Cisbio Bioassays recently developed an IP-One HTRF competitive assay based on the detection of the intracellular accumulation of a metabolite of IP3, D-myo-inositol monophosphate (IP1), thanks to lithium chloride (LiCl), a known inhibitor of inositol monophosphatase (IMPase) which degrades IP1 into inositol.[21] The IP-One HTRF assay takes advantage of measuring the final step in the IP cascade, allowing it to accumulate in the cell and to be measured without requiring a kinetic readout. An advantage of this assay for HTS over others is that it can be used with endogenously or heterologously expressed GPCRs, in either adherent or suspension cells, in the 1536-well plate format, to quantify the activity of agonists, antagonists, and inverse agonists. Of note, this last category of ligands cannot be seen via a calcium assay, for example, as increases in intracellular basal $Ca^{2+}$ concentration are not observed in cells expressing constitutively active Gq-coupled receptors.

### 3.2.5 Screening for G12/13 signaling pathway

The G12/13 family of G proteins activates Rho, a family of small GTPase, causing the downstream activation of the c-Jun N-terminal kinase (JNK) pathway. To our knowledge, no generic TR-FRET-based assay is actually available to screen for metabolites of this G12/13 pathway in an HTS format.

## 3.3. Later signaling events

### 3.3.1 Screening for βγ subunits activated pathways

Upon agonist GPCR activation, in parallel to the G protein α subunit-activated cascades, the βγ subunits of the G protein have been demonstrated to also modulate several effectors such as phospholipase or ion channels, but, again to our knowledge, no HTS-compatible TR-FRET-based assay is available to detect such events. In the case of a specific pathway, some metabolites can be titrated with an HTS-compatible TR-FRET assay, such as insulin, cortisol, or cyclic guanosine monophosphate (cGMP),[12] but this has to be implemented for a specific characterized GPCR target. Another possibility is to measure the GPCR-induced phosphorylation of some cellular substrates that occurs during the activation of a particular GPCR. These phosphorylation events could be analyzed in a sandwich format assay using a specific labeled antibody directed against the phosphorylated protein. However, such an assay is target specific and has to be developed for each GPCR. In this respect, for example, a Ras activation assay using time-resolved fluorescence was recently developed.[25] Ras is the prototypical member of the Ras superfamily of proteins, a guanosine-nucleotide-binding protein (small G protein also displaying a GTPase activity) closely related in structure to the Gα subunit of heterotrimeric G proteins (large GTPases). These proteins have been shown to be strongly involved in cancer, with mutated Ras stabilized into a constitutively active conformation, and they have also been shown to be involved in some GPCR signaling cascades. Similar to the assay previously described to measure binding of Eu-labeled GTP analog to a large G protein, in this assay the authors have used a terbium-labeled GTP (Tb-GTP).[25] However, no washing or filtration steps are required, as the assay is based on the homogeneous QRET. The Tb-GTP binding to Ras protects the fluorescence signal of terbium from quenching by a quencher in solution, whereas the signal of the non-bound fraction of Tb-GTP is quenched. Again, the time-resolved measurement of the fluorescence signal increases the sensitivity of the assay allowing HTS application.

### 3.3.2 Screening for G protein-independent pathways

After agonist stimulation, most GPCRs are phosphorylated by specific GPCR kinases, and the recruitment of β-arrestins to the phosphorylated GPCRs eventually terminates G protein signaling and leads to a coordinated process of receptor desensitization, inactivation, internalization, and, finally, either recycling to the membrane or lysosomal degradation. β-Arrestin proteins are also known to promote G protein-independent signaling of GPCRs, including those involving mitogen-activated protein kinases (MAPKs) such as ERK1/2, receptor and nonreceptor tyrosine kinases, phosphatidylinositol 3-kinases, and probably many others. A sandwich format TR-FRET assay designed to detect and quantify phosphorylation of ERK1/2 was recently marketed by Cisbio Bioassays.[17] It is based on the use of two labeled antibodies, one directed against ERK bound to the donor and the other one recognizing specifically the phosphorylated form of the protein and bound to the acceptor. It should be emphasized that MAPK signaling cascade events are temporally and spatially very well controlled in the cells; thus some optimization should be done before using this kit, as a single endpoint measurement of pERK could miss important events occurring in the cells. Of interest, a proof-of-concept study to measure recruitment of intracellular proteins to a GPCR has been performed.[33] This study relies on the insertion of small tags (HA, Flag, or Myc) at the carboxy-terminal tail of the receptor and in the protein of interest. After mild solubilization of the membrane allowing the antibodies labeled with HTRF®-compatible fluorophores to enter the cells, the authors could detect β-arrestin and G protein recruitment to PAR1 upon activation. If this format of assay is not compatible with HTS, it indicates nevertheless that HTRF® is perfectly compatible with the detection of intracellular events in living cells.

## 3.4. The future of HTS generic assays

In the past two decades, the complexity of GPCR pharmacology has dramatically increased with new concepts challenging the classic scheme of one receptor activating one specific G protein. Receptor homo- and/or heterodimerization, allosterism, biased signaling, agonist-dependent trafficking, and G protein-independent signaling are all relatively new concepts in the GPCR field, which have to be taken into consideration when looking for new ligands.[7,8] The issue would now be more a question of which pathway is the therapeutically pertinent one to look at when talking about a specific disease and, thus, which assay reflects the best what occurs in this pathway under physiological conditions. Biased signaling, for example, is a phenomenon in which certain agonists display better efficacies in activating one pathway over others. Thus, another important aspect to take into account is whether we can limit adverse effects by designing drugs activating only one specific pathway and not all the pathways the GPCR can couple to. For example, Lefkowitz and colleagues have demonstrated that the angiotensin II type 1 receptor ($AT_1$), the target of angiotensin-receptor blockers for the treatment of hypertension, can activate both Gq/11-type protein pathway and β-arrestin-induced pathway upon activation by the endogenous ligand angiotensin II.[34] On the contrary, biased ligands such as Sar1, Ile4, and Ile8-AngII (SII) stimulate β-arrestin signals, such as ERK1/2 activation, in the absence of detectable G protein agonism.[35] In cardiomyocytes, for example, in the absence of G protein activation, SII could stimulate proliferation, but not hypertrophy observed when the Gq/11 protein pathway is also activated.[36]

Thus, if a single functional assay capturing only one signaling pathway is selected for screening compound libraries, potentially valuable compounds could be missed if the compound does display biased activity. Therefore, multiplexing assays capable of detecting several signaling pathways may be the future methods to use in HTS. In TR-FRET-based assays, such a multiplexing may be possible (at least to detect simultaneously two secondary messengers). Indeed, terbium used as a donor can be associated with different acceptors, a fluorescein and a d2 near-infrared dye, for example, because they have well-separated emission wavelengths, thus allowing the detection of two metabolites from two distinct pathways, at the same time, in the same well, and upon the same agonist stimulation.[37]

Generic screening methods are invaluable strategies for HTS of large libraries of compounds and can be applied to several GPCR targets to determine the specificity and the signaling repertoire of each compound. However, it is not always possible to measure all the downstream events with HTS-compatible assays, and some hits could be missed. Therefore, complementary methods are needed. These methods are specific for the chosen target and range from binding assays to internalization assays and also to the complex analysis of the compound effects on dimeric and oligomeric GPCR entities.[38] Some target-specific TR-FRET-based assays are under development, and some are under validation for future application in HTS campaigns. Together with the generic methods, the target-specific methods contribute to the strategy to identify putative therapeutic molecules with reduced side effects.

## 4. SCREENING WITH TARGET-SPECIFIC METHODS

Two different target-specific TR-FRET methods to study GPCRs have been reported so far. The first one, a ligand-binding assay, has been validated for its use in HTS campaigns. The second one is based on the capability of GPCRs to form oligomers and the capability of the molecule to act on these complexes. This second method is still under development but would be very useful in the field of drug discovery because oligomers possess unique properties.

### 4.1. Drugs targeting a protein: Ligand-binding assay

As previously mentioned, screening for an active molecule for a given GPCR using functional assays may be rather difficult. Analyzing a single signaling pathway may lead to false negatives and, thus far, some signaling pathways remain difficult to screen in HTS format. Therefore, the identification of molecules that bind to a specific target, in a binding-like format assay, remains relevant, as they could potentially modulate the receptor signaling. There exist nowadays nonradioactive methods based on TR-FRET to perform ligand binding.

#### 4.1.1 TR-FRET-compatible ligands for binding assay

Like in classical radioactive binding assay, these are based on the displacement of a tracer compound that binds specifically and with a high affinity to the target. The tracer is often obtained by the derivatization of a known ligand with a fluorescent moiety. Depending on the receptor, this ligand can be a small molecule, a peptide, a small protein, or an antibody. The smaller the ligand, the more difficult its derivatization. Indeed, a major challenge is to keep high affinity after the fluorescent labeling of the ligand. Though small proteins and antibodies offer several possibilities to attach the fluorophore, as only part of the molecule interacts directly with the receptor, the same is not true for small ligands and peptides for which most of the molecule interacts directly with the receptor upon binding. In the latter case, a good knowledge of the binding pocket and of the pharmacophore is often necessary to synthesize high-affinity fluorescent ligands. This helps in the identification of the part of the molecule to which a linker can be added and further derivatized with a fluorophore.[39,40] As an example, Baker and colleagues have performed a systematic analysis of the derivatization of two adenosine A1 receptor ligands (one agonist and one antagonist) by different fluorophores and various linkers, and both appeared to affect the pharmacological and fluorescent properties.[41] Some of these ligands appeared to be powerful tools to monitor binding events in living cells using fluorescent correlation spectroscopy.[42] Over the past years, efforts have been made by several companies and academic laboratories to develop such ligands. By the end of 2011, fluorescent ligands targeting more than 70 GPCRs were commercially available from Cisbio Bioassays, Cellaura, Perkin Elmer, and Abcam.

While these fluorescent ligands can be used directly to perform binding assays (e.g., see Refs. 42,43 or DELFIA® technology by Perkin Elmer), these still require washing steps to separate the bound from the unbound fractions of the ligand. In the DELFIA® assays, for example, after incubating the cells with a europium-labeled ligand and the compounds to be screened until equilibrium is reached, the unbound fraction is washed away and the leftover fluorescence is read in a time-resolved manner after addition of an enhancing solution. These washing steps, which often prevent the miniaturization and automation of an assay, are not compatible with HTS, and solutions had to be found to circumvent this. HTRF represents a good alternative to the classic binding-assay format. Indeed, as already mentioned, the specific properties of the HTRF-compatible fluorophores allow highly specific signal detection in a homogeneous solution and prevent the detection of most of the background fluorescence, especially that coming from unspecific binding. Indeed, the HTRF binding assays are based on the measurement of the FRET signal between a fluorophore linked to the receptor and the fluorescent ligand (Fig. 7.3).

Upon addition of a competitive compound, the fluorescent ligand is displaced, and hence the FRET signal decreases. Accordingly, the ligand-binding HTRF® assay requires the labeling of the target GPCR specifically at the cell surface.

### 4.1.2 TR-FRET binding assays using GPCRs targeting antibodies

GPCR labeling with HTRF®-compatible fluorophores can be achieved using several strategies. The first strategy is to use a fluorescent antibody that recognizes the native receptor or a tag sequence inserted at the N-terminus of the receptor. Antibodies have the advantage of being easily labeled with any fluorophore. Such antibodies were successfully used to perform HTRF binding assays in cell lines expressing an HA-tagged vasopressin receptor V1a[44] or an HA-tagged complement 5A receptor C5AR,[45] for example. In both studies, anti-HA antibodies were labeled with an HTRF-compatible acceptor fluorophore, Alexa647 or AlexaFluor488, and applied on cells expressing the receptor of interest. The fluorescent ligands, either a europium cryptate derivative of a V1a antagonist or a terbium chelate-labeled C5R purified protein, and the compounds to be tested were then applied. Once the binding equilibrium is reached, the HTRF signal is directly recorded without any washing steps. The data reported support the robustness of the assay (compatible with 384-well plates) and its accuracy is comparable to that of radioactive assays.[44,45] Of interest, if antibodies directed against the native protein are available, the HTRF binding assay could be performed directly from native tissues, which would provide information regarding the pharmacological properties of the screened compounds in the native environment of the target (Fig. 7.3).

### 4.1.3 TR-FRET binding assays using innovative strategies to label GPCRs

A second strategy to label the target GPCR is to use self-labeling tags and, more efficiently, self-labeling proteins (for review, see Ref. 46). In short, these self-labeling techniques consist in fusing a suicide enzyme that will react with a fluorescent substrate leading to the formation of a covalent bond between the fluorescent moiety and the tag. These suicide enzymes include the SNAP-tag®,[47] the CLIP-tag®,[48] and the Halotag® technologies.[49] The ACP-tag technology is an additional self-labeling technique where the covalent transfer of fluorescent substituents from derivatized coenzyme A to the ACP-tagged fusion proteins is catalyzed by AcpSynthase.[50] While all these techniques could be used in theory to develop HTS-compatible HTRF binding assay, only assays based on the SNAP-tag®, CLIP-tag®, and Halotag® technologies have been optimized so far under the name Tag-lite® by Cisbio Bioassays. In the next paragraph, we will explain further the assay based on the SNAP-tag®, as the principle is the same for the different tags and only the nature of the substrate differs between the three labeling methods.

The SNAP-tag® is a mutant of a DNA repair protein $O^6$-alkylguanine DNA alkyltransferase which was optimized to react preferentially to benzylguanine derivatives. This reaction leads to irreversible covalent binding of the probe to the SNAP-tag® and release of guanine.[47] In the Tag-lite® binding assay,[19] the SNAP-tag® is fused to the N-terminus of a given GPCR and the benzylguanine derivatized with a TR-FRET-compatible fluorophore. This fusion is expressed in cell lines and labeled either with SNAP-Lumi4Tb substrate, an HTRF-compatible FRET donor, or with a SNAP-Red or SNAP-Green FRET acceptor. The cells can then be used immediately or stored in liquid nitrogen for later use. To proceed with the binding assay, the labeled cell suspension is dispensed in the assay plate together with the compound library to be screened, followed by the fluorescent ligand. Once equilibrium is reached, the TR-FRET signal is measured directly in an HTRF-compatible plate reader (Fig. 7.3). Note that the assay is also applicable on adherent cells and both with transiently transfected and stable cell lines.

Tag-lite® binding experiments have been performed on several GPCRs including class-A β2 adrenergic receptor, dopamine D2 receptor, opioid receptors, the class-B receptor VPAC1 for the vasoactive intestinal peptide, and the class-C GABA$_B$ receptor.[19,51] When compared to values from the literature, the $K_i$ values measured with the Tag-lite® assays were in the same range and the ranking order of the compounds was similar. In addition, the signal-to-noise ratio was found to be very high, and the specificity of the signal good enough for analysis in the 384-well plate format. A deeper analysis has been performed on the ghrelin receptor GSH-R1a.[51] Using the Snap-tagged receptor and a red fluorescent ghrelin ligand, 14 compounds were tested in a Tag-lite® binding experiment. The same compounds, including agonists, antagonists, and inverse agonists, were also tested in parallel in a radioactive binding assay. This study showed a good correlation for the compound potencies obtained in the two assays, and Tag-lite® binding assay was performed both on adherent cells and using cell suspension. In addition, Tag-lite® assay has been scaled down to the 384-well plate format, and the high signal-to-noise ratio makes it compatible with 1536-well plates.

## 4.2. Drugs targeting a protein complex

### 4.2.1 Oligomers as new therapeutic targets

Over the past two decades, the simple model for GPCR activation—one GPCR protomer activating one single G protein heterotrimer—has been challenged by an increasing amount of data showing that GPCR oligomerization, either homo- or heteromerization, is a common phenomenon.[10] Oligomerization has been shown to influence many different properties of GPCRs as illustrated below, and thus receptor dimers or oligomers, and especially heteromers, often possess unique characteristics totally different from those of their individual composing protomers and/or they promote regulation phenomena of one subunit by the other (e.g., allosteric modulation between units), making them unique and innovative targets to analyze. Thus, these oligomers represent innovative and challenging targets for drug discovery, and assays to identify them and to study their properties will certainly be developed in the near future.

#### 4.2.1.1 Effect of oligomerization on GPCR cell-surface expression

First, oligomerization has been shown to influence GPCR early biosynthesis in the endoplasmic reticulum and their maturation. For example, the $GABA_B$ receptor, a member of the class-C GPCR, was one of the first obligatory heteromers described in the literature.[52] It is composed of two subunits, $GABA_{B1}$ and $GABA_{B2}$, the first one being responsible for GABA binding and the second for G protein coupling.[53,54] $GABA_{B2}$ synthesis is necessary for the functional heterodimer to reach the cell surface. Indeed, the interaction between the two subunits is necessary to mask the retention motif contained in $GABA_{B1}$ C-terminus.[55,56] Other examples of obligatory dimers are found in the class-C GPCR, where the sweet and umami tastes are generated by functional responses from obligate heterodimeric pairs of T1 taste receptors,[57] and where the mGluRs, responsible for modulating fast glutamatergic transmission in the brain, are strict and obligatory homodimers linked by a disulfide bond.[26,58] GPCR oligomerization also influences other regulatory processes such as internalization, trafficking, and recycling. The CXCR4 and CCR5 receptors, for example, two members of the class-A GPCR and coreceptors for HIV entry in human cells, were shown to heteromerize, modulating T-cell function and generating specific internalization and degradation properties for the complex as compared to those of each protomer.[59]

## 4.2.1.2 Effect of oligomerization on GPCR activation and pharmacology

Besides modulating GPCR cell-surface expression, oligomerization is also a key factor for modifying the binding properties of the ligands and G protein coupling properties and thus the activated signaling pathways, opening a new level of complexity in the GPCR pharmacological properties.

First, GPCR heteromerization can generate specific binding properties either by modulating ligand binding through allostery or by promoting new binding pockets. For instance, CCR5 was shown to associate with another member of the chemokine receptor family, CCR2, resulting in a heterodimer capable of binding only a single chemokine molecule, and this was interpreted as negative binding cooperativity.[60] In the opioid receptor family, 6-guanidinonaltrindole (6-GNTI), an opioid agonist ligand used as an analgesic, was shown to preferentially activate heterodimers of κ-opioid (KOP) and δ opioid (DOP) receptors over the homomeric counterparts.[61] This example illustrates the possibility of finding ligands selective for a specific heterodimer. It means that, if the organization of heterodimers is tissue specific or even neuron specific in the central nervous system, considering that not all cells express the same repertoire of GPCRs, it should be possible to use heterodimer-selective ligands to target only the appropriate heterodimer in the appropriate cells, thus reducing adverse effects from activation of homomeric counterparts from the same cells or from other cells.

Second, heteromerization can induce a shift in the G protein family activated by the complex compared to individual components. Very strong evidence for a major role of heterodimers in key functional events in the cells is found in the dopaminergic receptor family. D1R and D2R have been shown to form heterodimers that could play a role in patients suffering from major depression, as their expression is increased in post-mortem brain of such patients.[62] Their heteromerization leads to a switch from a Gαs/olf (D1-like classical response) or Gαi (D2-like response) to a Gαq/11-mediated response,[63] and the compound SKF83959, previously known as a D1R agonist, is able to bind the D1–D2 heteromer and activate this specific Gq pathway in the brain. Moreover, D2R has also been shown to form heteromers with D5R, resulting in modified pharmacological properties when compared to activation of each protomer on its own.[64]

Third, GPCR heteromerization can induce a shift from a G protein-dependent pathway to a G protein-independent pathway. The heteromerization of the chemokine receptors CXCR4 and CXCR7 leads to a strong biased agonism, as it promotes the constitutive recruitment of β-arrestin and simultaneous impairment of the usual CXCR4 Gi-mediated signaling.[65] In the opioid receptor family, under normal physiological conditions, the μ-opioid receptor (MOP) is found mostly as homo-oligomers in the brain, and its trafficking is regulated by oligomerization.[66] Upon acute morphine stimulation, the G protein-mediated signaling pathway is activated and leads to analgesic effects. However, upon chronic morphine administration, the abundance of MOP-DOP heteromers is increased in key areas of the central nervous system which are implicated in pain processing, and they could impair the analgesic effects of morphine by recruiting β-arrestin and thus altering MOP homomer signaling.[67]

### 4.2.1.3 GPCR oligomers: New targets for HTS campaigns

Despite all these striking examples of the importance of oligomerization in GPCR functioning, and despite the fact that dimerization is a common phenomenon for other membrane proteins, only little thought has been given for a long time to the importance of quaternary structure in HTS campaigns and the priority was to design highly specific and selective ligands on one target protein. However, this was due to the lack of HTS-compatible methods to take into account the role of oligomerization in modulating pharmacological responses. Nowadays, new methods, and especially some based on TR-FRET, are emerging to allow first the identification of GPCR homo- or heteromers and then to study whether the association modifies the pharmacological and functional properties of the oligomer. Thus, it may be possible to screen compounds that will specifically target one GPCR heteromer and maybe achieve specific therapeutic potential with reduced adverse effects. So far, only a few homo- or heteromers examples have been validated for their potential as new therapeutic targets, and more will probably be needed before they become common targets for HTS screening. However, the development of specific HTS assays dedicated to study these complexes will help to stimulate screening campaigns in this direction.

### 4.2.2 TR-FRET screening for homo-/heteromer formation

The critical point before studying a peculiar GPCR complex and its functional properties is, of course, to be able to identify it, ideally in a native environment. And when such a homo- or heteromer is identified, all the generic methods mentioned above can be used to investigate its particular properties. However, another important point to consider is whether the modulation of this complex assembly can also modulate its properties. If this is the case, one can imagine screening for ligands favoring or disturbing the interactions. For example, drugs preventing the dissociation of transthyretin

(TTR) tetramers have been developed by Alhamadsheh and colleagues.[68] TTR is a tetrameric protein that transports thyroxine and retinol in the blood, but whose dissociation leads to amyloid cardiomyopathy. Using a fluorescence polarization-based high-throughput screen to detect small molecules that bind to the thyroxine-binding pocket of TTR under physiological conditions, the authors have identified molecules that stabilize the dimer–dimer interface and increase the activation energy for tetramer dissociation, thus rescuing cardiomyocytes from TTR proteotoxicity without causing cytotoxicity. TR-FRET-based technologies are under development to investigate these aspects of GPCR oligomerization.

### 4.2.2.1 TR-FRET detection of oligomers with tag insertion

Besides classical biochemical methods to identify oligomers or larger complexes, such as coimmunoprecipitation or functional complementation, which are hardly compatible with HTS and can induce biases in the results by promoting association, resonance energy transfer approaches offer many advantages to screen for GPCR-specific interactions. The Tag-lite® GPCR dimer assay from Cisbio combines many of these advantages, especially the possibilities to work on protein interactions occurring specifically at the cell surface, on living cells, and in a 96- or 384-well plate format. In this assay, studied GPCR subunits are labeled with cell-impermeable fluorophores compatible with TR-FRET, typically terbium cryptate as a donor on one subunit and a green or a red fluorescent molecule as an acceptor on the potential partner subunit (Fig. 7.3). Because energy transfer occurs only when the two labeled subunits are at a distance of less than 100 Å, which is about twice the diameter of a GPCR transmembrane domain, only two receptors in direct physical interaction are compatible with FRET emission, and especially with a high signal-to-noise ratio when measured in a time-resolved manner. This assay has been validated for both class-A and class-C GPCRs. For example, Maurel and colleagues used GPCR tagged at their N-terminus with a SNAP-tag, and showed that, whereas mGluRs assemble into strict dimers, the GABA$_B$ receptors spontaneously form dimers of heterodimers.[26] Alvarez-Curto and colleagues also studied cell-surface M3 muscarinic acetylcholine receptors fused with a SNAP-tag, and together with other techniques, TR-FRET allowed them to show that this receptor exist as dimeric or oligomeric complexes at the surface of cells and that this organization is regulated by ligand binding.[69]

This technology relies, of course, on the possibility to specifically label the GPCR target with fluorophores. As previously mentioned, several methods are now available, which include antibodies directed against the target itself or against a tag fused at the N-terminus of the GPCR target, self-labeling tags or proteins, or tags labeled by an added enzyme (for review, see Ref. 46). The use of orthogonal labeling methods, that is, the combination of two different tags such as SNAP- and CLIP-tags, or SNAP- and HALO-tags, is an interesting approach to specifically look at GPCR quaternary structures formed by the association of two different subunits, each carrying a specific tag labeled with one fluorophore (Fig. 7.3). For example, the SNAP- and CLIP-tags strategy was used to demonstrate that members of the mGluR family can form heteromers and that this association is limited to strict heterodimers, not dimers or homodimers.[27]

### 4.2.2.2 TR-FRET detection of oligomers of unmodified receptors

Alternatively, the receptor labeling with an antibody directed against the native protein is also a promising strategy. The main advantage is that no modification of the target GPCR is necessary, thus avoiding the impact of molecular modification of the receptor on its functionality and allowing association studies on primary cultures or native tissues (Fig. 7.3). For example, with the use of antibodies coupled to TR-FRET-compatible fluorophores, BILF1, an orphan GPCR involved in the infection of B lymphocytes by the Epstein-Barr virus, was shown to heterodimerize with various chemokine receptors, which may lead to an altered response of B lymphocytes to chemokines after infection.[70] However, this labeling strategy has also some drawbacks. Indeed, the use of bivalent antibodies can favor and constrain dimers or larger oligomers of GPCR, modifying the natural tendency of two GPCRs to associate. As an alternative method to circumvent this problem, the use of monovalent antibody fragments, typically just a Fab fragment, targeting GPCRs with similarly high affinity and specificity as classic full-length antibodies, would allow avoiding this bivalence-induced bias. Monoclonal Fab fragments recognizing several proteins are already commercially available and are coupled to fluorophores or enzyme such as horseradish peroxidase for use in ELISA or Western blot (see, e.g., the Web site of M-fold, a company that develops Fab on demand). As an example of a Fab targeting a GPCR, a monomeric Fab fragment was developed against the $\beta$2-adrenoceptor and exhibits peculiar pharmacological effects: it showed antagonist-like properties, whereas the same bivalent antibody exhibited agonist-like properties.[71]

Another labeling strategy to detect GPCR oligomers is to use fluorescent ligands, developed, for example, for the binding assays described previously. They can also help identify dimers or heteromers directly on native tissues, as no modification of the receptor is required. Albizu and colleagues used agonists and antagonists to the vasopressin and oxytocin receptors coupled to fluorophores compatible with TR-FRET and succeeded in demonstrating the presence of oxytocin receptor dimers or oligomers in native mammary glands.[72]

### 4.2.2.3 New perspectives for oligomerization TR-FRET assays

TR-FRET tools can also be used to study oligomer association/dissociation processes during internalization and recycling, for example, or other mechanism occurring inside the cells. For example, Rajapakse and colleagues were able, with TR-FRET microscopy, to study inside the cells protein–protein interactions that were hardly detectable with traditional microscopy imaging, with improved speed and sensitivity.[73] Unfortunately, contrary to all the above-mentioned technologies, this one is not yet transposable to HTS.

Of interest, screening for heterodimerization between distinct GPCRs appears to be promising to understand the role of orphan receptors. Note that roughly a third of the about 400 nonodorant GPCRs are still orphans. For example, the orphan receptor GPR50 was shown to heterodimerize constitutively and specifically with MT1 and MT2 melatonin receptors. While the association with MT2 had no effect on its function, GPR50 abolished agonist binding and G protein coupling to the MT1 protomer engaged in the heterodimer.[74]

With the increasing number of GPCRs shown to form heteromers in heterologous systems, the need for methods to attest their existence in native tissues has never been greater. Though microscopy techniques have been used to that aim, it appears now that TR-FRET methods could be an alternative with the advantage to potentially permit in the future the use of an HTS-compatible plate reader on native systems to study GPCR complexes (Fig. 7.3).

### 4.2.3 TR-FRET screening for specific pharmacological properties of oligomers

Once a specific heteromer or oligomer has been identified, all the assays presented previously and many more can be used to determine and characterize its potential, specific pharmacological properties. From a drug development point of view, the finding of tissues or pathologies in which

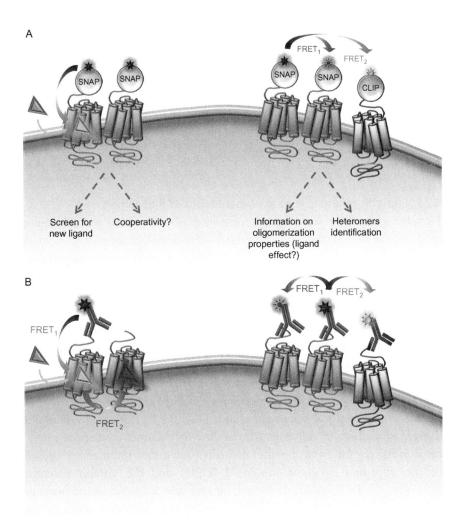

**Figure 7.3** GPCR target-specific assays compatible with TR-FRET. (A) Assays based on modified receptors. Binding experiments allow first screening for new ligands by monitoring the FRET signal decrease between a SNAP-tag fused GPCR labeled with a donor and a reference ligand labeled with an acceptor, for example. In the case of a GPCR dimer or oligomer, information about cooperativity between protomers, positive or negative, can also be obtained from this assay. In oligomerization assay, GPCR protomers are fused to a self-labeling tag and labeled with a mix of donor and acceptor substrates. If a specific FRET signal is detected, information about oligomerization properties can be obtained, and the effect of ligands on this oligomer can be investigated. In the case of heteromers, GPCR protomers are fused to different self-labeling tags and labeled specifically with a donor on one tag and an acceptor on the other. Heteromerization properties can again be studied if a specific FRET signal is detected between the two receptors. (B) Assays compatible with native systems. Most of the information described above can also be obtained in native systems using labeled antibodies recognizing the GPCR protomer or two fluorescent ligands binding to the protomers instead of self-labeling tags. The use of fluorescent ligands gives also access to information on the cooperativity within the oligomer. (See Color Insert.)

heterodimers or oligomers with peculiar pharmacological properties are specifically formed would constitute a unique target of great interest. So far, only a few examples have been reported in the literature to illustrate this concept, but the examples described below should stimulate the use of new assays to screen for molecules exhibiting specific pharmacological properties on a dimer or oligomer.

### 4.2.3.1 Specific pharmacological properties of GPCR oligomers

First, using the fluorescent ligands mentioned above, Albizu and colleagues were able to demonstrate an asymmetric relationship of the two protomers in the oxytocin receptor dimers or oligomers in native mammary glands (cooperative binding), confirming results obtained in cell lines.[72]

Second, adenosine A1 and A2A receptor were shown to heteromerize and their heteromerization allows adenosine to exert a fine-tuning modulation of glutamatergic transmission. Using radioligand binding, the authors showed the unique pharmacological properties for this heteromer, and they also showed that it is a unique target for caffeine and that chronic treatment leads to modification in the function of this heteromer that could underlie the strong tolerance to the psychomotor effects of caffeine.[75] A third example was reported in a study using confocal microscopy and TR-FRET. Kern and colleagues showed that subsets of native neurons in the hypothalamus (a structure implicated in the regulation of appetite) naturally present heterodimers composed of the dopamine receptor subtype-2 (D2R) and the ghrelin receptor (GHSR1a).[76] To that aim, they used, on one side, a ghrelin ligand fused to a red acceptor fluorophore and, on the other side, a specific antibody for D2R and a secondary antibody labeled with a TR-FRET-compatible donor. This association allosterically modifies the coupling properties of the dimer as compared to each receptor on its own, as well as desensitization mechanisms. Even without ghrelin activation, GHSR1a could be an allosteric modulator of dopamine-D2R signaling in response to endogenous dopamine, and thus it would allow a neuronal selective fine tuning of dopamine/D2R signaling, as neurons expressing D2R alone will be unaffected. This heteromerization has important therapeutic implications because extensive resources have been invested in developing GHSR1a antagonists as antiobesity agents, and polymorphisms in D2R impair D2R signaling and are associated with obesity in humans.[77] In this physiological context, the authors predicted that GHSR1a antagonists might exacerbate rather than prevent obesity. This new mechanism allowing modulation of specific neurons by GHSR1a-D2R heterodimer formation offers exciting opportunities for designing the next generation of drugs with improved side-effect profile for treating psychiatric disorders associated with dysregulation of dopamine signaling.

### 4.2.3.2 New strategies to specifically target GPCR heteromers in HTS TR-FRET assays

To be able to use the TR-FRET assays described above on a specific heterodimer, it will be first necessary to develop a method to specifically label the heterodimer. Of course, all the tags described above can be fused to the receptors to study the complex properties *in vitro*. But even more interesting, specific antibodies or ligand recognizing specifically the heterodimer can be developed and later labeled with TR-FRET-compatible probes. Indeed, Gupta and colleagues generated antibodies that selectively recognize the MOR-DOR heteromer and blocked the *in vitro* signaling, and could then analyze *in vivo* their increased abundance after chronic morphine treatment in pain-processing areas.[67]

Targeting oligomers represents one of the new challenges in drug discovery. Based on the current knowledge, this strategy could limit the adverse effects of the molecule when administered to animal models and later to patients. Indeed, since the oligomers seem to be tissue specific and opposite to what is observed for a receptor-specific drug, an oligomer-specific drug would then target the receptor present only in the specific tissue.

## 5. PERSPECTIVES

The TR-FRET-derived solutions available so far for the analysis of GPCR machinery cover the main events linked to their activation: ligand binding, G protein activation and other signaling cascades, oligomerization, and specific pharmacology of the oligomers. However, one could imagine extending further the offer to events such as conformational changes or internalization.

## 5.1. Detecting GPCR conformational changes upon binding

FRET- or BRET-based assays have been developed to monitor the conformational changes of GPCRs (for review, see Ref. 78). While these assays are often developed using microscopy detection, one could imagine developing a counterpart based on time-resolved FRET by using the lanthanide as a fluorescent donor and a compatible fluorescent acceptor, thus compatible with plate readers. Such implementation would be useful to detect a compound within a library that induces conformational changes of a given GPCR. This strategy has the advantage of detecting all molecules acting on a GPCR without requiring any knowledge on the signaling pathway activated by the molecules.

Eventually, a further development would be to link a specific conformational change to a specific property of the molecules (activator, inhibitor, modulator, etc.). Such a perspective could require the need to implement more local labeling and thus smaller tags, and a promising solution could be found in the non-natural amino acids.[79,80] So far, only nonpermeant TR-FRET substrates have been used for GPCR labeling, which did not permit the direct kinetic detection of all the intracellular events such as G protein and β-arrestin recruitment on intact cells as opposed to what is done with classic FRET or BRET assays. As already mentioned, a proof-of-concept study using antibodies reported intracellular events in real time by TR-FRET but hardly compatible with HTS.[33] However, cell-permeant substrates for SNAP and CLIP are now available (yet not derivatized with TR-FRET donors) and would allow the labeling of intracellular GPCR partners and even of the intracellular part of the receptor itself, which would permit the conception of new HTS-compatible TR-FRET-based assays.

## 5.2. Detecting GPCR internalization

A second axis for further development of tests would be to implement an internalization assay. Indeed, after activation, most GPCRs are internalized, and that property has already been used to screen for active compounds for a given receptor. At present, some solutions based on internalization are commercially available, but they rely on imaging and semiautomated analysis of the images and therefore not suitable for HTS campaigns (Innoprot). A recent article reports the use of SNAP and CLIP-tagged receptors to monitor GPCR internalization using time-resolved fluorescence measurement.[81] Cells expressing the receptor of interest are submitted to drug treatment at various concentrations and followed by labeling of the SNAP or CLIP with terbium. After a washing step, the amount of SNAP labeling is quantified relative to untreated cells. Once again, the format of this assay makes it rather ineligible for HTS. However, the development of a TR-FRET internalization assay that allows a direct measurement of the kinetics would surely complete the offer available so far for GPCR analysis.

## 5.3. Multiplexing: Simultaneous analysis of a compound effect in several signaling pathways

As mentioned previously, GPCR signaling is a complex phenomenon. Depending on the receptor, the ligand, the localization, and the oligomerization, the repertoire of activated signaling cascades differs. To better characterize this, the implementation of multiplexing assay represents a promising improvement. First of all, it would allow the detection of at least two events at the same time in the same cell, which would increase the reliability between the two tests. In addition, it would also decrease by twofold the amount of compound used and the time spend for screening the assays that were combined. HTRF has the peculiarity to be perfectly compatible with multiplexing. As already mentioned, the lanthanide emission spectra, and more particularly terbium-based FRET donors, show several peaks that are perfectly compatible with the excitation of several acceptors. In agreement, one could easily imagine combining both IP-One and cAMP measurements in the same well if, for example, the fluorescent IP is labeled with a red acceptor and the cAMP with a green one. Of note, TR-FRET multiplexing has already been reported to monitor the binding of an agonist, an antagonist, and a coactivator on the estrogen receptor.[82] The receptor is labeled with terbium, and the agonist and antagonist with fluorescein and semi-naphthalene fluorescein, respectively. The decrease of the FRET signal between the receptor and the antagonist could be recorded at the same time as the appearance of a FRET signal between the receptor and the agonist. Addition of a coactivator labeled with Cy5 gave rise to a concomitant third FRET signal that was also simultaneously recorded. Alternatively, another report of TR-FRET-based assay multiplexing has been reported by Horton and Vogel.[37] In this study, they combined two kinase assays, one using the terbium GFP FRET couple and the other one the europium AlexaFluor647 couple. As there was good separation between the different lanthanide emission and acceptor excitation spectra, the authors could show that both assays could be used together. Such technological solutions are perfectly conceivable in the field of HTS targeting GPCRs or other cell-surface receptors.

## 5.4. Toward further miniaturization

A further improvement of TR-FRET-based assays would be further miniaturization. While most of the reported assays were already proved to be compatible with a 384- and even a 1536-well plate format, a further improvement would be to scale down to the microarray format. A first step in that direction was made when detecting the interaction between lanthanide-labeled biotinylated BSA immobilized on a chip and Cy5-labeled streptavidin.[83] Another proof of concept for the use of receptor in microarrays has been provided by the analysis of the coactivator binding to the immobilized estrogen receptor in homogeneous format by FRET measurement on a microarray (for review, see Ref. 84). Altogether, this opens up new perspectives to develop TR-FRET-based microarrays to analyze GPCR properties.

## 6. CONCLUSION

As illustrated by the solutions available on GPCR, TR-FRET is currently a method of choice for HTS campaigns. Its high sensitivity makes it compatible with the use of the 384- or 1536-well plate format. Miniaturization allows reduction of the amount of compound used and therefore the price of the assay. In addition, most of the assays presented above do not require washing steps and thus are perfectly compatible with automation of the screening process.

More than GPCR, TR-FRET-based assays could be used to study most of the membrane proteins. Actually, activating a cell-surface protein results in modifications in the intracellular components. The kinase assays, for example, are also compatible to screen for compound on tyrosine kinases receptors. Tag-lite® strategy could be used on any cell-surface receptor both in its ligand-binding format and in its oligomerization assay. Though the fluorescent techniques are preferred to radioactive assays by pharmaceutical companies, the novel TR-FRET-based assays represent a further improvement and have the potential to gain more importance in the future screening campaigns.

## REFERENCES

1. Thorne N, Auld DS, Inglese J. Apparent activity in high-throughput screening: origins of compound-dependent assay interference. *Curr Opin Chem Biol* 2010;**14**:315–24.
2. Mathis G. Rare earth cryptates and homogeneous fluoroimmunoassays with human sera. *Clin Chem* 1993;**39**:1953–9.
3. Selvin PR. Principles and biophysical applications of lanthanide-based probes. *Annu Rev Biophys Biomol Struct* 2002;**31**:275–302.
4. Bockaert J, Pin JP. Molecular tinkering of G protein-coupled receptors: an evolutionary success. *EMBO J* 1999;**18**:1723–9.
5. Overington JP, Al-Lazikani B, Hopkins AL. How many drug targets are there? *Nat Rev Drug Discov* 2006;**5**:993–6.
6. Foord SM, Bonner TI, Neubig RR, Rosser EM, Pin JP, Davenport AP, et al. International Union of Pharmacology. XLVI. G protein-coupled receptor list. *Pharmacol Rev* 2005;**57**:279–88.
7. Gesty-Palmer D, Luttrell LM. Refining efficacy: exploiting functional selectivity for drug discovery. *Adv Pharmacol* 2011;**62**:79–107.
8. Luttrell LM, Kenakin TP. Refining efficacy: allosterism and bias in G protein-coupled receptor signaling. *Methods Mol Biol* 2011;**756**:3–35.

9. Wang L, Martin B, Brenneman R, Luttrell LM, Maudsley S. Allosteric modulators of G protein-coupled receptors: future therapeutics for complex physiological disorders. *J Pharmacol Exp Ther* 2009;**331**:340–8.

10. Terrillon S, Bouvier M. Roles of G-protein-coupled receptor dimerization. *EMBO Rep* 2004;**5**:30–4.

11. Bazin H, Trinquet E, Mathis G. Time resolved amplification of cryptate emission: a versatile technology to trace biomolecular interactions. *J Biotechnol* 2002;**82**:233–50.

12. Degorce F, Card A, Soh S, Trinquet E, Knapik GP, Xie B. HTRF: a technology tailored for drug discovery—a review of theoretical aspects and recent applications. *Curr Chem Genomics* 2009;**3**:22–32.

13. Stryer L. Fluorescence energy transfer as a spectroscopic ruler. *Annu Rev Biochem* 1978;**47**:819–46.

14. Mathis G. Probing molecular interactions with homogeneous techniques based on rare earth cryptates and fluorescence energy transfer. *Clin Chem* 1995;**41**:1391–7.

15. Frang H, Mukkala VM, Syysto R, Ollikka P, Hurskainen P, Scheinin M, et al. Nonradioactive GTP binding assay to monitor activation of g protein-coupled receptors. *Assay Drug Dev Technol* 2003;**1**:275–80.

16. Cabrera-Vera TM, Vanhauwe J, Thomas TO, Medkova M, Preininger A, Mazzoni MR, et al. Insights into G protein structure, function, and regulation. *Endocr Rev* 2003;**24**:765–81.

17. Thomsen AR, Hvidtfeldt M, Brauner-Osborne H. Biased agonism of the calcium-sensing receptor. *Cell Calcium* 2012;**51**:107–16.

18. van Borren MM, Verkerk AO, Wilders R, Hajji N, Zegers JG, Bourier J, et al. Effects of muscarinic receptor stimulation on Ca2 + transient, cAMP production and pacemaker frequency of rabbit sinoatrial node cells. *Basic Res Cardiol* 2010;**105**:73–87.

19. Zwier JM, Roux T, Cottet M, Durroux T, Douzon S, Bdioui S, et al. A fluorescent ligand-binding alternative using Tag-lite(R) technology. *J Biomol Screen* 2010;**15**:1248–59.

20. Handl HL, Vagner J, Yamamura HI, Hruby VJ, Gillies RJ. Development of a lanthanide-based assay for detection of receptor-ligand interactions at the delta-opioid receptor. *Anal Biochem* 2005;**343**:299–307.

21. Trinquet E, Fink M, Bazin H, Grillet F, Maurin F, Bourrier E, et al. D-myo-inositol 1-phosphate as a surrogate of D-myo-inositol 1,4,5-tris phosphate to monitor G protein-coupled receptor activation. *Anal Biochem* 2006;**358**:126–35.

22. Trinquet E, Bouhelal R, Dietz M. Monitoring Gq-coupled receptor response through inositol phosphate quantification with the IP-One assay. *Expert Opin Drug Discov* 2011;**6**:981–94.

23. Gabriel D, Vernier M, Pfeifer MJ, Dasen B, Tenaillon L, Bouhelal R. High throughput screening technologies for direct cyclic AMP measurement. *Assay Drug Dev Technol* 2003;**1**:291–303.

24. Martikkala E, Rozwandowicz-Jansen A, Hanninen P, Petaja-Repo U, Harma H. A homogeneous single-label time-resolved fluorescence cAMP assay. *J Biomol Screen* 2011;**16**:356–62.

25. Martikkala E, Veltel S, Kirjavainen J, Rozwandowicz-Jansen A, Lamminmaki U, Hanninen P, et al. Homogeneous single-label biochemical Ras activation assay using time-resolved luminescence. *Anal Chem* 2011;**83**:9230–3.

26. Maurel D, Comps-Agrar L, Brock C, Rives ML, Bourrier E, Ayoub MA, et al. Cell-surface protein-protein interaction analysis with time-resolved FRET and snap-tag technologies: application to GPCR oligomerization. *Nat Methods* 2008;**5**:561–7.

27. Doumazane E, Scholler P, Zwier JM, Eric T, Rondard P, Pin JP. A new approach to analyze cell surface protein complexes reveals specific heterodimeric metabotropic glutamate receptors. *FASEB J* 2011;**25**:66–77.

28. Ito S, Yoshimoto R, Miyamoto Y, Mitobe Y, Nakamura T, Ishihara A, et al. Detailed pharmacological characterization of GT-2331 for the rat histamine H3 receptor. *Eur J Pharmacol* 2006;**529**:40–6.

29. Pos Z, Wiener Z, Pocza P, Racz M, Toth S, Darvas Z, et al. Histamine suppresses fibulin-5 and insulin-like growth factor-II receptor expression in melanoma. *Cancer Res* 2008;**68**:1997–2005.

30. Williams C. cAMP detection methods in HTS: selecting the best from the rest. *Nat Rev Drug Discov* 2004;**3**:125–35.

31. Jean A, Conductier G, Manrique C, Bouras C, Berta P, Hen R, et al. Anorexia induced by activation of serotonin 5-HT4 receptors is mediated by increases in CART in the nucleus accumbens. *Proc Natl Acad Sci USA* 2007;**104**:16335–40.

32. Chambers C, Smith F, Williams C, Marcos S, Liu ZH, Hayter P, et al. Measuring intracellular calcium fluxes in high throughput mode. *Comb Chem High Throughput Screen* 2003;**6**:355–62.

33. Ayoub MA, Trinquet E, Pfleger KD, Pin JP. Differential association modes of the thrombin receptor PAR1 with Galphai1, Galpha12, and beta-arrestin 1. *FASEB J* 2010;**24**:3522–35.

34. Wei H, Ahn S, Shenoy SK, Karnik SS, Hunyady L, Luttrell LM, et al. Independent beta-arrestin 2 and G protein-mediated pathways for angiotensin II activation of extracellular signal-regulated kinases 1 and 2. *Proc Natl Acad Sci USA* 2003;**100**:10782–7.

35. Holloway AC, Qian H, Pipolo L, Ziogas J, Miura S, Karnik S, et al. Side-chain substitutions within angiotensin II reveal different requirements for signaling, internalization, and phosphorylation of type 1A angiotensin receptors. *Mol Pharmacol* 2002;**61**:768–77.

36. Violin JD, Lefkowitz RJ. Beta-arrestin-biased ligands at seven-transmembrane receptors. *Trends Pharmacol Sci* 2007;**28**:416–22.

37. Horton RA, Vogel KW. Multiplexing terbium- and europium-based TR-FRET readouts to increase kinase assay capacity. *J Biomol Screen* 2010;**15**:1008–15.

38. Zhang R, Xie X. Tools for GPCR drug discovery. *Acta Pharmacol Sin* 2012;**33**:372–84.

39. Leopoldo M, Lacivita E, Berardi F, Perrone R. Developments in fluorescent probes for receptor research. *Drug Discov Today* 2009;**14**:706–12.

40. Middleton RJ, Kellam B. Fluorophore-tagged GPCR ligands. *Curr Opin Chem Biol* 2005;**9**:517–25.

41. Baker JG, Middleton R, Adams L, May LT, Briddon SJ, Kellam B, et al. Influence of fluorophore and linker composition on the pharmacology of fluorescent adenosine A1 receptor ligands. *Br J Pharmacol* 2010;**159**:772–86.

42. Middleton RJ, Briddon SJ, Cordeaux Y, Yates AS, Dale CL, George MW, et al. New fluorescent adenosine A1-receptor agonists that allow quantification of ligand-receptor interactions in microdomains of single living cells. *J Med Chem* 2007;**50**:782–93.

43. Monnier C, Tu H, Bourrier E, Vol C, Lamarque L, Trinquet E, et al. Trans-activation between 7TM domains: implication in heterodimeric GABAB receptor activation. *EMBO J* 2011;**30**:32–42.

44. Albizu L, Teppaz G, Seyer R, Bazin H, Ansanay H, Manning M, et al. Toward efficient drug screening by homogeneous assays based on the development of new fluorescent vasopressin and oxytocin receptor ligands. *J Med Chem* 2007;**50**:4976–85.

45. Hu LA, Zhou T, Hamman BD, Liu Q. A homogeneous G protein–coupled receptor ligand binding assay based on time-resolved fluorescence resonance energy transfer. *Assay Drug Dev Technol* 2008;**6**:543–50.

46. Hinner MJ, Johnsson K. How to obtain labeled proteins and what to do with them. *Curr Opin Biotechnol* 2010;**21**:766–76.

47. Damoiseaux R, Keppler A, Johnsson K. Synthesis and applications of chemical probes for human O6-alkylguanine-DNA alkyltransferase. *Chembiochem* 2001;**2**:285–7.

48. Gautier A, Juillerat A, Heinis C, Correa Jr. IR, Kindermann M, Beaufils F, et al. An engineered protein tag for multiprotein labeling in living cells. *Chem Biol* 2008;**15**:128–36.

49. Zhang Y, So MK, Loening AM, Yao H, Gambhir SS, Rao J. HaloTag protein-mediated site-specific conjugation of bioluminescent proteins to quantum dots. *Angew Chem Int Ed Engl* 2006;**45**:4936–40.

50. Yin J, Liu F, Li X, Walsh CT. Labeling proteins with small molecules by site-specific posttranslational modification. *J Am Chem Soc* 2004;**126**:7754–5.

51. Leyris JP, Roux T, Trinquet E, Verdie P, Fehrentz JA, Ouedati N, et al. Homogeneous time-resolved fluorescence-based assay to screen for ligands targeting the growth hormone secretagogue receptor type 1a. *Anal Biochem* 2011;**408**:253–62.

52. Kaupmann K, Malitschek B, Schuler V, Heid J, Froestl W, Beck P, et al. GABA(B)-receptor subtypes assemble into functional heteromeric complexes. *Nature* 1998; **396**:683–7.

53. Galvez T, Duthey B, Kniazeff J, Blahos J, Rovelli G, Bettler B, et al. Allosteric interactions between GB1 and GB2 subunits are required for optimal GABA(B) receptor function. *EMBO J* 2001;**20**:2152–9.

54. Margeta-Mitrovic M, Jan YN, Jan LY. Function of GB1 and GB2 subunits in G protein coupling of GABA(B) receptors. *Proc Natl Acad Sci USA* 2001;**98**:14649–54.

55. Margeta-Mitrovic M, Jan YN, Jan LY. A trafficking checkpoint controls GABA(B) receptor heterodimerization. *Neuron* 2000;**27**:97–106.

56. Pagano A, Rovelli G, Mosbacher J, Lohmann T, Duthey B, Stauffer D, et al. C-terminal interaction is essential for surface trafficking but not for heteromeric assembly of GABA(b) receptors. *J Neurosci* 2001;**21**:1189–202.

57. Li X, Staszewski L, Xu H, Durick K, Zoller M, Adler E. Human receptors for sweet and umami taste. *Proc Natl Acad Sci USA* 2002;**99**:4692–6.

58. Romano C, Yang WL, O'Malley KL. Metabotropic glutamate receptor 5 is a disulfide-linked dimer. *J Biol Chem* 1996;**271**:28612–6.

59. Contento RL, Molon B, Boularan C, Pozzan T, Manes S, Marullo S, et al. CXCR4-CCR5: a couple modulating T cell functions. *Proc Natl Acad Sci USA* 2008;**105**:10101–6.

60. El-Asmar L, Springael JY, Ballet S, Andrieu EU, Vassart G, Parmentier M. Evidence for negative binding cooperativity within CCR5-CCR2b heterodimers. *Mol Pharmacol* 2005;**67**:460–9.

61. Waldhoer M, Fong J, Jones RM, Lunzer MM, Sharma SK, Kostenis E, et al. A heterodimer-selective agonist shows in vivo relevance of G protein-coupled receptor dimers. *Proc Natl Acad Sci USA* 2005;**102**:9050–5.

62. Pei L, Li S, Wang M, Diwan M, Anisman H, Fletcher PJ, et al. Uncoupling the dopamine D1-D2 receptor complex exerts antidepressant-like effects. *Nat Med* 2010;**16**:1393–5.

63. Rashid AJ, So CH, Kong MM, Furtak T, El-Ghundi M, Cheng R, et al. D1-D2 dopamine receptor heterooligomers with unique pharmacology are coupled to rapid activation of Gq/11 in the striatum. *Proc Natl Acad Sci USA* 2007;**104**:654–9.

64. So CH, Verma V, Alijaniaram M, Cheng R, Rashid AJ, O'Dowd BF, et al. Calcium signaling by dopamine D5 receptor and D5-D2 receptor hetero-oligomers occurs by a mechanism distinct from that for dopamine D1-D2 receptor hetero-oligomers. *Mol Pharmacol* 2009;**75**:843–54.

65. Decaillot FM, Kazmi MA, Lin Y, Ray-Saha S, Sakmar TP, Sachdev P. CXCR7/CXCR4 heterodimer constitutively recruits beta-arrestin to enhance cell migration. *J Biol Chem* 2011;**286**:32188–97.

66. He L, Fong J, von Zastrow M, Whistler JL. Regulation of opioid receptor trafficking and morphine tolerance by receptor oligomerization. *Cell* 2002;**108**:271–82.

67. Gupta A, Mulder J, Gomes I, Rozenfeld R, Bushlin I, Ong E, et al. Increased abundance of opioid receptor heteromers after chronic morphine administration. *Sci Signal* 2010;**3**:ra54.
68. Alhamadsheh MM, Connelly S, Cho A, Reixach N, Powers ET, Pan DW, et al. Potent kinetic stabilizers that prevent transthyretin-mediated cardiomyocyte proteotoxicity. *Sci Transl Med* 2011;**3**:97ra81.
69. Alvarez-Curto E, Ward RJ, Pediani JD, Milligan G. Ligand regulation of the quaternary organization of cell surface M3 muscarinic acetylcholine receptors analyzed by fluorescence resonance energy transfer (FRET) imaging and homogeneous time-resolved FRET. *J Biol Chem* 2010;**285**:23318–30.
70. Vischer HF, Nijmeijer S, Smit MJ, Leurs R. Viral hijacking of human receptors through heterodimerization. *Biochem Biophys Res Commun* 2008;**377**:93–7.
71. Mijares A, Lebesgue D, Wallukat G, Hoebeke J. From agonist to antagonist: Fab fragments of an agonist-like monoclonal anti-beta(2)-adrenoceptor antibody behave as antagonists. *Mol Pharmacol* 2000;**58**:373–9.
72. Albizu L, Cottet M, Kralikova M, Stoev S, Seyer R, Brabet I, et al. Time-resolved FRET between GPCR ligands reveals oligomers in native tissues. *Nat Chem Biol* 2010;**6**:587–94.
73. Rajapakse HE, Gahlaut N, Mohandessi S, Yu D, Turner JR, Miller LW. Time-resolved luminescence resonance energy transfer imaging of protein-protein interactions in living cells. *Proc Natl Acad Sci USA* 2010;**107**:13582–7.
74. Levoye A, Dam J, Ayoub MA, Guillaume JL, Couturier C, Delagrange P, et al. The orphan GPR50 receptor specifically inhibits MT1 melatonin receptor function through heterodimerization. *EMBO J* 2006;**25**:3012–23.
75. Ciruela F, Casado V, Rodrigues RJ, Lujan R, Burgueno J, Canals M, et al. Presynaptic control of striatal glutamatergic neurotransmission by adenosine A1-A2A receptor heteromers. *J Neurosci* 2006;**26**:2080–7.
76. Kern A, Albarran-Zeckler R, Walsh HE, Smith RG. Apo-ghrelin receptor forms heteromers with DRD2 in hypothalamic neurons and is essential for anorexigenic effects of DRD2 agonism. *Neuron* 2012;**73**:317–32.
77. Epstein LH, Temple JL, Neaderhiser BJ, Salis RJ, Erbe RW, Leddy JJ. Food reinforcement, the dopamine D2 receptor genotype, and energy intake in obese and nonobese humans. *Behav Neurosci* 2007;**121**:877–86.
78. Lohse MJ, Nuber S, Hoffmann C. Fluorescence/Bioluminescence resonance energy transfer techniques to study g-protein-coupled receptor activation and signaling. *Pharmacol Rev* 2012;**64**:299–336.
79. Liu CC, Schultz PG. Adding new chemistries to the genetic code. *Annu Rev Biochem* 2010;**79**:413–44.
80. Liu W, Brock A, Chen S, Schultz PG. Genetic incorporation of unnatural amino acids into proteins in mammalian cells. *Nat Methods* 2007;**4**:239–44.
81. Ward RJ, Pediani JD, Milligan G. Ligand-induced internalization of the orexin OX(1) and cannabinoid CB(1) receptors assessed via N-terminal SNAP and CLIP-tagging. *Br J Pharmacol* 2011;**162**:1439–52.
82. Kim SH, Gunther JR, Katzenellenbogen JA. Monitoring a coordinated exchange process in a four-component biological interaction system: development of a time-resolved terbium-based one-donor/three-acceptor multicolor FRET system. *J Am Chem Soc* 2010;**132**:4685–92.
83. Schaferling M, Nagl S. Optical technologies for the read out and quality control of DNA and protein microarrays. *Anal Bioanal Chem* 2006;**385**:500–17.
84. Schaferling M, Nagl S. Forster resonance energy transfer methods for quantification of protein-protein interactions on microarrays. *Methods Mol Biol* 2011;**723**:303–20.

CHAPTER EIGHT

# Fluorescent Protein-Based Biosensors and Their Clinical Applications

**Yusuke Ohba\*, Yoichiro Fujioka\*, Shigeyuki Nakada†, Masumi Tsuda\***

\*Laboratory of Pathophysiology and Signal Transduction, Hokkaido University Graduate School of Medicine, N15W7, Kita-ku, Sapporo, Japan
†Mitsui Engineering and Shipbuilding Co. Ltd., Tamano Technology Center, 3-16-1 Tamahara, Tamano, Okayama, Japan

## Contents

*Progress in Molecular Biology and Translational Science*, Volume 113
ISSN 1877-1173
http://dx.doi.org/10.1016/B978-0-12-386932-6.00008-9

## Abstract

Green fluorescent protein and its relatives have shed their light on a wide range of biological problems. To date, with a color palette consisting of fluorescent proteins with different spectra, researchers can "paint" living cells as they desire. Moreover, sophisticated biosensors engineered to contain single or multiple fluorescent proteins, including FRET-based biosensors, spatiotemporally unveil molecular mechanisms underlying physiological processes. Although such molecules have contributed considerably to basic research, their abilities to be used in applied life sciences have yet to be fully explored. Here, we review the molecular bases of fluorescent proteins and fluorescent protein-based biosensors and focus on approaches aimed at applying such proteins to the clinic.

## 1. INTRODUCTION

Green fluorescent protein (GFP) is a fluorescent protein that was originally isolated from the luminous organ of the jellyfish *Aequorea victoria* by Dr. Osamu Shimomura.[1] It is probably no exaggeration to say that GFP has continued to shed light on cell biology since its cDNA was isolated in 1992.[2] The main reason why GFP is so revolutionary is its ability to be easily introduced into cells by transfection, which is dependent on a key feature of GFP, that is, its ability to generate the intrinsic chromophore without cofactors or enzymatic components.[3]

It was not long before color variants of GFP were developed and fluorescent proteins were distributed worldwide.[3] Following these innovations, fluorescent proteins were isolated from organisms other than jellyfish (although mainly from *Cnidaria*)[4]; together, these proteins constitute a palette that can provide colorful decoration to living cells. Not only do these tools serve as markers of protein localization but they can also be used for monitoring intracellular environments under various physiological conditions. In addition, via use of the principles of reconstitution of protein fragments or a physicochemical energy transfer phenomenon, protein–protein interactions or conformational changes can be monitored, widening the potential applications of fluorescent proteins.

More recently, our research group has begun to further expand the potential of fluorescent proteins from basic biological research to the clinic.[5] As in basic research, biosensors engineered on the basis of fluorescent proteins have provided solutions for hitherto arduous tasks and may offer innovative technological advances in clinical laboratory examinations when used in combination with state-of-the-art detection devices. In this review, we focus on the molecular bases of fluorescent proteins, fluorescent protein-based biosensors, and methods to evaluate the signals from these sensors; we also discuss their potential for use in translational research or clinical situations.

## 2. FLUORESCENT PROTEINS

### 2.1. Chromophore formation of GFP and its derivatives

GFP is 238 amino acids long ($\sim$27 kDa) and folds into a cylindrical caged structure of 11 β-sheets (β-barrel) surrounding an internal distorted helix (Fig. 8.1A).[6] Protein folding provides the driving force for the formation of a chromophore (composed of three amino acids at positions 65–67, see Fig. 8.1B) in the internal helix, correctly orienting crucial residues to catalyze and direct chromophore synthesis pathways. The rigid β-barrel also comprises the protein matrix surrounding the chromophore, which shields it from the solvent and provides a chemically complex environment,

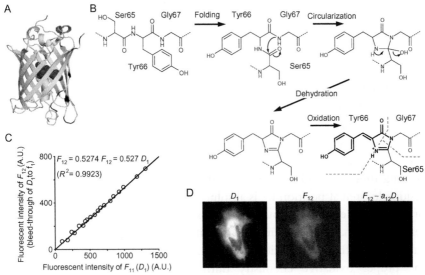

Figure 8.1 Basic properties of GFP. (A) Structure of *Aequorea victoria* GFP. Black balls within the β-barrel indicate the chromophore, whereas the main chains of amino acids from which circularly permutated GFPs are started are also highlighted. (B) Chemical reaction leading to chromophore formation. (C) Correction of emission bleed-through. Fluorescence intensities of $D_1$ obtained through the filter $f_2$ ($F_{12}$) are plotted against those obtained through $f_1$ ($F_{11}$). (D) Typical photographs before ($D_1$ and $F_{12}$) and after ($F_{12}-a_{12}D_1$) correction are shown.

thus ensuring the function of GFP as an autofluorescent protein.[7,8] The protein folding and chromophore formation processes are likely to be interdependent.[9] Indeed, the residues near the chromophore participate in determining the photochemical properties of each variant of GFP.[10,11] In addition, unfolding of GFP, which radically dismantles the β-barrel and repositions the amino acids surrounding the chromophore, completely abolishes fluorescence, although the covalent chemical structure of the central chromophore is maintained.[12]

Spectral tuning of fluorescent proteins is achieved via substitution of the amino acids composing the chromophore, which alters the extent of π-conjugation within the chromophore *per se* and also within the amino acids surrounding the chromophore. The β-barrel protects the chromophore from the environment and restricts its flexibility. Increasing evidence regarding the relationship between the structure and the optical properties of fluorescent proteins has led to the rational, direct manipulation of these properties. To date, a large number of mutants of the original GFP, as well as a range of fluorescent proteins originating from cnidarians other than jellyfish, have therefore become available and together constitute a wide palette of colors. Such proteins have been widely used to visualize protein dynamics and environmental changes in living cells, although until now their use has been restricted to basic research. For those interested in a more complete survey, there are a number of excellent reviews that cover the entire palette of fluorescent proteins currently available.[6,13–16]

The ability of fluorescent proteins to emit visible light derives from the posttranslational self-modification of three amino acids at positions 65–67 (Ser-Tyr-Gly in *A. victoria* GFP and Thr-Tyr-Gly in enhanced GFP) (Fig. 8.1B and Table 8.1), which results in chromophore formation.[3] Among these three chromophore-forming amino acids, only the glycine residue located at position 67 is absolutely conserved within all fluorescent proteins (Table 8.1). Although the tyrosine residue at position 66 is conserved in all naturally occurring GFP-like proteins (Table 8.1), it can be substituted with any aromatic amino acid. This residue is believed to provide the proper oxidative chemistry during chromophore maturation and prevent undesirable side reactions, such as hydrolysis and backbone fragmentation (Fig. 8.1B).[17,18]

Substitution of the tyrosine residue with an aromatic amino acid results in chemically distinct chromophores that emit light in the blue or cyan range[3]; for example, histidine is substituted in EBFP (enhanced blue fluorescent protein) and tryptophan in ECFP (enhanced cyan fluorescent

**Table 8.1** Amino acid sequences of the chromophores of fluorescent proteins

| | Amino acid position | | | | | |
|---|---|---|---|---|---|---|
| | 64 | 65 | 66 | 67 | 68 | 69 |
| *Aequorea* GFP | Phe | **Ser** | **Tyr** | **Gly** | Val | Gln |
| GFP-S65T | Phe | **Thr** | **Tyr** | **Gly** | Val | Gln |
| EGFP | Leu | **Thr** | **Tyr** | **Gly** | Val | Gln |
| EBFP | Leu | **Thr** | **His** | **Gly** | Val | Gln |
| ECFP | Leu | **Thr** | **Trp** | **Gly** | Val | Gln |
| EYFP | Phe | **Gly** | **Tyr** | **Gly** | Val | Gln |
| DsRed | Phe | **Gln** | **Tyr** | **Gly** | Ser | Lys |
| | (65) | (66) | (67) | (68) | (69) | (70) |

The actual amino acid numbers for DsRed are shown in parentheses. Amino acids at positions of 65–67 are directly involved in chromophore formation.

protein). Despite attempts to improve the photochemical characteristics of these blue- or cyan-emitting GFP homologs, blue-emitting fluorescent proteins (possessing histidine at position 66) exhibit low brightness in comparison with their green- or cyan-emitting counterparts. Recently, Dr. Nagai and colleagues succeeded in developing a bright ultramarine fluorescent protein, Sirius, by introducing Trp66Phe and several additional mutations into ECFP.[19] Sirius possesses the shortest emission wavelength among fluorescent proteins reported to date and is relatively bright and pH insensitive.

## 2.2. Red chromophore synthesis via blue intermediates

Amino acids at position 65 affect the resulting chemical structure of the chromophore and therefore vary widely among GFP-like proteins of various colors. Most GFP-derived fluorescent proteins have a threonine at this position, unlike *A. victoria* GFP (Table 8.1), which has a serine. This substitution results in proteins that have a single red-shifted excitation peak, fluoresce more intensely than wild type, and fold approximately four times faster than wild-type GFP.[20] The red chromophore of the *Discosoma* sp. (DsRed) fluorescent protein has an extended GFP-like core with an additional desaturated Cα—N bond at the Gln65 residue (position 66 in DsRed2) (Table 8.1). The red-shifted absorbance and emission are therefore provided by an extended π-electron conjugation.[4] An early hypothesis in which the maturation of the DsRed-like red chromophore was thought to proceed via a green GFP-like chromophore intermediate[21] has been recently revised. According to the current understanding, the maturation pathway of the DsRed-like red chromophore includes accumulation of a blue-emitting intermediate[22,23] possessing a structure thought to be similar to that of the mTagBFP chromophore, a blue derivative of DsRed. These properties have been utilized to generate fluorescent timers (see Section 3.1).

## 2.3. Circular permutation

The intricate posttranslational self-modification required for chromophore formation suggests that major rearrangements or insertions within GFP would prevent fluorescence emission. However, the three-dimensional structure of GFP (Fig. 8.1A) implies that there is still plenty of room for manipulation in the loop regions between the β-sheets. Indeed, Dr. Tsien and colleagues have succeeded in engineering several fluorescent rearrangements of GFP and other fluorescent proteins in which the amino and carboxyl portions are interchanged and rejoined, with a short spacer (GlyGlyThrGlyGlySer) connecting the original termini.[24] It is noteworthy that all interrupted positions are located in the original C-terminal half of enhanced GFP (Fig. 8.1A). In addition, some sites (His148, His169, and Ara227) are in β-strand segments, whereas others are in the loops, as expected (Fig. 8.1A). More importantly, certain residues within GFP, including Tyr145, tolerate insertion of other proteins, and conformational changes in the insert profoundly influence the fluorescence intensity.[24] These properties are utilized in the development of single-fluorescent protein-based biosensors (see Section 3). Circular permutations also alter the relative orientation of the chromophore to a fusion partner, and this property is exploited in the improvement of biosensors based on the principles of FRET (Förster resonance energy transfer; see Section 5).

## 2.4. Fluorescent proteins as tags

The discovery and cloning of GFP and its relatives have undoubtedly advanced our understanding of basic biology. For example, use of fluorescent proteins has facilitated routine monitoring of gene activation as well as the selective labeling and analysis of single proteins, cellular organelles, and even whole cells.[6,13]

Tracking cellular proteins *in vivo* with fluorescent tags was made routine by the development of GFP and its family members. Fluorescent proteins are usually appended to the amino or carboxyl terminus of the host protein. The choice of carboxyl versus amino terminal attachment can be guided by known properties of the protein; however, it is recommended that both types of tagged proteins be tested for functionality and subcellular localization. Some proteins are sensitive to the type of attachment of fluorescent proteins. For example, placing a fluorescent protein before a signal sequence in a transmembrane protein or secreted protein is likely to disrupt the subcellular targeting or expression of the protein. Small GTPases should be tagged on the amino termini because of the presence of "CAAX box," which is critical for the membrane localization of small GTPases.[25] In a few instances, fluorescent proteins have been added within the sequence of a protein. However, because of the size of a fluorescent protein, the resulting perturbations to the overall folding and function of the underlying protein can be significant. For internal fusions, careful considerations and planning are required to ensure that the GFP is added within loops or flexible domains. In one example, GFP was inserted randomly into a protein and a functional screen was used to test for effects of the inserted tag.[26]

When making a fluorescent protein fusion, linkers can be used to join the two sequences. This strategy helps to overcome the aforementioned concerns regarding protein folding and function, but the sequence itself is also critical for the proper functioning of tagged proteins. In general, it is prudent to add a flexible linker between the two partners. The standard flexible linkers consist of runs of glycines interspersed with serines or threonines (GlyGlyGlySerGlyGlyGlySer),[27] in which the glycines provide conformational flexibility and the hydrophilic residues allow hydrogen bonding with the solvent and prevent the linker from interacting with hydrophobic protein-binding interfaces. Similar linkers have been used to create flexible biosensors and tandem dimers of fluorescent proteins,[28,29] as well as circular permutations in GFP, as described elsewhere.[24] Hydrophilicity of the linker is prominently important when visualizing actin fibers using GFP-tagged $\beta$-actin. In this case, hydrophilic but flexible linkers (GlySerThrSerGly[30] or SerGlyLeuArgSerArgAla [the commercially available pEGFP-actin vector]) should be used, rather than normal linkers with high flexibility. For example, the GlyGlySerGlyGlySer linker substantially prevents incorporation of the resulting fusion protein into actin stress fibers or lamellipodial protrusions (data not shown).

## 2.5. Spectral unmixing (compensation) for multicolor imaging

Emergence of numerous colored variants of fluorescent proteins has encouraged researchers to launch multicolor imaging. Upon the expression of multiple fluorescent proteins in single cells or in spatially overlapping architectures, spectral bleed-through due to the large spectral overlap of fluorescent proteins prevents the detection of accurate, specific fluorescence intensities for every fluorescent protein. Spectral unmixing (or compensation) is therefore required to distinguish each specific signal from cells expressing multiple fluorescent proteins; in this method, a sample with a single fluorescent molecule is imaged onto several spectral channels to determine the extent of detector bleed-through.[31–34] Under the assumption that the spectrum of an individual chromophore linearly contaminates each of the other image channels (Fig. 8.1C), spectral unmixing can be executed by subtraction of the spectral bleed-through of each chromophore from each individual spectrum (Fig. 8.1D).[33]

Here is a practical example. Fluorescent dyes $D_1$ and $D_2$ are observed through optical filter sets $f_1$ and $f_2$ that are prepared for the observation of $D_1$ and $D_2$, respectively, but there is bleed-through of spectra of $D_1$ and $D_2$ to $f_2$ and $f_1$, respectively. To compensate for the bleed-through, cells expressing $D_1$ or $D_2$ alone are prepared and observed through both $f_1$ and $f_2$. When the fluorescence intensity of $D_i$ obtained through the filter $f_j$ is represented as $F_{ij}$, bleed-through of $D_1$ to $f_2$ is

$$F_{12} = a_{12}F_{11} = a_{12}D_1, \qquad [8.1]$$

where $a_{12}$ is an inclination of the linear function representing bleed-through of $D_1$ to the $D_2$ channel (through the $f_1$ filter; Fig. 8.1D). Similarly, bleed-through of $D_2$ into the $f_1$ channel can be expressed as follows:

$$F_{21} = a_{21}F_{22} = a_{21}D_2. \qquad [8.2]$$

In cells expressing both $D_1$ and $D_2$, the intensity obtained through the $f_1$ channel ($F_1$) contains the emission of $D_1$ and the bleed-through of $D_2$ (which is provided by Eq. 8.2); similarly, $F_2$ is the sum of $D_2$ and $a_{12} \cdot D_1$. Therefore, $F_1$ and $F_2$ are given by

$$F_1 = D_1 + a_{21} \cdot D_2; \qquad [8.3]$$
$$F_2 = a_{12} \cdot D_1 + D_2. \qquad [8.4]$$

By defining the matrix $A$, the equations can be rewritten as

$$\begin{pmatrix} F_1 \\ F_2 \end{pmatrix} = A \begin{pmatrix} D_1 \\ D_2 \end{pmatrix} \quad \left[ A = \begin{pmatrix} 1 & a_{21} \\ a_{12} & 1 \end{pmatrix} \right]. \qquad [8.5]$$

Thus, multiplying both sides of each equation by the inverse of the matrix ($A^{-1}$) from the left provides the corrected fluorescence intensities of $D_1$ and $D_2$.

$$\begin{pmatrix} D_1 \\ D_2 \end{pmatrix} = A^{-1} \begin{pmatrix} F_1 \\ F_2 \end{pmatrix} \left[ A^{-1} = \frac{1}{1 - a_{12}a_{21}} \begin{pmatrix} 1 & a_{21} \\ -a_{12} & 1 \end{pmatrix} \right]. \qquad [8.6]$$

Generally, when the number of fluorescent proteins to be compensated is $n$, spectral unmixing can be executed by considering an $n \times n$ matrix and its inverse as follows:

$$\begin{pmatrix} F_1 \\ F_2 \\ F_3 \\ \vdots \\ F_n \end{pmatrix} = A \begin{pmatrix} D_1 \\ D_2 \\ D_3 \\ \vdots \\ D_n \end{pmatrix}, \qquad [8.7]$$

$$\begin{pmatrix} D_1 \\ D_2 \\ D_3 \\ \vdots \\ D_n \end{pmatrix} = A^{-1} \begin{pmatrix} F_1 \\ F_2 \\ F_3 \\ \vdots \\ F_n \end{pmatrix}, \quad A = \begin{pmatrix} 1 & a_{21} & a_{31} & \cdots & a_{n1} \\ a_{12} & 1 & a_{32} & \cdots & a_{n2} \\ a_{13} & a_{23} & 1 & \cdots & a_{n3} \\ \vdots & \vdots & \vdots & \ddots & \vdots \\ a_{1n} & a_{2n} & a_{3n} & \cdots & 1 \end{pmatrix}. \qquad [8.8]$$

In practice, these operations can be performed using commercially available imaging software.

## 3. SINGLE-FLUORESCENT PROTEIN-BASED BIOSENSORS

In addition to monitoring protein dynamics or transcriptional activation (see Section 2.4), single-fluorescent protein-based biosensors have also been used to measure important characteristics of cellular environments, including pH, ion flux, and redox potential.[24,35–38]

### 3.1. Fluorescent timers

The processes of chromophore maturation and conversion (see Sections 2.1 and 2.2) can be used to create fluorescent proteins with new features. For example, a red fluorescent protein named drFP583 (E5)[39] changes its fluorescence from green to red and the recently developed mCherry-based fluorescent timers[40] convert their spectra from blue to red in a time-dependent manner. Changes in the emission spectra of these fluorescent timers are time dependent but concentration independent because of slow maturation of the chromophore. The predictable time course of the color change (from blue or green to red) enables quantitative analysis of temporal and spatial molecular events, thereby reporting the history of promoter activity and gene expression. Key amino acids in the chromophore environment critical for the timing properties of fluorescent timers have been determined via site-directed mutagenesis.[22]

## 3.2. Photoactivatable fluorescent proteins

Another recent development is the discovery of fluorescent proteins with light-modulated spectral properties, which are collectively termed "photo-activatable fluorescent proteins," and of fluorescent proteins that emit in the far-red region and possess large Stokes shifts. These molecules have a number of new applications, including multicolor super-resolution imaging in fixed and live cells.[41,42] In their resting states, photoactivatable fluorescent proteins (e.g., photoactivatable mCherries) are nonfluorescent, but they will emit bright red fluorescence in response to UV or violet light illumination.[41] The structure of the photoactivatable mCherry1 chromophore in the "dark state" is similar to that of the mTagBFP chromophore, but this chromophore is not fluorescent. Upon photon adsorption, the excited mTagBFP-like chromophore is converted into a red chromophore via a Kolbe-like radical reaction,[41,43] in a similar manner as the blue-to-red chromophore transition (see Section 2.2).

## 3.3. Sensor for ions and small molecules

The original *A. victoria* GFP has several weaknesses, including dual peaks in the excitation spectrum (395 and 475 nm) and pH-sensitive emission intensity; these properties were improved during the development of enhanced GFP. Using these properties, several fluorescent protein-based methods for pH measurement have been developed, however.[44-46] In order to monitor the acidic pH inside secretory vesicles, Dr. Rothman and colleagues developed two classes of pHluorins[44] by substitution of amino acids associated with the proton-relay network of Tyr66,[3,47-49] but not with the chromophore itself, in GFP. "Ratiometric" pHluorin contains a Ser202His mutation, along with eight additional substitutions, and displays a reversible excitation ratio change between pH 5.5 and 7.5 ($R_{395/475} = 0.35-0.95$), whereas "ecliptic" pHluorin gradually loses fluorescence as the pH is lowered, at least until the pH reaches $\sim 6.0$.[44] pHRed,[45] the first ratiometric, single-protein red fluorescent biosensor, was recently engineered by introducing an Ala213Ser mutation into mKeima, a red fluorescent protein with a long Stokes shift (excitation and emission wavelengths of 440 and 610 nm, respectively).[50] pHRed changes its excitation peak from 440 nm (protonated neutral chromophore of mKeima) to 585 nm (anionic chromophore) in response to acidification, thereby enabling pH monitoring by dual-excitation, single-emission observation.[45] Introduction of a His148Gln mutation into the yellow-emitting mutant of GFP (YFP) makes YFP a halide ion sensor.[51] The fluorescence intensity of this mutant YFP is sensitive to the concentrations of $F^-$, $Cl^-$, $Br^-$, $I^-$, $NO_3^-$, $SCN^-$, $ClO_4^-$, and formate, in addition to pH.[51] Voltage-sensitive fluorescent proteins (VSFPs) have also been developed to allow optical measurement of membrane potential.[52–55] By tagging Citrine, a variant of CFP, to the carboxyl terminus of a *Ciona intestinalis* voltage-sensor-containing phosphatase (Ci-VSP) via a short linker, a biosensor that reports fluorescence activation kinetics closely resembling the fast "gating" currents (VSFP3.1) has been developed, although the detailed mechanism through which the emission intensity is modulated is still lacking.

$Ca^{2+}$ sensors, which include camgaroo1,[24] pericam,[56] G-CaMP,[57] Case12,[38] and GECO, have also been well studied.[58] These biosensors, which result from the insertion of $Ca^{2+}$-sensing proteins into circular permutated fluorescent proteins, enable researchers to visualize elevation of intracellular $Ca^{2+}$ by increases in intensity of a single spectrum. Such biosensors have great advantages over FRET-based $Ca^{2+}$ biosensors such as cameleon[59] and FIP-$CB_{SM}$[60] because of their narrowed emission wavelength ranges and enhanced signal-to-noise ratios.

With the appropriate insertions, circularly permutated fluorescent proteins can detect changes in a range of parameters. Dr. Lukyanov and colleagues developed a fluorescent biosensor that is highly specific for hydrogen peroxide ($H_2O_2$) via insertion of circularly permuted YFP into the regulatory domain of the prokaryotic $H_2O_2$-sensing protein OxyR (HyPer).[61] A biosensor for adenylate nucleotides has also been generated from a circularly permuted variant of GFP and the ATP-binding bacterial protein GlnK1 (from *Methanococcus jannaschii*).[62]

## 4. BIMOLECULAR FLUORESCENCE COMPLEMENTATION

A range of proteins can be divided into two (or more) fragments that individually possess no function but that can be reconstituted as a functionally active complex when they are brought in proximity to each other. Fusing the fragments to partners that interact with each other enhances the efficacy of their assembly. Therefore, the activity of the complex can be used to monitor bimolecular or multimolecular interactions. After the original demonstration of the principle of fragment-based monitoring of protein–protein interaction, which employed fragments of ubiquitin,[63] fragments of a number of proteins, including β-galactosidase, dihydrofolate reductase, autofluorescent proteins, β-lactamase, and luciferases, have also been used.[64–71]

Of the available methods for the detection of protein interactions based on protein fragment association, the use of fragments of fluorescent proteins known as "bimolecular fluorescence complementation" (BiFC) might be the most widely used.[72,73] The BiFC assay is based on lessons learned from circular permutation of fluorescent proteins (see Section 2.3); in this assay, an interaction between proteins fused to nonfluorescent fragments of fluorescent protein facilitates fluorescence emission through the reconstitution of the fragments (Fig. 8.2A).[66] When used with a fluorescence microscope, the BiFC assay enables visualization of the intracellular distribution of protein interactions without the addition of other exogenous agents as substrates. BiFC was first used to monitor the noncovalent association of antiparallel

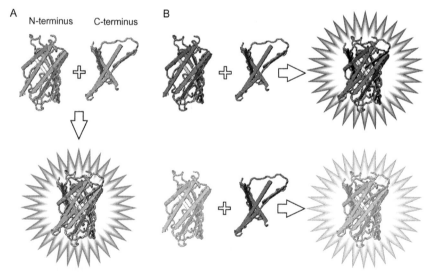

**Figure 8.2** Schematic representation of BiFC. (A) Although the N- or C-terminal fragments of fluorescent proteins are not fluorescent, they emit the proper fluorescence upon reconstitution. (B) Multicolor BiFC. As a result of the interaction between two fragments of different fluorescent proteins, the fluorescence spectrum is determined by the N-terminal fragment. This can be used to analyze the stoichiometry of the interaction.

leucine zippers[74] and subsequently utilized in mammalian cells to investigate interactions between basic leucine zipper (bZIP) and the nuclear factor-κB (NF-κB)/Rel family of transcription factors under physiological conditions.[66] Since then, BiFC has been successfully used in various model organisms, including mammalian cell lines, plants, and microorganisms.[75–77]

## 4.1. Advantages of BiFC

As described above, one important advantage of BiFC is the intrinsic fluorescence of the formed complex, which makes it possible to visualize protein–protein interactions without other reagents that are required in complementation methods using other protein fragments. In addition, multiple protein interactions can be simultaneously observed in a single cell using multicolor BiFC analysis.[78] In this approach, different interaction partners are fused to fragments of each different fluorescent protein, the binding of which therefore produces fluorescent complexes with distinct colors. In addition, thanks to the structural similarities among fluorescent proteins, some fluorescent protein fragments can associate with other different complementary fragments and produce BiFC complexes with distinct spectra (Fig. 8.2B).[78] This method allows visualization of competition for molecular binding among alternative interaction partners,[79] wherein the amino terminal fragments determine the spectra emitted by the formed fluorescent complexes. It is worth repeating that the interrupted positions for circular permutation are located in the carboxyl terminal halves of fluorescent proteins (see Section 2.3); therefore, the chromophores are present in the amino terminal fragments (Fig. 8.2B).

## 4.2. Disadvantages of BiFC

The intensity of fluorescence emanating from resulting BiFC complexes does not correspond to the kinetics of interactions between the fused proteins. First, the time required both for fluorescent protein fragment association and the chemical reactions that produce and mature the chromophore results in a delay between the beginning of the interaction and the beginning of fluorescence emission.[66,72] The extent of the delay is likely to depend on the kinetics of chromophore formation; therefore, BiFC complex fluorescence can often be detected much faster than expected (within minutes after fusion protein interaction) when quickly maturing fluorescent proteins are used. For instance, rapamycin-induced binding between FK506 binding protein (FKBP) and the FKBP-rapamycin binding (FRB) domain of mTOR becomes detectable by BiFC complex formation of YFP fragments within 10 min after rapamycin treatment.[80] Given that the signal continuously increases for at least 8 h, the rapidity of detection does not necessarily indicate a short half-time of fluorescent complex formation. Using fragments of Venus, a bright mutant of YFP with fast maturation, we have been able to monitor growth factor-induced Ras activation (which generally peaks at approximately 10 min), although this activation is slowed relative to activation of endogenous Ras.[81] Therefore, this rapid response might be due partially to the emergence of weak fluorescence signals emanating from a small fraction of associated fragments of a fluorescent protein followed by rapid chromophore maturation. Of course, detection of weak signals can be achieved by combining highly sensitive fluorescence detection with minimal background in both imaging systems and objects of observation. However, in general, BiFC analysis enables rapid detection of complex formation but does not allow instantaneous visualization of complexes or analysis of the kinetics of complex formation in real time.

Another limitation of BiFC is that it cannot directly identify novel interacting partners, because fluorescence complementation requires that the proteins be tagged with a fragment. The relative efficiencies of BiFC complex formation by different combinations of fluorescent protein fragments do not reflect the equilibriums among alternative interactions. However, they do reflect the relative efficiencies of complex formation among alternative interaction partners at the time of fusion protein synthesis.

## 5. FÖRSTER RESONANCE ENERGY TRANSFER

The abbreviation "FRET" might be better known than its expansion "Förster (or fluorescence) resonance energy transfer." FRET, first theorized in 1948,[82] is a phenomenon of radiationless excitation energy transfer from a donor chromophore to an acceptor chromophore.[83] FRET can be observed only when the donor and acceptor are very close to each other ($\sim 10$ nm; Fig. 8.3A) and when the following additional conditions are applied: (1) substantial overlap between donor emission and acceptor excitation spectra (Fig. 8.3B) and (2) optimal relative orientation between the donor and the acceptor (Fig. 8.3C). The rate of energy transfer $k_T$ ($s^{-1}$) is therefore given by

$$k_T = 8.71 \times 10^{23} \times \frac{J\kappa^2 \lambda_D}{n^4 r^6}. \qquad [8.9]$$

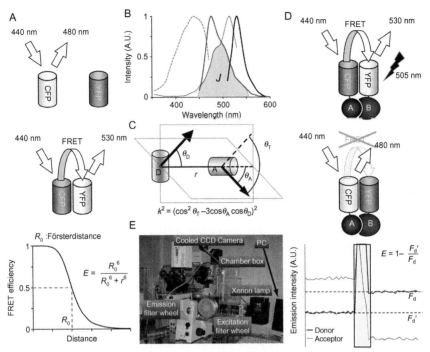

**Figure 8.3** The principle behind FRET. (A) When CFP exists alone, CFP excitation results in CFP emission. When CFP exists close to YFP, the excited energy of CFP is transferred to YFP; thus, YFP emission can be observed. In the bottom graph, the relative FRET efficiency is plotted against the distance between the chromophores (assuming that $\kappa^2 = 2/3$). (B) Spectral overlap ($J$) of donor emission and acceptor excitation spectra. Dashed and solid lines indicate the excitation and emission spectra, respectively, while gray and black lines indicate the spectra of CFP and YFP. (C) The relative orientations of chromophore dipoles and orientation coefficients. (D) Schematic representation of the photobleaching method. In the presence of FRET, disruption of YFP by photobleaching results in recovery of CFP emission. In the bottom graph, a typical photobleaching experiment is shown. From the donor emission intensities before ($F_d'$) and after ($F_d$) photobleaching, the FRET efficiency can be calculated using the equation indicated. (E) An epifluorescence microscopic system for FRET analysis.

Here, $r$ is the distance between the centers of the donor and acceptor chromophores, $\kappa^2$ is an orientation factor for dipole–dipole interaction (Fig. 8.3C), $\lambda_D$ is a rate constant for the donor fluorescence emission, and $J$ is the spectral overlap integral between the donor emission and the acceptor adsorption (Fig. 8.3B). The refractive index $n$ of the medium between the chromophores is generally taken as 1.33 in water; thus, $k_T$ is proportional to $1/r.^{6,84}$ When rotational diffusion prior to energy transfer randomizes the relative orientation of the two chromophores ($\kappa^2$ is assumed to be equal to 2/3), the efficiency of FRET ($E$) can be given by

$$E = \frac{R_0^6}{R_0^6 + r^6},$$  [8.10]

where the Förster distance $R_0$ is the distance between chromophores that provides the half-maximal FRET efficiency ($E/2$) (Fig. 8.3A). Initially, this technique was used by a limited number of biologists; with the explosion of fluorescent protein-based techniques since the cloning of GFP in 1992[2,85] and the discovery of GFP-like proteins from a variety of animals,[6,10,13,86,87] use of FRET has increased. In most GFP-based FRET experiments, CFP and YFP are used as the donor and the acceptor, respectively.

## 5.1. Intermolecular (or bimolecular) FRET

As described above, the GFP chromophore is fixed in the β-barreled structure, and the mobility of fluorescent proteins is restricted because of fusion to partner proteins. Hence, the efficiency of GFP-based FRET reporters largely depends on the orientation factor because of the time of rotation of such a large structure, which is longer than the excited-state lifetime (2–5 ns). Thus, GFP-based FRET cannot be a simple spectroscopic molecular ruler, and the efficiency of FRET must be determined by alternative methods. The FRET efficiency is also defined as the probability of energy transfer per donor excitation event. $E$ can therefore be obtained by measuring the intensity of donor emission in the absence ($F_d$) and presence ($F_d'$) of an acceptor[88,89]:

$$E = 1 - \frac{F_d'}{F_d}.$$  [8.11]

Although the actual measurements are far more complex, for intensity-based measurements of intermolecular FRET, in which the donor and acceptor are not linked covalently, their dissociations guarantee disappearance of FRET. In this case, the fluorescence intensity of the donor is measured to obtain FRET efficiency. A donor fluorescent protein (CFP) and an acceptor fluorescent protein (YFP) are fused to host proteins A and B, respectively, which can associate with each other (Fig. 8.3D). By exciting at 440 nm to a certain extent, $F_d'$ can be obtained (here, it is assumed to be 50). Next, YFP is specifically photobleached by a strong excitation at 510 nm, and $F_d$ (100) is measured with the same amount (as that for obtaining $F_d'$) of excitation at 440 nm. Here, $E$ is given by

$$E = 1 - \frac{50}{100} = 0.5. \qquad [8.12]$$

Therefore, 50% of the excited energy of CFP is transferred to YFP in the presence of A–B interaction. However, an increase in donor fluorescence after photobleaching can occur independently of the association between A and B. In addition, this method can therefore be executed only once because of the destructive nature of the photobleaching necessary to obtain $E$.

It is important to note that, in the aforementioned method, only the intensity of donor fluorescence is measured. Because acceptor molecules (such as YFP) not only absorb but also emit light in most FRET experiments, FRET can also be quantified by measuring the sensitized emission from the acceptor. When the sample is excited at the donor excitation wavelengths, fluorescence intensities from the acceptor and donor molecules are increased and decreased, respectively, in a manner that depends on the efficacy of energy transfer. Therefore, $E$ can be estimated using the ratio of fluorescence intensities emitted at donor versus acceptor emission wavelengths.[90] This dual-emission ratio method can repeatedly measure and reliably detect changes in FRET, regardless of the background FRET levels. However, the dual-emission ratio method is not as quantitative as donor fluorescence measurement because the ratio is affected by unavoidable factors deriving from optical instruments (specification of the used filters or detectors).[60] Thus, the acceptor/donor emission ratio can be used as an index of the degree of FRET.

The emission ratio, specifically the acceptor emission obtained by excitation of the donor, is also affected by factors other than the FRET efficiency. In the case of CFP–YFP FRET, excitation of the donor molecule (440 nm for CFP) also excites YFP, albeit less efficiently. Also, the CFP emission leaks into the FRET channel. Therefore, stoichiometric alterations in the amounts of donor and acceptor result in observed intensities of sensitized emission even in the absence of changes in FRET efficiency. To overcome this issue, researchers who wish to obtain reliable FRET indices must perform spectral unmixing. As described in Section 2.5, cells expressing only the donor or the acceptor are imaged through three sets of optical filters (donor excitation–donor emission, donor excitation–acceptor emission, and acceptor excitation–acceptor emission) under the same imaging conditions used for FRET observation. This experiment provides slopes that represent the bleed-through of CFP and YFP into the FRET channel (i.e., $CFP_{bt} = a \times CFP$; $YFP_{bt} = b \times YFP$; bt, bleed-through). In the case of FRET with CFP and YFP, the level of cross talk between the donor and the acceptor is negligible. Cells expressing both CFP and YFP are then observed. Reliable FRET indices are provided by calculation of[91]

$$FRET_c = FRET - a \times CFP - b$$
$$\times YFP(= FRET - CFP_{bt} - YFP_{bt}) \qquad [8.13]$$

and

$$\text{Emission ratio} = \frac{FRET_c}{CFP}. \qquad [8.14]$$

Various methods introducing different observation strategies for FRET indices and efficiency have been also reported.[92–94]

When intermolecular FRET is utilized *in vitro*, such arithmetic processing is no longer necessary to obtain accurate FRET indices. For example, if donor- and acceptor-labeled proteins are synthesized in bacteria or insect cells or via *in vitro* translation and mixed together at a 1:1 ratio in a solution in which the proteins can freely diffuse, accurate FRET indices can be obtained by measuring the emission ratio because the invariable concentrations of donor and acceptor keep their bleed-throughs constant. In addition, their dissociation results in a close-to-infinite distance between them and guarantees the absence of FRET in the absence of interaction. Thus, *in vitro*, intermolecular FRET systems generally produce larger responses than do intramolecular FRET systems, in which researchers must use trial and error to optimize the relative orientations among the donor, acceptor, and host proteins to gain a greater increase or decrease in FRET in response to interactions or conformational changes (see next section).

## 5.2. Intramolecular (or unimolecular) FRET

While intermolecular FRET is useful for monitoring protein–protein interactions, intramolecular FRET is well suited for the analysis of conformational changes. However, intramolecular FRET constructs have a great advantage over intermolecular FRET *in vivo*, with respect to the stoichiometric issues described above. For example, expression levels of donor- or acceptor-tagged proteins are influenced by the properties of the host proteins, making it difficult to control relative expression levels of the donor and the acceptor. In addition, exogenous proteins can bind to endogenous partners, which dampens an increase in FRET in response to protein interaction. In contrast, intramolecular FRET constructs are fairly insensitive to endogenous molecules and contain the same amounts of donor and acceptor (in the absence of protein cleavage or degradation), making the emission ratio a reliable FRET index. In other words,

$$YFP = c \times CFP. \qquad [8.15]$$

When combined with Eq. (8.14), this becomes

$$\frac{FRET}{CFP} = \frac{[FRET_c + a \times CFP + b \times (c \times CFP)]}{CFP}$$

$$= \frac{FRET_c}{CFP} + (a + bc). \qquad [8.16]$$

On the other hand, obtaining "good" intramolecular FRET-based biosensors that are well suited for practical use is a major barrier to the use of this method. A number of biosensors that enable visualization of cellular events under physiological conditions have been developed and have been summarized in several excellent reviews[83,95]; thus, we will refrain from discussing each individual biosensor. Much work has been done in an effort to construct better biosensors than current biosensor versions. A rational trial to improve the properties of biosensors involves introduction of circular permutations into one or both of the donor/acceptor fluorescent proteins. Dr. Miyawaki and colleagues, who originally developed the FRET-based $Ca^{2+}$ biosensor Cameleon and its derivate yellow Cameleon (YC), have succeeded in generating an improved YC by conducting circular permutation on Venus, the acceptor of YC.[96] The improved YC3.60, which possesses cp173 Venus (starting its new amino terminus at Asp173), is equally bright but shows a five- to sixfold larger dynamic range compared with YC3.12 with "wild-type" Venus. We have also improved the CrkL-based biosensor[5] using a similar strategy in which circular permutation was conducted on the donor fluorescent protein, ECFP (see Section 7). In addition, a similar approach using multiple cpCFPs and cpYFPs was used to enhance the dynamic range of a troponin C (TnC)-based $Ca^{2+}$ indicator, TN-L15.[97,98] Despite these achievements, it remains unclear whether it is the distance between or the relative orientation of the chromophores that is responsible for the improved dynamic ranges. Thus, an enormous number of experiments to test such constructs are still required to generate improved combinations of donors and acceptors for each FRET biosensor.

Another rational strategy for the improvement of FRET constructs is fine-tuning of the linker regions within the biosensors. For example,

Dr. Nagai and colleagues recently developed YCs with high $Ca^{2+}$ affinity ($K_d' = 15$–$140$ nM, Cameleon-Nano) and 1450% signal changes by replacing the polyglycine linker with a longer form.[99] More recently, Dr. Matsuda and colleagues reported an optimized backbone for FRET-based biosensors (the extension for enhanced visualization by evading extra-FRET (Eevee) system).[100] In this system, introduction of a long, flexible linker ranging from 116 to 244 amino acids in length makes FRET biosensors almost completely dependent on the distance between the donor (ECFP) and the acceptor (Ypet). The Eevee system has succeeded in improving a variety of biosensors for tyrosine kinases, serine/threonine kinases, and GTPases.[100]

## 6. METHODS FOR EVALUATING FRET EFFICIENCY

The basis for obtaining FRET efficiency by the sensitized emission method was described in the preceding section (see Section 5.1). Although fluorescence plate readers, fluorescence spectrometers, or similar equipment may be used, fluorescence microscopy is the most popular method for evaluation; Using microscopy, dynamic cellular events, including protein interaction and activation, can be analyzed in living cells or organisms. General requirements for such experiments include an epifluorescence microscope, a cooled charge-coupled device (CCD) for image acquisition, and filters suited for the excitation and emission spectra of the donor and the acceptor (Fig. 8.3E). Mechanical shutter(s) and filter wheels (in which filters are set) under the control of imaging software may be essential in time-lapse experiments to avoid excess exposure of cells to light during the periods of acquisition and to acquire two or three images for FRET evaluation, respectively. After image acquisition, image-processing software may help to calculate FRET indices for each pixel. As described above, in intramolecular FRET systems, FRET indices are estimated by simply calculating the ratio of acceptor emission to donor emission (that are obtained by donor excitation); in such systems, it is possible to obtain FRET indices from simultaneous acquisition of emission images with the use of a spectral separator. Such equipment is commercially available (Dual-View from Photometrics or W-View from Hamamatsu Photonics) and enables imaging with high temporal resolution. Details of the practical experimental procedures are provided in an excellent review.[101]

## 6.1. Donor fluorescence lifetime as a measure of FRET efficiency

Measurement of the lifetime of donor fluorescence is the most reliable method to "quantify" FRET efficiency.[102] Lifetime measurement is independent of chromophore concentrations and thus quantitatively comparable from sample to sample. In this approach, $E$ is given by

$$E = 1 - \frac{\tau_d'}{\tau_d}, \qquad [8.17]$$

where $\tau_d'$ and $\tau_d$ are the donor lifetimes in the presence and absence of the acceptor, respectively. After excitation by a short laser pulse, the excited-state donor (when it contains a single component of lifetime) decays with a single exponential time course ($\tau_d$). FRET results in a short lifetime ($\tau_d'$) (see below and Fig. 8.4A).

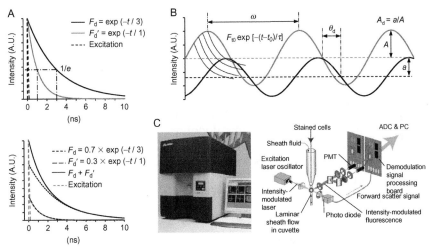

**Figure 8.4** FRET evaluation by fluorescence lifetime. (A) Upper panel: time-dependent decay of fluorescence intensity with lifetimes of 3 ns (black line) and 1 ns (gray) in response to pulsed excitation (dashed). Lower panel: the resulting averaged fluorescence lifetime (solid line) when the bound fraction (with a fluorescent lifetime of 1 ns) and unbound fraction (3 ns) are mixed in the ratio 3:7. (B) Principle of the frequency-domain method for the measurement of fluorescence lifetime. Modulated fluorescent emissions (black), in response to modulated excitations (gray) with a circular frequency of $\omega$, are shown, along with phase differences and amplitude ratios. (C) A cytometer for lifetime measurement, Flicyme. Exterior view (left) and interior optical paths with electronic circuits (right) are shown.

## 6.2. Time domain

Changes in the fluorescence intensity (or the number of photons emitted) of the donor chromophore [$F_d(t)$ (photons $cm^{-3}s^{-1}$)] excited by the ideal pulsed light source at time point 0 can be given by

$$F_d(t) = F_0 \exp\left(\frac{-t}{\tau_d}\right) \left(\text{or simply } F_d(t) \propto \exp\left(\frac{-t}{\tau_d}\right)\right), \qquad [8.18]$$

where $F_0$ (photons $cm^{-3}$ $s^{-1}$) represents the fluorescence intensity at time point 0 (s) (i.e., the product of the number of excited electrons of fluorescence molecules (electrons $cm^{-3}$) and the rate constant of the radiative transition state $(s^{-1})$).[103] Thus, the observed fluorescence decays exponentially (Fig. 8.4A). Also, the change in fluorescence intensity of the donor in the presence of acceptor $F_d'(t)$ is given by

$$F_d'(t) = F_0 \exp\left(-\frac{t}{\tau_d'}\right) \quad \text{or} \quad F_d'(t) \propto \exp\left(-\frac{t}{\tau_d'}\right). \qquad [8.19]$$

When the proportion of the donor-bound acceptor is given by $a$ (unbound form, $1 - a$), the time-dependent change in donor fluorescence intensity is therefore given by

$$F_d(t) \propto a \cdot \exp\left(-\frac{t}{\tau_d'}\right) + (1 - a) \cdot \exp\left(-\frac{t}{\tau_d}\right). \qquad [8.20]$$

Examples of $F_d(t)$ ($\tau_d = 3$ [ns]) and $F_d(t)'$ ($\tau_d' = 1$ [ns]), as well as $F(t)$ when $a$ is 0.3, are shown in Fig. 8.3A. Thus, donor occupancy ($a$) can be calculated by fitting the fluorescence decay obtained by fluorescence lifetime imaging microscopy (FLIM). The FLIM method of time domain is often combined with two-photon excitation microscopy, which is coupled with femtosecond-pulsed lasers.

## 6.3. Frequency domain

In contrast, in the frequency-domain method, time-dependent changes in fluorescence intensities in response to modulated excitation power are continuously acquired. Therefore, this method can be used with a wide-field fluorescence microscope equipped with faster image acquisition devices such as an image intensifier or electron multiplying CCD (EMCCD). When fluorescent proteins (or dyes) are excited by a sine wave with an angular frequency of $\omega$ $(s^{-1})$, the fluorescence signal output is elicited as a sine curve with the same frequency but a distinct phase and amplitude (difference in phases and ratio in amplitudes are $\theta_d$ and $A_d$, respectively; Fig. 8.5B).[104,105] From the difference in phases, a fluorescence lifetime is given by

$$\tau_{d,\theta} = \frac{\tan\theta_d}{\omega}, \qquad [8.21]$$

$$\tau_{d,A} = \frac{\sqrt{(1/A_d)^2 - 1}}{\omega}. \qquad [8.22]$$

where $\tau_{d,\theta}$ and $\tau_{d,A}$ represent phase and modulation lifetimes, which are a unique value in the case of fluorescent proteins with a single lifetime component (Fig. 8.5B). One of the problems inherent in the frequency-domain method is that the excited state of CFP derived from *Aequorea* GFP decays with multiple (three or four) time constants, making frequency-domain lifetime measurements of FRET using CFP as a donor complicated. Several fluorescent proteins that can be substituted for CFP, including Cerulean and mTurquoise, are reported to possess a single lifetime component.[106,107]

## 6.4. Lifetime measurement using a flow cytometer

Although fluorescence microscopy has been and continues to be the main tool used to perform FRET experiments, the number of cells that can be observed within a period is restricted. For quantitative measurement, techniques aimed at obtaining FRET indices with high throughput may be necessary. Fluorescence flow cytometry is a good candidate for handling a vast number of cells in a short period; however, subtle studies regarding the successful application of this technique are available. We would therefore like to introduce a unique instrument, the world's first flow cytometer designed to detect fluorescence lifetime along with fluorescence intensity, which we are now developing. This machine, named "Flicyme," for fluorescence lifetime cytometer (Fig. 8.5C), possesses great potential for the highly quantitative observation of FRET efficiency. In addition, it can be easily utilized for high-throughput drug screening, especially for molecular targeted drugs.

Flicyme determines the fluorescence lifetime and fluorescence intensity for each individual living cell that passes through a flow cell at high speed. Key specifications of Flicyme are listed in Table 8.2. Cells are illuminated by the diode laser (407 or 440 nm, 60 mW) with a modulation frequency of 28 MHz, which results in the acquisition of lifetime with a precision of 0.02 ns and at a maximum rate of 10,000 cells/s. These achievements are highly advantageous for facilitating FRET experiments in many situations,

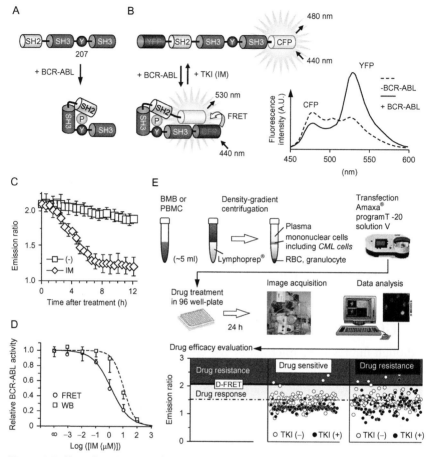

**Figure 8.5** Clinical application of the FRET biosensor Pickles. (A) Structure of CrkL and its change in response to Tyr207 phosphorylation by BCR-ABL. (B) Structure of Pickles. CrkL is sandwiched by YFP and CFP. Phosphorylation on Y207 and its reversal, respectively, increase and decrease FRET efficiency. Typical spectra of Pickles in the presence (solid) and absence (gray) of BCR-ABL are also shown (right panel). (C) Time- and IM-dependent decrease in the FRET efficiency of Pickles. CML cells were treated with vehicle (square) or IM (rhombus) at time point 0. Emission ratios (FRET/CFP) were obtained using fluorescent microscopy, and their ratios are plotted, along with SD. (D) Comparison of the dose responsiveness of the biosensor and of Western blotting. Emission ratio of Pickles (FRET) and levels of phosphorylated CrkL (WB) at the indicated concentrations were quantitated, normalized to those observed in the absence of IM, and plotted, along with SD. (E) Flowchart of FRET-based evaluation of drug efficacy in patient samples. Finally, the patient is judged as resistant when cells with FRET > 2.04 remain after IM treatment (right), while cells with emission ratios between 1.5 and 2.02 are used to evaluate drug responsiveness.

**Table 8.2** Specifications of Flicyme

| Light source | 407 or 445 nm laser diode (60 mW) |
|---|---|
| Detection parameter and detector | Scattering light, 2 ch: forward scatter, photodiode; side scatter, PMT<br>Fluorescence intensity and lifetime, 3 ch: PMT |
| Detection fluorescence wavelength | Ch1: 482/35 nm (e.g., CFP or AG)<br>Ch2: 542/30 nm (e.g., YFP) or 579/34 nm (e.g., KO)<br>Ch3: 650-long pass (e.g., Keima) |
| Sample flow rate | 6 m/s |
| Sample flow volume ($\mu l\ min^{-1}$) | Lo: 40, mid: 80, hi: 160 |
| Maximum acquisition rate | 10,000 events/s |
| Modulation frequency | 28 MHz |
| Size (mm) | 500 (*W*) × 675 (*D*) × 700 (*H*) |

PMT, photomultiplier; AG, Azami Green; KO, Kusabira Orange.

including initial screening of FRET constructs (to select one with greater gains), drug screening in which comparison of FRET efficiencies among samples is required, and detection of a small subpopulation within a heterogeneous population via observation of a large number of cells (see below).

## 7. CLINICAL APPLICATION OF FRET

In the latter part of this review, we would like to introduce a novel diagnostic method based on bioimaging technology and fluorescent proteins, using chronic myeloid leukemia (CML) as a model. CML, a hematological malignancy involving the transformation of hematopoietic stem cells in the bone marrow, is characterized by the formation of an abnormal chromosome (Philadelphia chromosome, Ph1) and the expression of its transcript BCR-ABL.[108] About 8000 people in Japan and 63,000 worldwide, representing about one-fifth to one-fourth of all leukemia patients, suffer from this disease. BCR-ABL encodes a constitutively active tyrosine kinase and causatively contributes to malignant transformation of the leukemia cells by activating a range of signaling pathways through tyrosine phosphorylation of its substrates, including CrkL and signal transducer and activator of transcription.[109,110]

In 2001, the tyrosine kinase inhibitor imatinib mesylate (IM) was approved as a first-line therapy for CML by the U.S. Food and Drug Administration (FDA). Emergence of this drug, the therapeutic outcomes of which are better than those of other options, has radically improved therapy for CML and has provided a strategy to control the disease, other than allogeneic hematopoietic stem cell transplantation (allo-HSCT).[111,112] On the other hand, for a considerable number of patients, either IM fails to work from the beginning or resistance to IM develops during treatment. The mechanisms of IM resistance can be dependent or independent of mutations in the Abl kinase domain; such mutations occur in ~50% of IM-resistant patients. In the former case, because the second-generation drugs nilotinib (NL) and dasatinib are now available for the treatment of IM-resistant patients, as well as for first-line treatment, the effectiveness of these medicines may be predictable and dependent on the types of mutation present.[113] When there is no mutation present, however, the only way in which to evaluate drug efficacy is to actually prescribe the drug and follow the course of treatment by blood and bone marrow tests for several months to a year or more. Given that treatment with an ineffective medicine cannot be expected to hamper progression of the disease, the most suitable medication should, ideally, be used from the start.[114] To solve this problem and to optimally select the most effective drug for each individual patient, it is necessary to develop a technique(s) that evaluates the effect of every available medication on each patient's leukemic cells and detects the presence of any resistant cells before starting therapy.

## 7.1. Biosensor design and its specification

To address the aforementioned issues, we developed a novel biosensor to monitor BCR-ABL activity in living cells.[5] CrkL,[115] an authorized representative of the substrates of BCR-ABL,[116–118] was used for the backbone of the biosensor. CrkL harbors one Src homology 2 (SH2) domain and two SH3 domains in addition to the tyrosine residue that is phosphorylated by BCR-ABL (Fig. 8.5A). Of these, the SH2 domain possesses the ability to bind to phosphorylated tyrosine, leading to structural changes upon phosphorylation by BCR-ABL or the inhibition of BCR-ABL activity in response to drug treatment (Fig. 8.5A). The biosensor for BCR-ABL activity was designed on the basis of this characteristic of CrkL; in our construct, CrkL is sandwiched between CFP and YFP.

CrkL phosphorylation by BCR-ABL triggers a structural change in CrkL, leading to an increase in FRET efficiency due to the close proximity of CFP and YFP; conversely, inhibition of BCR-ABL kinase activity by inhibitors results in a decrease in FRET (Fig. 8.5B). The final version of the biosensor, named "Pickles" for phosphorylation indicator of CrkL *en* substrate, required fine-tuning beyond merely adding CFP and YFP to either end of CrkL. As in the case of development of other FRET biosensors, certain modifications, including truncation of CrkL, circular permutations of CFP, and monomeric mutation in of YFP, were required for its development and clinical application, and the detailed procedures are described in our previously published research paper.[5]

The final product of Pickles (version 2.31; Pickles_2.31) displayed an increase in FRET efficiency of approximately 80–100% upon phosphorylation by BCR-ABL; this increase was specific for BCR-ABL (and its cellular counterpart c-Abl) over other nonreceptor-type tyrosine kinases tested. Because our biosensor was designed based on bioimaging techniques, a faithful time-dependent decrease in FRET efficiency was observed in the drug-treated cells compared to the control cells using time-lapse microscopy in tumor cells (Fig. 8.5C). In addition, when we tested its dose-dependent response to IM, Pickles could detect the drug effect at lower concentrations and in a wider measureable range compared to other existing techniques, for example, Western blotting (Fig. 8.5D). Moreover, we noticed that the time required for evaluation is ∼12 h; BCR-ABL activity reached a nadir 3–6 h after drug treatment (Fig. 8.5C).

## 7.2. Special features of Pickles

As described above, GFP-based FRET technology was originally used to observe the spatiotemporal activation of proteins in living cells. Therefore, the most prominent feature of our Pickles system is that we can perform evaluation of drug responsiveness at the single-cell level using FRET technology. In fact, a drug-resistant cell population of 1% or less can be distinguished from other drug-sensitive cells. In our system, in which a researcher manually observes a few hundreds of cells with a microscope, 1% is the highest sensitivity. To improve sensitivity, we attempted similar analyses using a combination of immunofluorescence with an antibody specific for phosphorylated CrkL and a flow cytometer, which can simultaneously analyze significantly higher cell numbers; however, unexpectedly, the detectability of resistant cells was the same as that in our system. This highlights the low noise and high potential of our system, indicating that when we are able to increase the number of analyzed cells, we will able to identify less resistant cells. Therefore, we are currently intensely engaged in increasing the analyzed cell number by (1) automating image acquisition with computer-controlled architecture and (2) measuring lifetime using Flicyme (see Section 6.4).

The single-cell analysis made possible by Pickles also increases our understanding of the mechanisms underlying the disease. The common mutations in BCR-ABL and their relationship to drug susceptibility have been extensively examined; thus, if the mutation observed in a given patient is a common one, it is relatively straightforward to determine the most effective drug. However, the effectiveness of the second-generation drug NL on cells containing the Gly250Glu BCR-ABL mutant (hereafter referred to as G250E) remains controversial.[119,120] We thus attempted to investigate this issue using Pickles. Each individual cell expressing the G250E mutant of BCR-ABL showed diverse NL sensitivities, resulting in a large standard deviation. When we examined the dose-dependent responses of BCR-ABL, we found that drug responsiveness decreased in a manner that was dependent on the expression level of BCR-ABL only for the G250E mutant and NL combination. Therefore, the expression level of BCR-ABL with the G250E mutation in the patients' tumor cells might have been low in the study that described this mutation as being sensitive to NL but high in the report that concluded that G250E is resistant to NL. Although it has previously been reported that drug susceptibility decreases with an increase in BCR-ABL expression,[121] this tendency is especially important for the G250E mutant. In either case, it is necessary to take the expression level of BCR-ABL into consideration, specifically when this mutation is detected, and to note that drug options may vary depending on BCR-ABL expression levels.

## 7.3. Evaluation of drug efficacy in patient-derived cells

Approval from the Internal Review Board in our research institute has already been obtained for this system, and drug efficacy is being evaluated using patient-derived tumor cells. More than 50 cases have been examined over the past 4 years. This project is still ongoing and includes many cases that will require careful follow-up. However, from the patients who were incorporated into the study at earlier stages, we have learned that our system is able to predict patients' drug susceptibilities and detect the existence of drug-resistant cells by examination prior to starting treatment. The examination uses peripheral blood or bone marrow samples obtained by blood specimen collection or bone marrow aspiration; although these sampling techniques are invasive, they are included in conventional tests for CML, so our analyses result in no additional patient burdens. For each sample, after the separation of mononuclear cells (including the tumor cells) by density gradient centrifugation, cDNA encoding Pickles is introduced into the cells by electroporation. After confirming fluorescence emission at 24 h, drug treatment is begun. The FRET efficiency in individual cells is then calculated from the image data obtained using fluorescence microscopy (Fig. 8.5E).

Given that tumor cells in the chronic phase of CML contain cells at each step of differentiation, it is not necessarily the case that BCR–ABL activity is high in all leukemic cells. This phenomenon is reflected in the range of FRET efficiencies observed. For this reason, a threshold value (D-FRET) was used to identify cells in which BCR–ABL activity is high. This value was mathematically derived from the mean and the standard deviation of the FRET efficiency in cells that did not express BCR–ABL. Given that the probability for cells with low BCR–ABL activity to exceed D-FRET is less than 0.003, the BCR–ABL activity of cells that exhibit higher FRET efficiencies than this value is significantly high.

We are currently engaged in drug efficacy evaluation using the aforementioned processes and criteria. If cells from both peripheral blood and bone marrow with values above D-FRET disappear upon drug treatment, the patient is considered sensitive, whereas if they remain above this value, we conclude that the patient is resistant (Fig. 8.5E). In addition, when the distribution pattern of FRET efficiency was carefully compared with the clinical course, we saw that we were able to anticipate the patients' responses to drugs, albeit only partially, using this analysis, which was performed at the initial stage of the disease. For instance, we experienced a case in which drug-resistant cells existed only in the bone marrow sample, while the peripheral blood cells were susceptible to drug treatment. At the beginning of treatment, a decrease in the number of tumor cells was observed; however, the patient had achieved neither a complete cytogenetic response nor a major molecular response after 12 months of therapy (a suboptimal response) and had to be put on a different dosage and ultimately different drugs. One can speculate that the resistant cells that had been detected during the initial examination remained and caused relapse.

## 7.4. Future tasks

In our system, the biosensor is introduced into tumor cells via transfection, and its expression level depends on the ability of the tumor cell to synthesize the protein. An outstanding issue, therefore, is that the success or failure of this diagnosis is influenced by the gene introduction efficiency. From our knowledge, different transfection efficiencies are observed in different patients, even among those whose leukemia stage or leukemic cell populations are similar. The development of a suitable congenic method is unquestionably needed, and we are looking into this issue.

Another problem is the distribution of this technology to a broad range of clinical sites. As for the analysis, the processes have already been automated by the image-processing software MetaMorph (Universal Imaging); most of the image acquisition processes depend on manual work with a microscope and thus requires skilled researchers. A system through which anyone can easily analyze samples, for example, automated image acquisition by a computer-controlled microscope, is also being developed. This will also help to reduce the time required for the overall diagnostic processes. Alternatively, there is a plan to build up a center to collect samples and analyze them. In this case, because fresh samples are better for gene transfer, it would also be necessary to overcome the problem of preservation and transportation of the samples. After overcoming these issues, we aim to establish this technology for use in leukemia diagnostics as soon as possible.

## 8. CLOSING REMARKS

In recent years, the use of fluorescent proteins has literally thrown much light on biological research. GFP-based FRET is an epochal imaging technology; events occurring in living cells can now be captured as a color change. However, FRET has historically been limited to the basic research field. The research introduced here is the first step to widen GFP's appeal to clinicians and is a "pioneer study" for tailor-made medicines.[122] However, we believe that, for our work to be clinically useful, it must become widespread, and we will therefore make a zealous effort to distribute these techniques. The advantages of this technology include not only therapeutic improvement but also a decrease in the time required to obtain the maximum treatment effect. Moreover, given the high cost of molecular targeted drugs, the spread of this technology to clinical sites will definitely lead to the practice of efficient and economical medicine. In addition to the clinical benefits, the specific advantage of this system, that is, analysis of living cells, may contribute to basic research into the mechanisms of drug resistance in CML cells.

## REFERENCES

1. Morise H, Shimomura O, Johnson F, Winant J. Intermolecular energy transfer in the bioluminescent system of Aequorea. *Biochemistry* 1974;**13**:2656–62.
2. Prasher D, Eckenrode V, Ward W, Prendergast F, Cormier M. Primary structure of the Aequorea victoria green-fluorescent protein. *Gene* 1992;**111**:229–33.
3. Heim R, Prasher DC, Tsien RY. Wavelength mutations and posttranslational autoxidation of green fluorescent protein. *Proc Natl Acad Sci USA* 1994;**91**:12501–4.
4. Yarbrough D, Wachter RM, Kallio K, Matz MV, Remington SJ. Refined crystal structure of DsRed, a red fluorescent protein from coral, at 2.0-A resolution. *Proc Natl Acad Sci USA* 2001;**98**:462–7.
5. Mizutani T, Kondo T, Darmanin S, Tsuda M, Tanaka S, Tobiume M, et al. A novel FRET-based biosensor for the measurement of BCR-ABL activity and its response to drugs in living cells. *Clin Cancer Res* 2010;**16**:3964–75.
6. Chudakov DM, Matz MV, Lukyanov S, Lukyanov KA. Fluorescent proteins and their applications in imaging living cells and tissues. *Physiol Rev* 2010;**90**:1103–63.
7. Pakhomov AA, Martynov VI. GFP family: structural insights into spectral tuning. *Chem Biol* 2008;**15**:755–64.
8. Remington SJ. Fluorescent proteins: maturation, photochemistry and photophysics. *Curr Opin Struct Biol* 2006;**16**:714–21.
9. Remington SJ. Negotiating the speed bumps to fluorescence. *Nat Biotechnol* 2002;**20**:28–9.
10. Shaner NC, Steinbach PA, Tsien RY. A guide to choosing fluorescent proteins. *Nat Methods* 2005;**2**:905–9.
11. Shu X, Shaner NC, Yarbrough CA, Tsien RY, Remington SJ. Novel chromophores and buried charges control color in mFruits. *Biochemistry* 2006;**45**:9639–47.
12. Baldini G, Cannone F, Chirico G. Pre-unfolding resonant oscillations of single green fluorescent protein molecules. *Science* 2005;**309**:1096–100.
13. Day RN, Davidson MW. The fluorescent protein palette: tools for cellular imaging. *Chem Soc Rev* 2009;**38**:2887–921.
14. Kremers G-J, Gilbert SG, Cranfill PJ, Davidson MW, Piston DW. Fluorescent proteins at a glance. *J Cell Sci* 2011;**124**:157–60.
15. Davidson MW, Campbell RE. Engineered fluorescent proteins: innovations and applications. *Nat Methods* 2009;**6**:713–7.
16. Shaner NC, Patterson GH, Davidson MW. Advances in fluorescent protein technology. *J Cell Sci* 2007;**120**:4247–60.
17. Barondeau DP, Kassmann CJ, Tainer JA, Getzoff ED. Understanding GFP posttranslational chemistry: structures of designed variants that achieve backbone fragmentation, hydrolysis, and decarboxylation. *J Am Chem Soc* 2006;**128**:4685–93.
18. Barondeau DP, Kassmann CJ, Tainer JA, Getzoff ED. The case of the missing ring: radical cleavage of a carbon-carbon bond and implications for GFP chromophore biosynthesis. *J Am Chem Soc* 2007;**129**:3118–26.
19. Tomosugi W, Matsuda T, Tani T, Nemoto T, Kotera I, Saito K, et al. An ultramarine fluorescent protein with increased photostability and pH insensitivity. *Nat Methods* 2009;**6**:351–3.
20. Heim R, Cubitt AB, Tsien RY. Improved green fluorescence. *Nature* 1995;**373**:663–4.

21. Gross LA, Baird GS, Hoffman RC, Baldridge KK, Tsien RY. The structure of the chromophore within DsRed, a red fluorescent protein from coral. *Proc Natl Acad Sci USA* 2000;**97**:11990–5.

22. Pletnev S, Subach FV, Dauter Z, Wlodawer A, Verkhusha VV. Understanding blue-to-red conversion in monomeric fluorescent timers and hydrolytic degradation of their chromophores. *J Am Chem Soc* 2010;**132**:2243–53.

23. Strack RL, Strongin DE, Mets L, Glick BS, Keenan RJ. Chromophore formation in DsRed occurs by a branched pathway. *J Am Chem Soc* 2010;**132**:8496–505.

24. Baird G, Zacharias D, Tsien R. Circular permutation and receptor insertion within green fluorescent proteins. *Proc Natl Acad Sci USA* 1999;**96**:11241–6.

25. Willumsen BM, Norris K, Papageorge AG, Hubbert NL, Lowy DR. Harvey murine sarcoma virus p21 ras protein: biological and biochemical significance of the cysteine nearest the carboxy terminus. *EMBO J* 1984;**3**:2581–5.

26. Giraldez T, Hughes TE, Sigworth FJ. Generation of functional fluorescent BK channels by random insertion of GFP variants. *J Gen Physiol* 2005;**126**:429–38.

27. Huston JS, Levinson D, Mudgett-Hunter M, Tai MS, Novotný J, Margolies MN, et al. Protein engineering of antibody binding sites: recovery of specific activity in an anti-digoxin single-chain Fv analogue produced in Escherichia coli. *Proc Natl Acad Sci USA* 1988;**85**:5879–83.

28. Campbell R, Tour O, Palmer A, Steinbach P, Baird G, Zacharias D, et al. A monomeric red fluorescent protein. *Proc Natl Acad Sci USA* 2002;**99**:7877–82.

29. Alfthan K, Takkinen K, Sizmann D, Söderlund H, Teeri TT. Properties of a single-chain antibody containing different linker peptides. *Protein Eng* 1995;**8**:725–31.

30. Ballestrem C, Wehrle-Haller B, Imhof BA. Actin dynamics in living mammalian cells. *J Cell Sci* 1998;**111**(Pt 12):1649–58.

31. Garini Y, Young IT, McNamara G. Spectral imaging: principles and applications. *Cytometry A* 2006;**69**:735–47.

32. Harris AT. Spectral mapping tools from the earth sciences applied to spectral microscopy data. *Cytometry A* 2006;**69**:872–9.

33. Zimmermann T. Spectral imaging and linear unmixing in light microscopy. *Adv Biochem Eng Biotechnol* 2005;**95**:245–65.

34. Zimmermann T, Rietdorf J, Girod A, Georget V, Pepperkok R. Spectral imaging and linear un-mixing enables improved FRET efficiency with a novel GFP2-YFP FRET pair. *FEBS Lett* 2002;**531**:245–9.

35. Bizzarri R, Serresi M, Luin S, Beltram F. Green fluorescent protein based pH indicators for in vivo use: a review. *Anal Bioanal Chem* 2009;**393**:1107–22.

36. Hanson GT, Aggeler R, Oglesbee D, Cannon M, Capaldi RA, Tsien RY, et al. Investigating mitochondrial redox potential with redox-sensitive green fluorescent protein indicators. *J Biol Chem* 2004;**279**:13044–53.

37. Mizuno T, Murao K, Tanabe Y, Oda M, Tanaka T. Metal-ion-dependent GFP emission in vivo by combining a circularly permutated green fluorescent protein with an engineered metal-ion-binding coiled-coil. *J Am Chem Soc* 2007;**129**:11378–83.

38. Souslova EA, Belousov VV, Lock JG, Strömblad S, Kasparov S, Bolshakov AP, et al. Single fluorescent protein-based Ca2+ sensors with increased dynamic range. *BMC Biotechnol* 2007;**7**:37.

39. Terskikh A. Fluorescent timer protein that changes color with time. *Science* 2000;**290**:1585–8.

40. Subach FV, Subach OM, Gundorov IS, Morozova KS, Piatkevich KD, Cuervo AM, et al. Monomeric fluorescent timers that change color from blue to red report on cellular trafficking. *Nat Chem Biol* 2009;**5**:118–26.

41. Subach FV, Patterson GH, Manley S, Gillette JM, Lippincott-Schwartz J, Verkhusha VV. Photoactivatable mCherry for high-resolution two-color fluorescence microscopy. *Nat Methods* 2009;**6**:153–9.

42. Subach FV, Patterson GH, Renz M, Lippincott-Schwartz J, Verkhusha VV. Bright monomeric photoactivatable red fluorescent protein for two-color super-resolution sptPALM of live cells. *J Am Chem Soc* 2010;**132**:6481–91.

43. van Thor JJ, Gensch T, Hellingwerf KJ, Johnson LN. Phototransformation of green fluorescent protein with UV and visible light leads to decarboxylation of glutamate 222. *Nat Struct Mol Biol* 2002;**9**:37–41.

44. Miesenböck G, De Angelis DA, Rothman JE. Visualizing secretion and synaptic transmission with pH-sensitive green fluorescent proteins. *Nature* 1998;**394**:192–5.

45. Tantama M, Hung YP, Yellen G. Imaging intracellular pH in live cells with a genetically encoded red fluorescent protein sensor. *J Am Chem Soc* 2011;**133**:10034–7.

46. Hanakam F, Albrecht R, Eckerskorn C, Matzner M, Gerisch G. Myristoylated and non-myristoylated forms of the pH sensor protein hisactophilin II: intracellular shuttling to plasma membrane and nucleus monitored in real time by a fusion with green fluorescent protein. *EMBO J* 1996;**15**:2935–43.

47. Ormo M, Cubitt A, Kallio K, Gross L, Tsien R, Remington S. Crystal structure of the Aequorea victoria green fluorescent protein. *Science* 1996;**273**:1392–5.

48. Ehrig T, O'Kane DJ, Prendergast FG. Green-fluorescent protein mutants with altered fluorescence excitation spectra. *FEBS Lett* 1995;**367**:163–6.

49. Heim R, Tsien RY. Engineering green fluorescent protein for improved brightness, longer wavelengths and fluorescence resonance energy transfer. *Curr Biol* 1996;**6**:178–82.

50. Kogure T, Karasawa S, Araki T, Saito K, Kinjo M, Miyawaki A. A fluorescent variant of a protein from the stony coral Montipora facilitates dual-color single-laser fluorescence cross-correlation spectroscopy. *Nat Biotechnol* 2006;**24**:577–81.

51. Jayaraman S, Haggie P, Wachter RM, Remington SJ, Verkman AS. Mechanism and cellular applications of a green fluorescent protein-based halide sensor. *J Biol Chem* 2000;**275**:6047–50.

52. Dimitrov D, He Y, Mutoh H, Baker BJ, Cohen L, Akemann W, et al. Engineering and characterization of an enhanced fluorescent protein voltage sensor. *PLoS ONE* 2007;**2**: e440.

53. Siegel MS, Isacoff EY. A genetically encoded optical probe of membrane voltage. *Neuron* 1997;**19**:735–41.

54. Knöpfel T, Diez-Garcia J, Akemann W. Optical probing of neuronal circuit dynamics: genetically encoded versus classical fluorescent sensors. *Trends Neurosci* 2006;**29**:160–6.

55. Lundby A, Mutoh H, Dimitrov D, Akemann W, Knöpfel T. Engineering of a genetically encodable fluorescent voltage sensor exploiting fast Ci-VSP voltage-sensing movements. *PLoS ONE* 2008;**3**:e2514.

56. Nagai T, Sawano A, Park ES, Miyawaki A. Circularly permuted green fluorescent proteins engineered to sense Ca2+. *Proc Natl Acad Sci USA* 2001;**98**:3197–202.

57. Nakai J, Ohkura M, Imoto K. A high signal-to-noise Ca(2+) probe composed of a single green fluorescent protein. *Nat Biotechnol* 2001;**19**:137–41.

58. Zhao Y, Araki S, Wu J, Teramoto T, Chang Y-F, Nakano M, et al. An expanded palette of genetically encoded Ca$^+$ indicators. *Science* 2011;**333**:1888–91.

59. Miyawaki A, Llopis J, Heim R, McCaffery J, Adams J, Ikura M, et al. Fluorescent indicators for Ca2+ based on green fluorescent proteins and calmodulin. *Nature* 1997;**388**:882–7.

60. Romoser V, Hinkle P, Persechini A. Detection in living cells of Ca2+-dependent changes in the fluorescence emission of an indicator composed of two green fluorescent protein variants linked by a calmodulin-binding sequence. A new class of fluorescent indicators. *J Biol Chem* 1997;**272**:13270–4.

61. Belousov VV, Fradkov AF, Lukyanov KA, Staroverov DB, Shakhbazov KS, Terskikh AV, et al. Genetically encoded fluorescent indicator for intracellular hydrogen peroxide. *Nat Methods* 2006;**3**:281–6.

62. Berg J, Hung YP, Yellen G. A genetically encoded fluorescent reporter of ATP:ADP ratio. *Nat Methods* 2009;**6**:161–6.

63. Johnsson N, Varshavsky A. Split ubiquitin as a sensor of protein interactions in vivo. *Proc Natl Acad Sci USA* 1994;**91**:10340–4.

64. Pelletier JP, Mineau F, Fernandes JC, Duval N, Martel-Pelletier J. Diacerhein and rhein reduce the interleukin 1beta stimulated inducible nitric oxide synthesis level and activity while stimulating cyclooxygenase-2 synthesis in human osteoarthritic chondrocytes. *J Rheumatol* 1998;**25**:2417–24.

65. Rossi F, Charlton CA, Blau HM. Monitoring protein-protein interactions in intact eukaryotic cells by beta-galactosidase complementation. *Proc Natl Acad Sci USA* 1997;**94**:8405–10.

66. Hu C, Chinenov Y, Kerppola T. Visualization of interactions among bZIP and Rel family proteins in living cells using bimolecular fluorescence complementation. *Mol Cell* 2002;**9**:789–98.

67. Wehrman T, Kleaveland B, Her J-H, Balint RF, Blau HM. Protein-protein interactions monitored in mammalian cells via complementation of beta-lactamase enzyme fragments. *Proc Natl Acad Sci USA* 2002;**99**:3469–74.

68. Galarneau A, Primeau M, Trudeau L-E, Michnick SW. Beta-lactamase protein fragment complementation assays as in vivo and in vitro sensors of protein protein interactions. *Nat Biotechnol* 2002;**20**:619–22.

69. Spotts JM, Dolmetsch RE, Greenberg ME. Time-lapse imaging of a dynamic phosphorylation-dependent protein-protein interaction in mammalian cells. *Proc Natl Acad Sci USA* 2002;**99**:15142–7.

70. Paulmurugan R, Umezawa Y, Gambhir SS. Noninvasive imaging of protein-protein interactions in living subjects by using reporter protein complementation and reconstitution strategies. *Proc Natl Acad Sci USA* 2002;**99**:15608–13.

71. Remy I, Michnick SW. A highly sensitive protein-protein interaction assay based on Gaussia luciferase. *Nat Methods* 2006;**3**:977–9.

72. Kerppola T. Visualization of molecular interactions by fluorescence complementation. *Nat Rev Mol Cell Biol* 2006;**7**:449–56.

73. Kerppola TK. Bimolecular fluorescence complementation (BiFC) analysis as a probe of protein interactions in living cells. *Annu Rev Biophys* 2008;**37**:465–87.

74. Ghosh I, Hamilton AD, Regan L. Antiparallel Leucine Zipper-Directed Protein Reassembly: Application to the Green Fluorescent Protein. *J Am Chem Soc* 2000;**122**:5658–9.

75. Bracha-Drori K, Shichrur K, Katz A, Oliva M, Angelovici R, Yalovsky S, et al. Detection of protein-protein interactions in plants using bimolecular fluorescence complementation. *Plant J.* 2004;**40**:419–27.

76. Chen B, Liu Q, Ge Q, Xie J, Wang Z-W. UNC-1 regulates gap junctions important to locomotion in C. elegans. *Curr Biol* 2007;**17**:1334–9.

77. Fujioka Y, Utsumi M, Ohba Y, Watanabe Y. Location of a possible miRNA processing site in SmD3/SmB nuclear bodies in Arabidopsis. *Plant Cell Physiol* 2007;**48**:1243–53.

78. Hu C-D, Kerppola TK. Simultaneous visualization of multiple protein interactions in living cells using multicolor fluorescence complementation analysis. *Nat Biotechnol* 2003;**21**:539–45.

79. Grinberg AV, Hu C-D, Kerppola TK. Visualization of Myc/Max/Mad family dimers and the competition for dimerization in living cells. *Mol Cell Biol* 2004;**24**:4294–308.

80. Robida AM, Kerppola TK. Bimolecular fluorescence complementation analysis of inducible protein interactions: effects of factors affecting protein folding on fluorescent protein fragment association. *J Mol Biol* 2009;**394**:391–409.

81. Tsutsumi K, Fujioka Y, Tsuda M, Kawaguchi H, Ohba Y. Visualization of Ras-PI3K interaction in the endosome using BiFC. *Cell Signal* 2009;**21**:1672–9.

82. Förster T. Intermolecular energy transference and fluorescence. *Ann Physik* 1948;**437**:55–75.
83. Miyawaki A. Visualization of the spatial and temporal dynamics of intracellular signaling. *Dev Cell* 2003;**4**:295–305.
84. Stryer L. Vision: from photon to perception. *Proc Natl Acad Sci USA* 1996;**93**:557–9.
85. Tsien RY. The green fluorescent protein. *Annu Rev Biochem* 1998;**67**:509–44.
86. Matz MV, Fradkov AF, Labas YA, Savitsky AP, Zaraisky AG, Markelov ML, et al. Fluorescent proteins from nonbioluminescent Anthozoa species. *Nat Biotechnol* 1999;**17**:969–73.
87. Miyawaki A. Innovations in the imaging of brain functions using fluorescent proteins. *Neuron* 2005;**48**:189–99.
88. Bastiaens P, Majoul I, Verveer P, Soling H, Jovin T. Imaging the intracellular trafficking and state of the AB5 quaternary structure of cholera toxin. *EMBO J* 1996; **15**:4246–53.
89. Miyawaki A, Tsien R. Monitoring protein conformations and interactions by fluorescence resonance energy transfer between mutants of green fluorescent protein. *Methods Enzymol* 2000;**327**:472–500.
90. Adams S, Harootunian A, Buechler Y, Taylor S, Tsien R. Fluorescence ratio imaging of cyclic AMP in single cells. *Nature* 1991;**349**:694–7.
91. Gordon G, Berry G, Liang X, Levine B, Herman B. Quantitative fluorescence resonance energy transfer measurements using fluorescence microscopy. *Biophys J* 1998;**74**:2702–13.
92. Erickson MG, Liang H, Mori MX, Yue DT. FRET two-hybrid mapping reveals function and location of L-type Ca2+ channel CaM preassociation. *Neuron* 2003;**39**:97–107.
93. Berney C, Danuser G. FRET or No FRET: a quantitative comparison. *Biophys J* 2003;**84**:3992–4010.
94. Zal T, Gascoigne NRJ. Photobleaching-corrected FRET efficiency imaging of live cells. *Biophys J* 2004;**86**:3923–39.
95. Aoki K, Kiyokawa E, Nakamura T, Matsuda M. Visualization of growth signal transduction cascades in living cells with genetically encoded probes based on Förster resonance energy transfer. *Philos Trans R Soc Lond B Biol Sci* 2008;**363**:2143–51.
96. Nagai T, Yamada S, Tominaga T, Ichikawa M, Miyawaki A. Expanded dynamic range of fluorescent indicators for Ca(2+) by circularly permuted yellow fluorescent proteins. *Proc Natl Acad Sci USA* 2004;**101**:10554–9.
97. Heim N, Griesbeck O. Genetically encoded indicators of cellular calcium dynamics based on troponin C and green fluorescent protein. *J Biol Chem* 2004;**279**:14280–6.
98. Mank M, Reiff DF, Heim N, Friedrich MW, Borst A, Griesbeck O. A FRET-based calcium biosensor with fast signal kinetics and high fluorescence change. *Biophys J* 2006;**90**:1790–6.
99. Horikawa K, Yamada Y, Matsuda T, Kobayashi K, Hashimoto M, Matsu-ura T, et al. Spontaneous network activity visualized by ultrasensitive Ca(2+) indicators, yellow Cameleon-Nano. *Nat Methods* 2010;**7**:729–32.
100. Komatsu N, Aoki K, Yamada M, Yukinaga H, Fujita Y, Kamioka Y, et al. Development of an optimized backbone of FRET biosensors for kinases and GTPases. *Mol Biol Cell* 2011;**22**:4647–56.
101. Aoki K, Matsuda M. Visualization of small GTPase activity with fluorescence resonance energy transfer-based biosensors. *Nat Protoc* 2009;**4**:1623–31.
102. Yasuda R. Imaging spatiotemporal dynamics of neuronal signaling using fluorescence resonance energy transfer and fluorescence lifetime imaging microscopy. *Curr Opin Neurobiol* 2006;**16**:551–61.
103. Selvin PR. Fluorescence resonance energy transfer. *Methods Enzymol* 1995;**246**:300–34.

104. Murata S, Herman P, Lakowicz JR. Texture analysis of fluorescence lifetime images of AT- and GC-rich regions in nuclei. *J Histochem Cytochem* 2001;**49**:1443–51.

105. Murata S, Herman P, Lakowicz JR. Texture analysis of fluorescence lifetime images of nuclear DNA with effect of fluorescence resonance energy transfer. *Cytometry* 2001;**43**:94–100.

106. Rizzo MA, Springer GH, Granada B, Piston DW. An improved cyan fluorescent protein variant useful for FRET. *Nat Biotechnol* 2004;**22**:445–9.

107. Goedhart J, van Weeren L, Hink MA, Vischer NOE, Jalink K, Gadella TWJ. Bright cyan fluorescent protein variants identified by fluorescence lifetime screening. *Nat Methods* 2010;**7**:137–9.

108. Kurzrock R, Estrov Z, Kantarjian H, Talpaz M. Conversion of interferon-induced, long-term cytogenetic remissions in chronic myelogenous leukemia to polymerase chain reaction negativity. *J Clin Oncol* 1998;**16**:1526–31.

109. Groffen J, Stephenson JR, Heisterkamp N, de Klein A, Bartram CR, Grosveld G. Philadelphia chromosomal breakpoints are clustered within a limited region, bcr, on chromosome 22. *Cell* 1984;**36**:93–9.

110. Heisterkamp N, Groffen J, Stephenson JR, Spurr NK, Goodfellow PN, Solomon E, et al. Chromosomal localization of human cellular homologues of two viral oncogenes. *Nature* 1982;**299**:747–9.

111. Druker BJ. Translation of the Philadelphia chromosome into therapy for CML. *Blood* 2008;**112**:4808–17.

112. Druker BJ, Tamura S, Buchdunger E, Ohno S, Segal GM, Fanning S, et al. Effects of a selective inhibitor of the Abl tyrosine kinase on the growth of Bcr-Abl positive cells. *Nat Med* 1996;**2**:561–6.

113. Weisberg E, Manley PW, Cowan-Jacob SW, Hochhaus A, Griffin JD. Second generation inhibitors of BCR-ABL for the treatment of imatinib-resistant chronic myeloid leukaemia. *Nat Rev Cancer* 2007;**7**:345–56.

114. Roy L, Guilhot J, Krahnke T, Guerci-Bresler A, Druker BJ, Larson RA, et al. Survival advantage from imatinib compared with the combination interferon-alpha plus cytarabine in chronic-phase chronic myelogenous leukemia: historical comparison between two phase 3 trials. *Blood* 2006;**108**:1478–84.

115. Feller SM. Crk family adaptors-signalling complex formation and biological roles. *Oncogene* 2001;**20**:6348–71.

116. Nichols GL, Raines MA, Vera JC, Lacomis L, Tempst P, Golde DW. Identification of CRKL as the constitutively phosphorylated 39-kD tyrosine phosphoprotein in chronic myelogenous leukemia cells. *Blood* 1994;**84**:2912–8.

117. Oda T, Heaney C, Hagopian JR, Okuda K, Griffin JD, Druker BJ. Crkl is the major tyrosine-phosphorylated protein in neutrophils from patients with chronic myelogenous leukemia. *J Biol Chem* 1994;**269**:22925–8.

118. ten Hoeve J, Arlinghaus RB, Guo JQ, Heisterkamp N, Groffen J. Tyrosine phosphorylation of CRKL in Philadelphia+ leukemia. *Blood* 1994;**84**:1731–6.

119. O'Hare T, Walters DK, Stoffregen EP, Jia T, Manley PW, Mestan J, et al. In vitro activity of Bcr-Abl inhibitors AMN107 and BMS-354825 against clinically relevant imatinib-resistant Abl kinase domain mutants. *Cancer Res* 2005;**65**:4500–5.

120. Redaelli S, Piazza R, Rostagno R, Magistroni V, Perini P, Marega M, et al. Activity of bosutinib, dasatinib, and nilotinib against 18 imatinib-resistant BCR/ABL mutants. *J Clin Oncol* 2009;**27**:469–71.

121. Gorre ME, Mohammed M, Ellwood K, Hsu N, Paquette R, Rao PN, et al. Clinical resistance to STI-571 cancer therapy caused by BCR-ABL gene mutation or amplification. *Science* 2001;**293**:876–80.

122. Lu S, Wang Y. Fluorescence resonance energy transfer biosensors for cancer detection and evaluation of drug efficacy. *Clin Cancer Res* 2010;**16**:3822–4.

CHAPTER NINE

# Fluorescent Macromolecular Sensors of Enzymatic Activity for *In Vivo* Imaging

**Alexei A. Bogdanov, Mary L. Mazzanti**
Laboratory of Molecular Imaging Probes, Department of Radiology, University of Massachusetts Medical School, Worcester, Massachusetts, USA

## Contents

## Abstract

Macromolecular imaging probes (or sensors) of enzymatic activity have a unique place in the armamentarium of modern optical imaging techniques. Such probes were initially developed by attaching optically "silent" fluorophores via enzyme-sensitive linkers to large copolymers of biocompatible poly(ethylene glycol) and poly(amino acids). In diseased tissue, where the concentration of enzymes is high, the fluorophores are freed from the macromolecular carrier and regain their initial ability to fluoresce, thus allowing *in vivo* optical localization of the diseased tissue. This chapter describes the design and application of these probes and their alternatives in various areas of experimental medicine and gives an overview of currently available techniques that allow imaging of animals using visible and near-infrared light.

*Progress in Molecular Biology and Translational Science*, Volume 113
ISSN 1877-1173
http://dx.doi.org/10.1016/B978-0-12-386932-6.00009-0

## 1. INTRODUCTION

Fluorescence-based imaging and detection strategies have been a mainstay of basic and preclinical research for over four decades. This is in large part due to the cost-effectiveness, safety, and sensitivity of these techniques (usually requiring only nanomolar concentrations of dyes while yielding micrometer to nanometer microscopic resolution in single cells as well as in tissue samples).

In parallel, an increasing number of fluorescent and bioluminescent molecules (optical imaging probes) have become available and these are allowing investigation of ever more complex biological phenomena, such as genomic regulation, signal transduction, cell excitability, and stem cell generation and differentiation. Although many of the currently available fluorescent probes were primarily developed for *in vitro* and *in situ* studies using fluorescence microscopy and fluorescence-based screening assays, a growing number are being developed specifically for *in vivo* applications such as cell tracking and monitoring both normal and pathological processes in animal models. Success in these endeavors has spurred interest in the use of fluorescence imaging for detecting human disease. In particular, enzyme-sensing fluorescent reporters have been used for detecting a number of different pathologies in animal models and hold great promise in the detection and treatment of human diseases, especially those exhibiting an inflammatory component (i.e., atherosclerosis, cancer, and osteoarthritis). The continued development and improvement of optical detectors, along with the increasing availability of high-specificity, biologically safe fluorescent probes, has made transition of fluorescence-based imaging of enzyme activity to the clinical arena a plausible endeavor.

This chapter discusses some of the important research that has brought this experimental modality closer to clinical application and describes the compounds that are most likely to have relevance and feasibility for transition into clinical practice with emphasis given to fluorescent reporters of enzymatic activity.

## 2. ENZYME ACTIVITY AND DYSREGULATION AS A PREDICTOR OF DISEASE

Strategies aimed at imaging and detecting enzyme activity have substantial potential for aiding in the diagnosis and prognosis of a wide range of diseases. Ample evidence exist showing that dysregulation of enzymatic activities play a role in the etiology and/or progression of cancer, atherosclerosis, stroke and heart disease, diabetes, arthritis, multiple sclerosis, Alzheimer's disease,[1] HIV, and other infections. For example, a number of proteases have been implicated in cancer, including urokinase-type plasminogen activator (uPA),[2] matrix metalloproteinases (MMPs),[3] and cysteine proteinases.[4] These enzymes are believed to facilitate extracellular matrix breakdown, allowing cancer cells as well as activated stroma cells to invade neighboring tissue.[5] MMPs are expressed in higher levels in some forms of cancer, and their expression level correlates to tumor stage,[6] invasiveness,[7,8] metastasis,[9] and degree of tumor vascularization.[10] Increased expression and activity of cysteine cathepsins and uPA have also been found in some tumors,[11–14] and elevated levels of cathepsin B, D, and L are found in mammary adenocarcinomas and are prognostic of poor outcome.[15–18] More recently, major roles have been identified for cyclooxygenases, lipoxygenases, integrin-linked kinase 1, and glutaminase in the progression of various malignancies.[19–21]

Atherosclerosis is another clinically prevalent disease that is now thought to arise, in part, from damage incurred though the activation of inflammatory pathways and associated enzymes. Lipoprotein-associated phospholipase A2 (Lp-PLA2) activity generates proinflammatory lipids that participate in the formation of atherosclerotic necrotic cores,[22] which in turn recruit macrophages and granulocytes that release myeloperoxidase (MPO). The resultant MPO-mediated generation of hypochlorite anion leads to matrilysin (MMP7) activation and contributes to plaque rupture as a result of fibrous cap degradation.[23] Increases in Lp-PLA2 have been linked to an increased risk of cardiac death, myocardial infarction, acute coronary syndromes, and ischemic stroke.[24]

The critical role of enzyme activity in many different diseases has also been deduced from beneficial clinical outcomes obtained following therapeutic interventions that alter specific enzyme activities. Examples include the use of angiotensin converting enzyme inhibitors for the treatment of hypertension,[25] elastase inhibitors for the treatment of cardiovascular disease,[26] HIV aspartyl protease inhibitors to prevent development of AIDS,[27] proteasome inhibitors for the treatment of multiple myeloma,[28] and leukoprotease inhibitors for the treatment of cystic fibrosis.[29] Development of enzyme inhibitors for the treatment of cancer is currently an intensive area of research, and therapeutic benefit has been shown for inhibitors of AMP-activated protein kinase,[30] MMP9 and MMP14,[31,32] cathepsin S,[33], and cathepsins B, L, and S.[34]

The diagnostic value of determining enzyme levels in patients has been appreciated for years in clinical medicine and is the basis for a number of simple and common serum blood tests. Included in these is the determination of the serum serine protease known as prostate-specific antigen and of creatine kinase for diagnosing prostate cancer and heart attack, respectively. Although less specific for differential diagnosis, elevated serum levels of aspartate aminotransferase (AST or SGOT), alanine aminotransferase (ALT or SGPT), alkaline phosphatase, $5'$ nucleotidase, and gamma-glutamyl transpeptidase are routinely used as indicators of liver dysfunction. Given the cost-effectiveness and almost universal availability of these assays, the need for a method to anatomically localize and quantitate enzyme levels in patients may not be readily apparent. However, in diseases such as cancer, atherosclerosis, and infection, interventional procedures used for further diagnosis (i.e., biopsy, exploratory surgery) and treatment (site-directed radiation, tumor dissection, thrombectomy/embolectomy) can be accomplished with greater efficacy and safety when the exact location of diseased tissue can be predetermined. Additionally, the choice of treatment may depend on the extent of disease spread (such as metastatic tumors), in which case an anatomical map of disease burden is extremely useful.

## 3. FLUORESCENCE DETECTION *IN VIVO*

Although more established clinical imaging modalities such as MRI and PET have been applied to noninvasive detection of enzyme activity, there is a growing interest in the use of fluorescence-based methods for this purpose because of the inherent safety and cost-effectiveness of this imaging modality. Fluorescence-based enzyme sensors are particularly attractive for evaluating treatment effects after therapy, as the patient can be imaged multiple times without the fear of radiation-induced side effects. This is especially true, given the ever-growing number of therapeutics that target enzyme activity. In addition, fluorescence-based methods are attractive because they offer the capacity to monitor multiple (in some cases up to 10) targets (fluorescent molecules having different spectral parameters) simultaneously during the same exam/procedure. For these reasons, there has been concerted effort in recent years to translate the wealth of knowledge gained from the development of preclinical fluorescence-based imaging and tomography methods to the clinical arena.[35–37] Currently, indocyanine green is the only near-infrared (NIR) fluorophore that has been approved by the FDA for clinical use, and there are numerous studies that have shown its utility as an i.v.-administered, nonspecific fluorescent contrast agent for applications in intraoperative angiography and tumor/sentinel lymph node mapping and staging (reviewed in Refs. 36,37). In addition, 5-aminolevulinic acid (which is not itself fluorescent but leads to the accumulation of fluorescent porphyrins in malignant tissue) has also been used to demarcate tumor margins in neurosurgical procedures and has been judged to contribute to positive clinical outcomes.[38,39] Improvements in clinical fluorescence detection systems[40,41] and the use of more advanced imaging methods such as fluorescence lifetime imaging (FLI)[42–45] should further strengthen the standing of fluorescence imaging in the clinical diagnosis and treatment of disease.

Fluorescence is a property of organic and inorganic materials in which electromagnetic energy (usually in the form of visible or ultraviolet light) is absorbed by the material and results in the emission of light (photons). In general, the excitation light is of shorter wavelength (higher energy) than the emission light. Certain tissues and molecules of the body naturally exhibit fluorescence, and a growing number of clinical applications harness this property to aid in interventions such as endoscopy and surgery. The most routine use of fluorescence for clinical detection and diagnosis is in the application of autofluorescence imaging (AFI) as a component of endoscopic exams. AFI detects the natural tissue fluorescence that is emitted by biomolecules such as collagen, flavins, and porphyrins. The emission wavelength is altered by changes in the metabolic state of the tissue, which can be used as an indicator of malignancy. Also under current development are methods that rely on the administration of fluorescent dyes and fluorescently tagged macromolecules which then are distributed throughout the body. The success of the latter approach depends on the ability of the fluorescent probe to relay accurate and precise information about physiological processes and disease. This, in turn, is largely dependent on the ability of the probe to target the tissue or disease process of interest and to then render a signal that has sufficient intensity to be detected. Although these characteristics vary for the individual probes developed to date, generalizations can be made in terms of what characteristics are necessary for fluorescent probes to be useful in a clinical setting. As described in Ref. 46, an ideal fluorescence sensor for *in vivo* imaging of disease would be characterized by (1) a peak fluorescence close to 700–900 nm; (2) high quantum yield; (3) narrow excitation/emission spectrum; (4) high chemical and photostability; (5) nontoxicity; (6) excellent biocompatibility, biodegradability, or excretability; (7) availability of monofunctional derivatives for conjugations; and (8) commercial viability and capacity for high-volume synthesis to support the large-scale production required for human use.

The primary limitation of fluorescence imaging *in vivo* is the inability to detect the signal in deep tissues. This limit is imposed by the depth to which both the excitation light from the illumination source and emission light from the fluorophore can penetrate the tissue under investigation. Almost all biological tissues absorb light in the visible to UV range, which limits penetration to only a few hundred micrometers. However, longer excitation wavelengths (far-red to NIR, i.e., 700–1500 nm) allow the deepest penetration, that is, up to several centimeters.[47]

One of the potential means of circumventing the problems associated with illumination of tissue is the use of bioluminescent reporters. Bioluminescent proteins such as firefly luciferase (*Photinus pyralis*) occur naturally in a number of species and emit photos upon catalyzing the oxidation of a substrate and therefore do not rely on an external excitation source. Bioluminescence generally generates a strong signal and high sensitivity in animal studies; however, because the organism or tissue under study must express luciferase, these agents are generally restricted to experimental studies in which a transgenic animal that expresses the enzyme can be generated. However, examples do exist in which bioluminescent probes have been used in nontransgenic animals. In this case, the substrate (luciferin) is administered to the animal (i.p. or i.v.) at a time point after the bioluminescent probe-carrying luciferase has reached the target tissue. After luciferin administration, the bioluminescent signal is generated within a short period (1–2 min), peaks at around 10–12 min, and returns to baseline within an hour.

Regardless of the excitation source, once excited, photons emitted from the fluorescent molecule have to reach the detector, which is, in most cases, located outside the body. As is the case of the excitation source, dyes emitting lower energy photons (i.e., NIR 700–900 nm) generally perform better in terms of absorption by molecules in the body and can be detected at a depth of several millimeters to a centimeter. A secondary, although extremely important issue is that of light scattering in the tissue, which reduces the amount of signal returned and also makes spatial localization of returned signal problematic. The uncertainty of the signal source becomes greater at larger tissue depths.

A tertiary issue affecting the quality of information that can be obtained from *in vivo* fluorescence detection is the autofluorescence of biological tissue. Biomolecules, such as hemoglobin, tryptophan, NADH, pyridoxine, collagen, elastin, flavins, porphyrins, nucleotides, and even water, can absorb light across the UV–vis range, and some of these molecules emit fluorescence that results in attenuation of the fluorophore signal of interest. Although the natural fluorescence of biological tissue has been put to use for some clinical applications (as previously described), it is considered a nuisance variable when attempting to detect the fluorescence signal from an exogenously administered dye. This problem is reasonably dealt with by using NIF dyes (790–830 nm emission) because signals emitted from biological tissue is lower in this wavelength range, or alternatively, by using special detection methods (time-resolved measurements[48–51] or spectral unmixing techniques[52,53].

The inherent intensity of the signal from the fluorophore is also a consideration for *in vivo* imaging, as it directly affects sensitivity and the SNR. High signal intensity is an obvious benefit for *in vivo* applications because it reduces the amount of fluorophore that must be administered and also reduces the amount of energy deposition needed for dye excitation, thus limiting potential phototoxicity. The signal from a single-fluorophore molecule is a product of the molecule's quantum yield (QY, i.e., number of photons emitted per photon absorbed) and its molar extinction coefficient. The signal will also be affected by optical path length, pH, and the polarity of the surrounding molecules. In addition, the emitted photons may not contribute to the detected signal as a result of a nonradiative dipole–dipole coupling mechanism such as Förster resonance energy transfer or FRET, with the surrounding tissue macromolecules, decreasing photon intensity 10-fold for each centimeter of tissue depth.[54] Some newer classes of engineered probes such as fluorescent semiconductor nanocrystals (quantum dots, QDs) can greatly exceed this limit and have demonstrated superior performance for *in vivo* applications.[55] In addition, QDs are more resistant to photobleaching and have the added flexibility of being tunable to any given emission frequency. Enthusiasm is somewhat tempered for these agents, however, given the concerns about their biological safety and their ability to be targeted to specific tissues.[55] Nonetheless, development of QDs is an active area of research, and expectations are high that QDs with safer toxicity profiles will become available in time.

Gains in fluorescent signal detection have also been made by advancements in detector technology and computer algorithms for separation and amplification of the signal. Typically a charge-coupled device (CCD) camera, which is sensitive to a wide range of spectral frequencies, is used to capture the emitted photons from fluorescence and convert these to a digital signal. Cooling the camera reduces stray noise in the electronics and improves the SNR. A secondary image of the tissue or anatomy of interest allows the 2D fluorescent signal to be localized within the animal. One obvious limitation of 2D imaging is its inability to discriminate signals from shallow and deep tissue. Fluorescence molecular tomography (FMT) is a developing technique that offers greater anatomical localization, spatial resolution, and quantitation[56] (Fig. 9.1). In FMT, the emitted fluorescence is measured from a matrix of multiple excitation sources and is normalized by the excitation signal transmitted through the tissue. Further

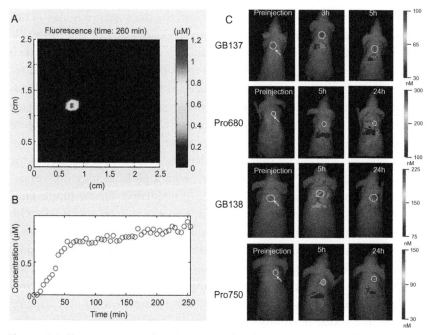

**Figure 9.1** Fluorescence-mediated tomography (FMT) in imaging of near-infrared probe activation *in vitro* and *in vivo*: (A) FMT reconstruction image of an NIR fluorescence source, a 2.5-mm tube filled with a mixture of PGC–Cy5.5 and a model protease (pancreatic trypsin) resulting in liberation of fluorescent Cy5.5 dye (1 μM); (B) Time course of the reaction measured by time-lapse FMT, demonstrating the reaction kinetics resulting in fluorescence signal shown in (A) (courtesy of Dr. Vasilis Ntziachristos (Technical University of Munich, Germany)); and (C) FMT imaging of the time-dependent accumulation of small molecular weight and macromolecular probes in a mouse xenograft model of human cancer. Shown are FMT-reconstructed fluorescent signals in subcutaneous adenocarcinomas of GB137 (quenched activity-based probe); activity-based probe GB138 and macromolecular probes ProSense 680 and ProSense 750. Tumors are shown using circular regions of interest. Adapted from Ref. 57. (See Color Insert.)

computational analysis based on the absorption and scattering properties of the tissue allows determinations to be made of the point of origin of the fluorescence signal and a 3D image to be reconstructed, even from tissue several centimeters deep. However, FMT also necessitates longer image acquisition times, and care must be taken in the interpretation of data, as the reconstruction algorithms are not infallible. This technology is now offered in commercial animal imaging systems and has allowed quantitation and distribution of tumor-targeted fluorescent probes throughout the entire depth of the animal.[58] FMT borrows much of its methodology from diffraction tomography and diffuse optical tomography in which a NIR light source on the body surface delivers NIR range photons several centimeters into the tissue (usually brain or breast) and the emitted photonic energy is used to acquire chemical and anatomical information about the tissue. These systems have been tested clinically for the detection of shallow tumors in breast and other subdermal locations.[40,41] Newer detectors also allow the extraction of information from the emitted fluorescence based on properties other than intensity, such as FLI which yields information about chemical processes taking place in the vicinity of the dye molecule or probe. FLI is also desirable in some applications because the measured signals are independent of the fluorophore concentration.

Last, the ability of a fluorescent signal to relay usable information about a disease process will be largely determined by the targeting capability of the probe, that is, the ability to be selectively taken up by the tissue of interest and/or to target a specific biomolecule. The latter is no small feat for enzyme targets, given the large number of proteins with similar or redundant functions and associated enzymatic activities found in the body. For example, roughly 600 proteases have been identified by bioinformatic analysis of the mouse and human genome.[59] The delivery of fluorescent probes to target sites is also challenging, given the efficiency of the reticuloendothelial system at sequestering and eliminating foreign molecules. However, as we discuss in a later section, a number of innovative strategies have been developed to circumvent these problems.

## 4. PRINCIPLES OF ENZYMATIC ACTIVITY SENSORS BASED ON FLUORESCENCE CHANGES

As previously discussed, aberrant enzyme activity has been implicated in a number of diseases, and therefore its detection *in vivo* could substantially impact clinical outcomes and aid in research studies of the disease. Although measures of mRNA and proteomic screens give some information about specific enzyme levels, many enzymes are activated during posttranslational processing. Thus, a more direct measure is needed to discern actual enzyme activity. Fluorescent indicators are uniquely suitable for this task because of their ability to undergo rapid (milliseconds to seconds) transition from a "silent" nonfluorescing to a fluorescing state, or from a shorter to a longer emission wavelength. The goal of designing an enzyme-sensitive probe is to link the change in the fluorescence state to the activity of the enzyme, and this has been accomplished over the years using a number of different macromolecular chemistries and configurations. In some configurations, a single enzyme molecule can activate multiple fluorophore-linked substrates within the same microenvironment and thereby produce an amplification of the fluorescent signal in the diseased tissue (the basis for use of enzymes such as $\beta$-galactosidase, $\beta$-lactamase, etc., as reporter genes). As we discuss below, a number of different chemistries have been used in the development of fluorescent probes targeting different diseases/enzyme systems. The overall difference is in the way the fluorescence-generating substrate is targeted to the enzyme (e.g., covalently vs. noncovalently) and the way in which the precursor (substrate) molecule is made silent (quenched) in the nonactivated state. For example, fluorophores may be quenched by other molecules as a result of collisional interaction (dynamic quenching), by interaction with surface plasmons (such as a gold nanoparticle surface), or by closely positioned secondary fluorophores that quench emission because of formation of noncovalent complexes (close quenching). The efficacy of FRET is determined by the energy transfer rate, which is a function of the overlap of the donor emission and acceptor excitation spectra, the relative orientation of the transition dipoles, and the distance between the donor and the acceptor.[60] The latter is usually measured in Forster's distance which is usually limited to 10 nm.[61] Fluorophores can also be switched on and off by modulation of their electronic properties through oxidation–reduction and other chemical reactions.[62]

## 5. MACROMOLECULAR FLUORESCENT PROBES

The most frequently used amino-reactive fluorescent dye FITC (fluorescein isothiocyanate) reacts both with $\alpha$-N-terminal amino groups and N-$\epsilon$-amino groups of lysine at high pH values with the formation of the corresponding thiocarbamylates.[63] This fluorochrome-activated analog is frequently used for tagging proteins with readily detectable green fluorescent tags. The first detailed description of *in vitro* imaging of cells using a macromolecule carrying multiple fluorescein fluorophore molecules was provided by French et al.[64] This study established that, upon labeling of BSA with several FITC molecules, the latter were clearly subject to quenching with a resultant decrease of fluorescence intensity and lifetime. The increase of fluorescein emission intensity and a drastic sixfold lengthening of fluorescence lifetime (FL) from 0.5 to 3 ns were observed over time in cells after dye uptake. Experiments with FITC-labeled L- and D-polylysine lead to the conclusion that proteolysis was responsible for the fluorescence increase, as D-polylysine (unlike L-polylysine), which is noncleavable by cellular proteases, did not result in fluorescence changes.[64] In later studies, the same FITC-labeled probe was used for tracking migrating cells in a 3D matrix.[65]

These initial studies were important in that they provided inroads for future development of *in vivo* probes that report on proteolysis. Fluorescence intensity and FL changes resulting from cleavage of the enzymatic macromolecular substrate arise from conformational and chemical changes occurring in the cleaved fluorescent products. Most polypeptides have secondary and tertiary structures defined by hydrogen-bonding and hydrophobic interaction of peptide bonds and amino acid side residues. If reactive groups within a polypeptide chain (primarily $N$-$\varepsilon$-amino groups of lysines) are located spatially close to each other, the covalent linking of fluorophores to these groups will result in low fluorescence. Natural globular proteins such as albumins have low segmental mobility as opposed to their synthetic poly- and oligoamino acid counterparts. For example, polylysines exist in extended coil conformation at a pH above 8.5. The conformational flexibility of linear or branched synthetic poly(amino acids) can potentially result in higher numbers of interacting dyes because of (1) higher overall number of accessible reactive groups and (2) their mobility that facilitates the formation of transient or stable dye dimers (or stacks). The probability of fluorescent dye quenching in the above case is high because of the ease of collisional interaction and, more importantly, the formation of nonfluorescent dye dimers (H- or J-dimers) or higher order aggregates (also known as "self-quenching"). The formation of nonfluorescent aggregates is more common for fluorochromes that are electroneutral and/or are hydrophobic such as NIR cyanine dyes (Fig. 9.2). Sulfation of cyanines results in better water solubility but does not prevent the formation of nonfluorescent H-aggregates. Therefore, the design of macromolecular imaging substrates can benefit from using sulfated cyanines and other NIR fluorochromes because of the overall increase of solubility of macromolecules with fluorophore-linked

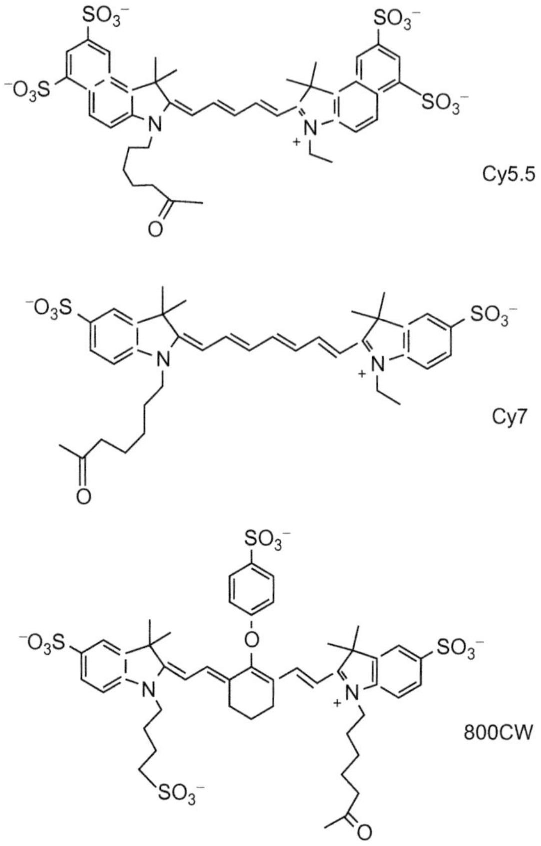

**Figure 9.2** Examples of near-infrared cyanine fluorophores commonly used for labeling macromolecules: Cy5.5, Cy7 (GE-Healthcare Life Sciences, Piscataway NJ), and IRDye 800CW (Li-COR Corp., Lincoln NE).

side chains. Some sulfated dyes display inherently low self-quenching even after covalent linking to polypeptides, for example, green fluorescent Alexa Fluor 488. Such dyes have many advantages over traditionally used fluorescein analogs for microscopy applications. However, for the reasons that were described in previous sections of this review, *in vivo* fluorescence imaging currently relies on fluorophores with far-red to NIR emission.

The linking of NIR fluorophores to plasma proteins has advantages and pitfalls: the advantages include availability of purified or recombinant blood plasma proteins, near-uniform mass, near-constant number of reactive groups, and low immunogenicity in a native form. The disadvantages include the limited number of above reactive groups, insufficiently long circulation time, nonspecific uptake, and the known ability of dyes to induce immune response if they are linked to protein-based scaffolds. Linear synthetic poly(amino acids) are also known to have short blood circulations times, and strong residual negative or positive charge can result in a high level of kidney uptake and toxicity. However, the immunogenicity of poly(amino acids) with high degrees of side group modification with various ligands is usually lower than in the case of proteins. To circumvent the problems associated with very rapid elimination and immunogenicity of poly-L-lysine, we envisioned using a partial covalent modification of N-ε-amino groups of lysine with activated methoxypoly(ethylene glycol) (MPEG) esters for increasing the hydrodynamic radius and, consequently, increasing circulation times of imaging sensors *in vivo*. The resulting graft copolymer (termed protected graft copolymer, PGC, Fig. 9.3) has since been used successfully for many imaging and drug delivery applications (reviewed in Ref. 66) (see Table 9.1). One of the benefits of PEGylation (i.e., covalent linking of poly(ethylene glycol) (PEG) to other molecules or surfaces) is that, because of the very high level of water hydration (three molecules of water coordinate with a single ethylene oxide monomer within PEG polymer) and very high segmental flexibility, MPEG molecules create a shell of "soft matter" around the central backbone of polylysine. As a consequence, the backbone and the fluorochromes covalently linked to it are protected from rapid elimination from the bloodstream. Another important property of PEG is the ability to protect biomacromolecules from

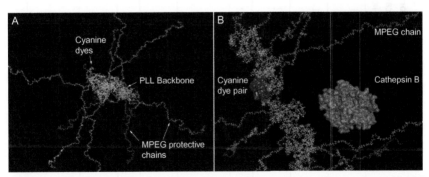

**Figure 9.3** Molecular operating environment (MOE) modeling the structure of PGC fragment ($n = 20$ lysine monomers) with a backbone PLL conjugated to IRDye 800CW cyanine dyes. (A) Main structural elements and (B) the same model showing the backbone with more detail. The molecular interacting surface of a pair of interacting IRDye 800CW dyes (H-dimer) is rendered in blue. A molecule of cathepsin B is shown for comparison. (See Color Insert.)

**Table 9.1** Enzyme-activated NIR-dye-labeled macromolecular imaging probes and their use for optical imaging *in vivo*

| Macromolecule, type | Specificity | Fluorophore, mol/mol carrier macromolecule | Animal model | Target/background ratio[a] | Imaging method | Reference |
|---|---|---|---|---|---|---|
| PGC[b] | Cathepsin B/H/L/S | Cy5.5, 11 | Murine breast adenocarcinoma | 12 | PE[c] | 67 |
| PGC | Cathepsin B/H/L/S | Cy5.5, 11 | Murine breast tumor | 9.4 (250 pM dye) and 27.5 (2.5 nM dye) | FMT | 68 |
| PGC | Cathepsin D | Cy5.5, 22 | Murine hCaD-expressing tumors | 22 | PE | 69 |
| PGC | MMP2 | Cy5.5, 12 | Murine fibrosarcoma | 3 | FRI | 70 |
| PGC | Cathepsin B/H/L/S | Cy5.5 | Murine atherosclerosis (NO and ApoE knockouts) | | | 71 |
| PGC | Cathepsin B/H/L/S | Cy5.5 | Murine breast tumor | 5 | FMT | 72 |
| PGC | Thrombin | Cy5.5 | Murine thrombosis | 2 | FRI | 73 |
| PGC | Cathepsin B/H/L/S | Cy5.5 | Murine dysplastic intestinal adenomas | 2–4.5 | FRI | 74 |
| PGC | Cathepsin B/H/L/S | Cy5.5 | Murine breast cancer | 1.5 times higher signal in invasive versus noninvasive tumors | FRI | 75 |

| | | | | | | |
|---|---|---|---|---|---|---|
| PGC | Cathepsin B/H/L/S | Cy5.5 | Murine lymph node metastasis | 4.8 (i.v. injection) 6.6 (SC injection) | FRI | 76 |
| PGC | HIV-1PR | Cy5.5 | Murine glioma as abdominal xenograft | 4 | PE | 77 |
| PGC | Cathepsin B/H/L/S | Cy5.5 | Murine osteoarthritis | 50–70% signal reduction post therapy | FRI | 78 |
| PGC | Caspase-1 | Cy5.5, 18 | Murine glioblastome-expressing caspase-1 | 1.7 | PE | 79 |
| PGC | Cathepsin B/H/L/S | Cy5.5 | Murine osteoarthritis | 3 | FRI | 80 |
| PGC | MMP2 | Cy5.5 | Murine glioma | Twofold FMT signal decrease after MMP3 inhibition | FMT | 81 |
| PGC | MMP2 and 9 | Cy5.5 | Murine myocardial infarction *ex vivo* | 7 | IM | 82 |
| ProSense 680[d] | Cathepsin B/H/L/S | VivoTag 680 | Murine lung carcinoma | | FMT | 83 |
| PGC | uPA | Cy5.5 or Cy7 15–19 | Murine fibrosarcoma and adenocarcinoma | 3.2 and 3.1 | FRI | 84 |
| PGC | MMP2 and 9 | Cy5.5 | Murine atherosclerosis (ApoE ko) | 5 (*ex vivo*) | FRI, FMT | 85 |
| ProSense 680 | Cathepsin B/H/L/S | VivoTag 680, 6 | Murine myocardial infarction | 19 | FMT | 86 |

Continued

**Table 9.1** Enzyme-activated NIR-dye-labeled macromolecular imaging probes and their use for optical imaging *in vivo*—cont'd

| Macromolecule, type | Specificity | Fluorophore, mol/mol carrier macromolecule | Animal model | Target/background ratio | Imaging method | Reference |
|---|---|---|---|---|---|---|
| PGC | Cathepsin K | Cy5.5, 18 | Murine atherosclerosis, surgically exposed | 1.7 | IM | 87 |
| ProSense 680 | Cathepsin B/H/L/S | VivoTag 680 | Colonic adenocarcinomas | 8.9 | Microendoscopy | 88 |
| Noncovalently bound homotetrameric avidin | Lectin receptor | Rhodamine X, 3 | Murine peritoneal ovarian metastases | | IM | 89 |
| Galactosamine-conjugated serum albumin (GmSA) | D-galactose | Rhodamine X, 20 | Murine peritoneal ovarian metastases | | IM | 90 |
| PGC; ProSense 750 | Cathepsin B/L/S | VivoTag 750 | Murine pulmonary inflammation | 3 | FMT | 91 |
| ProSense 680, MMPSense680 | Cathepsin B/H/L/S and MMP2/3/9/13 | VivoTag 680 | Murine pancreatic cancer | Allowed detection of tumors 1–2 mm | PE and intravital FCFM | 92 |
| PAMAM-4 dendrimer + peptide | MMP7 | Cy5.5, 8 | Murine intestinal adenoma | 2.2 *in vivo* 6 *ex vivo* | PE | 93 |
| PGC; ProSense 750 | Cathepsin B/L/S | VivoTag 750 | Murine ovarian cancer metastases to peritoneal cavity | 3.5 | LP | 94 |

| | | | | | | |
|---|---|---|---|---|---|---|
| PGC; MMPSense | MMP9 and 2; 3, 7–10; and 12–16 | Cy5.5 | Murine stroke | 1.3 | PE | 95 |
| ProSense 680 | Cathepsin B/H/L/S | VivoTag 680 | Rodent pancreatic cancer | Detection rate with 100% sensitivity and 92.2% specificity | Intravital FCFM | 96 |
| Peptides (PS-5, PS-25, PS-40) | Cathepsin B | VivoTag 680 | Murine atherosclerosis | PS5 *ex vivo* = 4.76, PS40 *ex vivo* = 7.13, PS40 *in vivo* ~2.75 | FRI and FMT | 97 |
| ProSense 750 | MMP 2/3/7/9/12/13 | VivoTag 750 | Murine esophageal adenocarcinoma | 4.5 | FEN | 98 |
| PGC-800 | Cathepsin B/H/L/S | IRDye 800CW | Murine myocardial infarction | | FLI | 45 |

[a]Calculated using data from the original reports referenced in this table.
[b]PGC, protected graft copolymer, that is, MPEG-bPLL.
[c]PE, planar excitation; FRI, fluorescence reflectance imaging; FMT, fluorescence molecular tomography; LP, laparoscopy; IM, intravital microscopy; FCFM, fiberoptic confocal fluorescence microscopy; PAMAM, PEGylated polyamidoamine; FEN, fluorescence endoscopy; FLI, Fluorescence lifetime imaging.
[d]Commercially available probes based on PGC (ProSense™, MMPSense™, and Angiosense™, Perkin-Elmer, Waltham MA).

complement activation and activation of immune response. During the course of optimization experiments, we established that only MPEG with an average mass of 5000 was sufficient for providing single-exponential blood elimination kinetics and no antibody formation to acylated poly-L-lysine and attached small molecular weight molecules was observed after repeated i.v. injections of PGC.

The resultant PGC copolymers used in initial *in vivo* experiments (which in the fluorophore-free state are composed of MPEG-grafted PLL) have an approximate molecular mass of 350–450 kD and carry approximately 50–55 amino groups per molecule[99] which can be readily covalently modified. In the presence of an excess of free activated amino-reactive ester of cyanine dyes, because of steric constraints, not all of the available amino groups can be efficiently modified in the water milieu: it is usually feasible to link approximately 30–45 molecules of large cyanine dyes per molecule of PGC, depending on the structure of the dye, and produce a macromolecule in which most of the fluorophores end up in a quenched nonfluorescent state. The background fluorescence of dye-conjugated PGC protease sensors is usually low, while the activation efficiency of the sensors depends on several factors: for example, (1) the ability of the dyes to maintain a quenched state, that is, the strength of dye dimerization; (2) quantum yield of fluorescence in the nonquenched state; and (3) the ease of backbone fragmentation by protease. All the above parameters vary, though not very widely, among various NIR fluorochromes. Spectral properties of the dye-conjugated PGC depend on loading density. If the PGC carrier is linked to less than 12 dyes/molecule, their absorbance spectra indicate very little close interaction between individual dyes (blue-shifted absorbance peak is very small; Fig. 9.4). If the loading is high, that is, $> 35$ dyes/molecule, there is a pronounced blue-shifted absorbance peak which is characteristic of H-dimer formation. This peak is much less prominent in samples after the proteolytic cleavage of the probes. Fluorescent properties change dramatically after the cleavage with multifold increase of fluorescence intensity (see Fig. 9.4 and Table 9.2).

At this time, the backbone-linked heptamethine IRDye 800CW® has provided the best performance in terms of the robustness of fluorescence intensity increase as a function of proteolytic activation (fragmentation) of dye-linked PGC (Table 9.2). The cleavage of PGC probes by model protease depends on whether the cyanine dyes are linked at high or low density to the backbone and whether MPEG protective chains are linked to the backbone via ester (semistable) linkers or not. Semistable linkers are

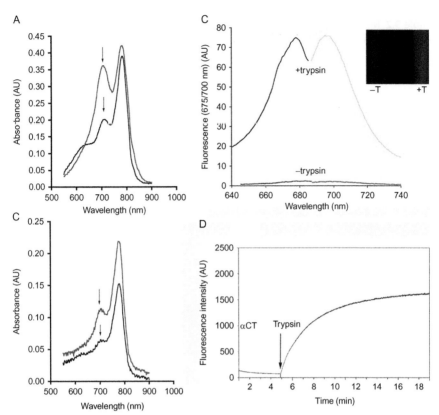

**Figure 9.4** Spectral properties of PGC-based enzyme-sensitive probes. (A) Absorbance spectra of IRDye 800CW dye linked to PGC at high conjugation density (gray trace) and low conjugation density (black trace). (B) Same as A but after 30 min incubation in the presence of trypsin. Arrows are pointing at the blue-shifted peak of H-dimers of IRDye 800CW dye. (C) The change of excitation (ex) and emission (em) spectra of PGC–Cy5.5 before and after the addition of trypsin. The corresponding effect on detectable red fluorescence of Cy.5.5 is shown in the inset. (D) Experimental demonstration of PLL-backbone encoded specificity: the addition of α-chymotrypsin (αCT) does not result in PGC-Cy5.5 cleavage, whereas trypsin results in rapid degradation of PGC-Cy5.5 molecules and fluorescence intensity increase. (For color version of this figure, the reader is referred to the online version of this chapter.)

degraded by nonspecific esterases and also quickly fall apart in alkaline or acidic media. They usually result in efficient and complete degradation of PGC probes into small fragments and result in the highest relative changes of fluorescence intensity and lifetime. Overall, FL changes are greater for cyanine dyes that fluoresce in the far-red range of the spectrum; that is, Cy5.5, IRDye 680RD produce a very profound change of FL (up to fivefold)

**Table 9.2** Macromolecular backbone–fluorophore conjugated probes and their activation with a model protease *in vitro*

| NIR dye (m.w.) | Excitation maximum (nm PBS) | Emission maximum (nm PBS) | Molar extinction coefficient $\varepsilon$ (l/mol cm) | Number dyes/mol[a] | Fluorescence Intensity increase after trypsin[b] activation times at 37° C, 30 min | FL change, before and after trypsin activation (ns) |
|---|---|---|---|---|---|---|
| Cy5.5 mono NHS ester (1128) | 674 | 688 | 250,000 | 35 | $53 \pm 6$ | 0.21/1.1 |
| IRDye 680RD (1003.5) | 672 | 694 | 150,000 | 30 | $52 \pm 7$ | 0.14/0.92 |
| IRDye 35 (1151) | 682 | 696 | 200,000 | 30 | $33 \pm 7$ | ND |
| IRDye 800CWNHS ester (1166) | 774 | 789 | 240,000 | 40 | $108 \pm 4$ | 0.27/0.50 |

[a]Molecular mass of PGC was calculated as 350 kD.
[b]0.01 mg/ml trypsin (130 U/mg).
IRDye 35 is a research dye provided by LI-COR Biosciences and is not commercially available.

and FL is longer than for NIR dyes such as Cy7 and IRDye 800CW. However, *in vivo* imaging experiments suggest that heptamethine dyes (such as IRDye 800CW®) have advantages that are not limited to the spectral range (i.e., red-shifted excitation and emission) of these fluorophores. It has been suggested that MPEG protective chains could theoretically interfere with enzymatic lysis of the bonds within PGC sensors. The assumption in this case is that the sensitivity of the probe would suffer. However, such interference is unlikely, given the relatively small hydrodynamic diameters of lysosomal hydrolases. The interference is of greater importance for larger proteins that are secreted and activated in the extracellular space, for example, in the extracellular matrix. In any case, the extent of PGC sensor fragmentation will depend on whether the protease and the sensor are allowed to react in the same compartment for a sufficient length of time. While model serine proteases have high catalytic rates, real *in vivo* target's catalytic constants may be much lower, which undoubtedly will influence the sensitivity of any sensor, including macromolecular dye-conjugated PGC.

The use of polylysine backbones with "encoded" cathepsin-sensing specificity within the backbone has proved feasible in numerous *in vivo* applications (see Table 9.1 and Sections 7 and 8). However, there are numerous hydrolases, including cathepsins, that can cleave backbones with either free or acylated N-ε-amino group-conjugated PGC sensors. These sensors were not initially designed with the intent of providing high selectivity and specificity for given hydrolases. Such specificity was eventually provided by using iodoacetylated PGC as a "template" for linking synthetic peptides encoding specific peptide sequences and for linking cysteines as well as free amino groups[99] (Fig. 9.5). The latter were modified with NIR dyes (e.g., Cy5.5). This design resulted in probes that rendered large fluorescence intensity increase (350-fold) after reacting with hydrolases that were capable of cleaving the linker. The addition of specific linkers may assist in solving the problem of potential lack of accessibility to NIR fluorophores if the latter are linked directly to the backbone N-ε-amino groups of lysine. However, the caveat is that even the above more sophisticated design of PGC sensor is susceptible to an attack of multiple extracellular proteases, including the secreted cathepsin B (reviewed in Ref. 100) which is still capable of hydrolyzing both the backbone and the linkers. One potential solution is to use a poly-D-lysine backbone, which is not cleavable by naturally occurring proteases. This approach has been used to manufacture a true noncleavable long-circulating imaging agent (Angiosense) that is useful for monitoring tumoral blood volume using NIR imaging.[101]

## 6. PHARMACOKINETICS OF PGC AND IMAGING OF ACTIVATION

Macromolecular sensors based on long-circulating PGC graft copolymers with built-in enzyme-mediated fluorophore activation are large molecules that as a rule have low uptake in the organs of the reticuloendothelial system (liver, spleen). Only PGC based on MPEG-bPLL with low degree of PEGylation or a very low degree of backbone modification of the backbone (10% of amino groups) is usually removed from the circulation rapidly (half-life in blood < 1 h). The biodistribution of fluorophore-linked PGC in animals (rats, rabbits, and nonhuman primates) was found to approximate that of GdDTPA-labeled PGC, suggesting that blood half-life of PEGylated enzyme-sensitive macromolecular sensors ranges in various species between 24 (e.g., in rats) and up to 32 h (in rabbits). Long circulation times in the

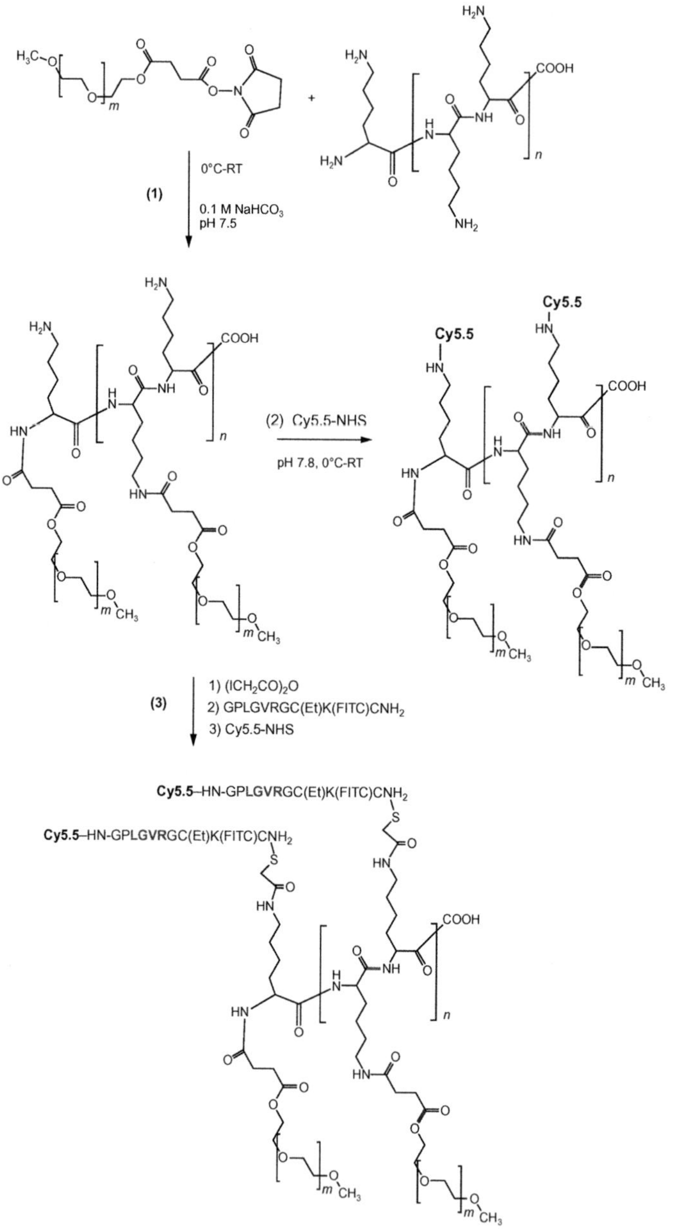

**Figure 9.5** A schema of PGC-based probe synthesis: **(1)** synthesis of MPEG-bPLL; **(2)** synthesis of cathepsin B/H/L-cleavable PGC-Cy5.5; and **(3)** multistep synthesis of MMP-cleavable PGC carrying peptide linkers with MMP-specific LGVR amino acid sequences (shown in italics). Peptide bonds cleavable by cathepsins are shown with an arrow. Esterase-cleavable bonds are shown with an arrowhead. (For color version of this figure, the reader is referred to the online version of this chapter.)

bloodstream enables slow buildup of PGC sensor at the sites of enhanced vascular permeability. The presence of vascular barrier, which is usually compromised in solid tumors and inflammatory lesions, is not freely permeable even in the case of relatively small (5–10 nm sized) molecules such as PGC carriers.

According to the Kedem–Katchalsky equation,[102] the transport rate of a solute, that is, in our case a macromolecular sensor ($J_s$) across the blood vessel wall, can be described as

$$J_s = \omega RT(C_v - C_e) + J_v(1 - \sigma)\bar{C}, \qquad [9.1]$$

where $\bar{C} = 1/2(C_v + C_e)$, that is, the average concentration of the sensor in vascular and extravascular spaces; $J_v$ is the fluid volume flux; $\sigma$ is the solute reflection; $\omega$ are the solute permeability coefficients; and $RT$ is the gas constant–thermodynamic temperature product. The first term takes into account the diffusive transport, and the second term describes the rate of convection. Assuming the temperature to be constant,

$$\omega RT = PS,$$

where $PS$ is the permeability–surface area product.

The vast majority of all fluorescence imaging methods are based on measuring the change of fluorescence intensity in the target organ over time. In general, the total fluorescence intensity measured in the organ is

$$I_o(t) = I_e(t) + I_v(t), \qquad [9.2]$$

where $I_o$ is the total measured fluorescence, $I_e$ is the fluorescence intensity of extravascular compartment, and $I_v$ is the intravascular fluorescence intensity. However, for quenched long-circulating macromolecular sensors, $I_v \sim 0$, and therefore Eq. (9.2) is simplified as

$$I_o(t) \cong I_e(t).$$

The change of extravascular, that is, target organ, fluorescence ($I_e$) in the absence of the intracompartmental self-quenching of enzyme-liberated fluorescent products will be proportional to the buildup of the concentration of free fluorophore ($L$) that is cleaved off a macromolecular carrier. In case the buildup proceeds at a linear rate, that is, the release rate of the fluorophore mediated by the enzyme from the carrier is linear, the fluorescence intensity change rate can be written as

$$\frac{dI_o}{dT} = \frac{dI_e}{dt} = A\left(\frac{dL}{dT}\right)$$

and the rate of fluorophore liberation by the enzyme can be rewritten as

$$\frac{dL}{dT} = acD_t, \qquad [9.3]$$

where $D_t$ is the concentration of the sensor in the target organ at any moment of time, $c$ is the dimensionless concentration of the fluorophore in the carrier macromolecule, and $a$ is the proportionality coefficient.

Thus, in the presence of nonzero blood flow $J_v$ through the target organ, in general, extravascular fluorescence intensity measured in the target organ can be expressed as

$$I_o \cong I_e = AD_T c, \qquad [9.4]$$

where $D_T$ can be calculated using the initial known i.v. injected dose of the macromolecular imaging sensor $D_o$:

$$D_T = D_o(1 - \exp(-Bt))$$

and

$$B = \left(\frac{PSJ_v}{V_e K_{av}}\right),$$

where $PS$ is the permeability–surface product of vasculature through which the macromolecular carrier is being transported, $J_v$ is the blood flow rate, $V_e$ is the extravascular volume, and $K_{av}$ is the available volume fraction in the extravascular compartment. Some of the above parameters can be estimated experimentally using intravital microscopy and whole-body optical imaging.[103,104]

It should be noted that, in general, the kinetics of fluorescent product release or formation is nonlinear and Eq. (9.3) is written in a simplified form. Equation (9.3), which affects Eq. (9.4), ideally has to reflect the order of enzymatic reaction, which unfortunately is unknown for many lysosomal hydrolases and other enzymes that are involved in proteolysis of "quenched" macromolecular carriers of fluorophores in the extracellular matrix.

# 7. APPLICATIONS OF MACROMOLECULAR FLUORESCENT SENSORS IN CANCER IMAGING

The potential for optical imaging of disease-related enzyme activation has been most widely studied in animal models of cancer. The events that take place at the interface between proliferating tumor and surrounding tissue, that is, tumor progression, stromal cell activation, tissue invasion, and metastasis, are always associated with coordinated activation of enzymatic cascades. Enzymes such as MMPs, cathepsins, and uPA have been linked to the accelerated breakdown of the extracellular matrix and induction of endothelial and supporting cell migration and tumor angiogenesis.[2,105–107] For example, it is known that tumor cells trigger protease-dependent extracellular matrix remodeling in response to interactions with neighboring mesenchymal and hematopoietic stromal cells, which in turn allow malignant cells to infiltrate the basement membrane and invade other tissues.[105,108] A number of different fluorescence-based sensors have been developed and shown to be capable of tumor detection in animal models of cancer. These differ most significantly in the way in which the sensor targets the tumor, thus giving specificity to the generated signal.

Our laboratory achieved one of the initial successes in enzyme-mediated tumor detection using a polymer-based NIRF sensor initially designed for imaging tumor proteolytic activity[66] (Fig. 9.1). This PGC, which was described in more detail in Section 5, is biocompatible and consists of a high-molecular-weight carrier of fluorophores with protease sensitivity and specificity "built in" the structure elements of the macromolecular sensor. The unique chemistry of the PGC–fluorophore sensor allows it to be retained in the bloodstream (i.e., have long circulation times) and to pass through highly permeable tumor vasculature after which it accumulates in tumors. Other inherent strengths of this design are that (1) the "off" and "on" states of fluorescence emission are directly linked to the activity of the targeted enzyme; (2) the enzyme is not inactivated by the probe and thus can continue to free more cleavable Cy5.5 products, thus amplifying the fluorescent signal over time (> 24 h); and (3) when used to target intracellular enzymes, the cleaved fluorescent products are sequestered in the cells that comprise the solid tumor (cancer cells and activated stromal cells), thereby allowing the target-to-background ratio (TBR) to increase as the free probe (and its associated background signal) is eliminated from the body. PGC-based probes have also been developed that target extracellar enzymes such as MMPs, uPA, and thrombin.

Injection of PGC-based fluorescent sensors into tumor-bearing animals usually resulted in tumor-associated fluorescence intensity which was on average 12 times higher than the surrounding background and allowed detection of tumor xenografts *in vivo*.[67,68] Subsequent studies have determined that, although taken up and broken down in some tumor cells, the macromolecular sensor is largely taken up by stromal cells that are recruited to the site of tumor formation (fibroblasts, monocytes/

macrophages, and dendritic cells[109]). In these cells, cysteine proteases (possibly cathepsins B, H, or L) are believed to be responsible for the activation of the PGC probe. Such PGC-based sensors have been designed to target a number of proteases including those expressing cathepsin D,[69,74] cathepsin B,[72] MMP,[70,81,110] urokinase plasminogen activator,[84,111] caspase-1,[79] and coagulation enzymes.[112] Both commercially available (ProSense[TM], MMPSense[TM]) and laboratory-synthesized PGC-based fluorescent probes have demonstrated the ability to image a variety of cancers in animal models including ovarian,[94] colon,[88] pancreatic,[92,96] lung,[83] as well as fibrosarcoma,[84] glioma,[77] and metastatic disease.[76,113] High-molecular-weight enzyme-activatable sensors should have a number of advantages in the detection of tumors accessible to potential clinical optical imaging applications (i.e., those within 7–10 mm of the skin or endoluminal surface such as breast, prostate, and colon cancers), especially given the demonstrated sensitivity of these sensors, that is, in detecting subnanomole levels of tumor-associated enzymes. They may also prove valuable for the differentiation and staging of certain cancers. For example, a well-differentiated human breast cancer (BT20) and a highly invasive metastatic human breast cancer (DU4475) selectively express cathepsin B, which can be readily detected *in vivo* with PGC-based fluorescent sensors.[75] Moreover, in tumors of equal size, there was a 1.5-fold higher fluorescence signal in the highly invasive breast cancer, which correlated with a higher cathepsin B protein content. PGC-based metalloproteinase-3 sensing probes have also been used to assess therapeutic efficacy after treatment of tumors with oncolytic adenovirus.[81] A summary of the main examples of PGC-assisted imaging in animal models, including models of cancer, is given in Table 9.1.

## 8. ALTERNATIVE ENZYME-TARGETED STRATEGIES IN IMAGING CANCER

Although results from *in vitro* and *in vivo* testing of PGC-based probes have been overall highly promising, some limitations do remain. One is that enzymes that are released and act in the extracellular space (such as MMPs, uPA, and thrombin) will produce fluorescent cleaved fragments that are susceptible to rapid diffusion and elimination from the target tissue. To address this issue, Tsien and colleagues[114] used nonquenched NIRF (Cy5.5)-tagged polyarginine cell-penetrating peptides (CPPs) that bear a positive charge and linked them through a protease-cleavable linker to a negatively charged peptide. These activatable CPPs will accumulate in the interstitial space but cannot enter cells until the linker is cleaved by extracellular MMP which then frees the CPP and allows it to enter the cell. In this way, the enzyme-activated probe can be retained in the tumor cells, while nonactivated probes are, over time, eliminated from the tissue/animal, thus creating detectable contrast between anatomical regions with high versus low protease activity. Tumor cells secreting MMPs (2 and 9) in mice and biopsied human squamous cell carcinomas could be identified by a two- to threefold increase in NIR fluorescence, compared to surrounding tissue. Subsequent studies showed that the probe could detect several different cancer types (including breast cancer) and that probe accumulation was highest at the tumor–stromal interface.[115]

A similar strategy uses a protease-targeted reactive group (typically an electrophile) to covalently attach a quenched fluorescent dye to the active site of the enzyme. Upon covalent modification, the fluorophore is unquenched and thus becomes activatable. In this way, the fluorescence signal can be directly linked to the activity of the enzyme. Such activity-based probes (ABPs)[116,117] have been developed for *in vitro* detection of proteases, kinases, and phosphatases.[116] When injected into tumor-bearing athymic mice, quenched near-infrared fluorescent activity-based probes (qNIRF-ABPs) targeting lysosomal cysteine proteases allowed *in vivo* detection of cathepsin-producing tumors (SNR of tumor was roughly ninefold higher than surrounding tissue)[118] and determination of enzyme activity attenuation after therapeutic intervention. Covalent attachment to the enzyme allows the dye to be sequestered in cathepsin-producing tissue, thus permitting *ex vivo* microscopic analysis subsequent to *in vivo* imaging. An important difference between ABP-based enzyme probes and PGC-based sensors is that the former inactivates the target enzyme once it has been covalently modified, whereas the latter leaves the enzyme active and thus capable of cleaving and activating additional dye-linked molecules in the tissue (amplification effect). Because of the difference in time required in each case for generating optical signal of acceptable magnitude, direct comparisons of the two methods (i.e., long-circulating "quenched" macromolecular sensors and qNIRF-ABPs) are difficult. However, a recent study suggests that the amplification effect does not invariably result in an overall higher signal generation.[57] Low-molecular-mass qNIRF-ABPs sensors of cathepsin B showed overall high reconstructed volume-averaged fluorescent intensity signals in tumors and low backgrounds in liver and spleen when animals bearing subcutaneous xenografts were imaged using fluorescence-mediated tomography (FMT). The comparison of the overall FMT-reconstructed fluorophore concentration in the subcutaneous (ectopic) mammary tumors showed that, in the case of qNIRF-ABPs, the probe concentration was higher. However, high-molecular-mass commercial PGC analogs (ProSense[TM]) showed much better tumor/lung ratios and overall better target volume/ background ratios, with a much brighter signal achieved in tumors at late time points (8 h, see Fig. 9.1C).[57] In all likelihood, the observed differences between qNIRF-ABPs and PGC-based sensors were due to differences in both the kinetics and retention of the respective dye molecules. ABPs are small and have rapid distribution in extravascular space and clearance, whereas ProSense shows slow diffusion into the extravascular space and slow uptake by cells, where the liberated fluorophores have much longer resident times than qNIRF-ABPs. The latter undergo rapid degradation and exocytosis or compaction in lysosomes after covalent or noncovalent interaction with the target enzymes.

Other non-PGC-based macromolecular NIRF probes have also been developed. Scherer and colleagues showed detection of intestinal neoplasia using a PEGylated polyamidoamine (PAMAM)-Generation 4 dendrimer core covalently coupled to a Cy5.5-labeled peptide which acted as a selective substrate for MMP7. The probe is "self-quenched" (optically silent) in the absence of MMP7 but leads to a 2.2-fold increase in fluorescence in MMP7-expressing tumors[93] at 3 h postinjection. *Ex vivo* imaging yielded even higher TBR (sixfold) and allowed identification of tumors as small as 0.01 cm.[2] However, unlike polyamino acid-based PGCs, PAMAM dendrimers are not biodegradable.

A well-appreciated difficulty in imaging enzyme activity, especially for *in vivo* applications, is the somewhat broad specificity of most biological enzymes (especially proteolytic enzymes). Thus, the resulting fluorescence signal will lack strict specificity for a single enzyme and, in addition, may lack strict specificity for diseased tissue, as even healthy tissues express enzyme activity. One approach that has been implemented to overcome this limitation is to use higher specificity disease targets, such as membrane receptors, as the targeted moiety (reviewed in Ref. 119). Kobayashi *et al.* have described the use of a fluorescent sensor (Av-3ROX) that is composed of a self-quenching avidin–rhodamine X conjugate, which has affinity for lectin on certain cancer cells (such as ovarian, colon, gastric, and pancreatic cancer, which have high metastatic potential). Upon binding to the lectin receptor, the probe is internalized and fluorescence is activated after the probe is cleaved into fragments (dissociation of avidin tetramers to monomers) by lysosomal proteases. This scheme takes advantage of both the broad specificity of lysosomal enzymes and the relative selectivity of the lectin membrane receptor. In addition, because the fluorescent fragments are sequestered inside the cells, there may be greater opportunity for buildup of signal as the non-reacted probe is eliminated from the tissue.[89] Using intravital imaging in murine models of peritoneal ovarian metastases, Av-3ROX allowed imaging of submillimeter cancer nodules (sensitivity of 92% and specificity of 98%) with minimal contamination by background signal. Higher resolution is likely if the probe is detected using current state-of-the-art animal imaging techniques. In a subsequent study, a nonimmunogenic alternative of avidin, galactosamine-conjugated serum albumin (GmSA), was used with similar results.[90]

## 9. MACROMOLECULAR SENSORS IN FLUORESCENT IMAGING OF INFLAMMATION AND VASCULAR DISEASE

Macromolecular fluorescence-based enzyme sensors have also shown utility in detecting pathological changes occurring in cardiovascular disease (reviewed in Ref. 120). Vascular wall inflammation precedes atherosclerosis, and one of the early pathological signs of the disease, the fatty streak, is characterized by the accumulation of monocyte-derived macrophages and T lymphocytes and their associated proteases.[121] Proteolytic enzymes from highly activated macrophages also hasten plaque erosion and rupture, and their levels are correlated with clinical outcomes in patients with cardiovascular disease.[101,122–124] Elevated levels of MMP9 have been found in the brain tissue of stroke patients and also correlate with clinical outcome.[125–128]

Experimentally induced atherosclerotic lesions in mice were successfully imaged using FMT after the injection of PGC-based NIRF probes for cathepsin B[71] and MMPs (MMP2 and 9).[85] Histological analysis showed colocalization of the probes with the targeted enzymes and with activated macrophages. In addition, the probe-mediated signal increase could be diminished after treatment with protease inhibitors.[85] PGC-based MMP-sensitive probes (MMPSense[TM]) have also been used to image increased MMP levels occurring after middle cerebral artery occlusion (MCAO) in mice.[95] In this study, TBRs were 1.3 times higher in the affected hemisphere, and both the fluorescent signal and lesion volume could be diminished by MMP inhibitors. When repeated in MMP9-deficient mice, a major portion of the probe signal could be attributed to MMP9-specific activity. MMP-sensitive probes have also been used to detect inflammatory responses occurring after myocardial infarction in mice.[82] Smaller fluorescence molecular probes have also been developed to image inflammation and enzyme-related vascular pathologies.[86,129–131]

PGC-based NIRF probes designed to target inflammation-specific molecules are also well suited for detecting osteo- and rheumatoid arthritis, as the disease is characterized by early inflammatory responses in which release of MMPs, cysteine proteases, and cathepsin B into the synovial fluid causes the destruction of arthritic joints due to the degeneration of proteoglycan and type II collagen. Because these events occur before morphological changes in joint structure can be detected, there is a strong impetus for finding noninvasive imaging methods to assess treatment options that may prevent further joint damage. In two separate studies, a cathepsin B–activated NIR probe (PGC-Cy5.5) injected 24 h prior to imaging resulted in a three-fold increased fluorescence signal intensity in the affected joints of animals with experimentally induced arthritis as compared to control animals.[78,80] In arthritic animals treated with the anti-inflammatory drug methotrexate (35 mg of MTX/kg 48 h prior to probe injection), a significantly lower fluorescence signal was observed as compared with untreated arthritic animals.[78] In a more recent study, an altered form of the probe was used in which poly-D-lysine replaced poly-L-lysine and a cathepsin K–sensitive peptide linker (GHPG-GPQGKC) was used for attachment of Cy5.5. This configuration resulted in a probe that was activatable only by cathepsin K[132] and was subsequently shown to detect osteoclast-derived cathepsin K–mediated bone degradation, which is largely responsible for bone resorption.

The use of macromolecular enzyme-sensing probes has limitations in certain cases where the extent of probe extravasation from the bloodstream into the extravascular space of the affected organ is very limited. For example, in the heart, which is located near the lungs and liver (the latter have shown high levels of protease-sensing probe activation (see Section 8)[57]), the resultant levels of fluorescent signal (TBR) can be rather small. Small-molecular-weight probes such as qNIRF-ABPs can potentially accumulate in the heart rapidly. However, high levels of accumulation can result in cardiotoxicity. Therefore, we tested the approach that used the effect of cyanine dye FL change in response to probe cleavage by proteases. Although the far-red cyanine dye Cy5.5 gives the most pronounced increase in lifetime after PGC proteolysis (approximately 5×, see Table 9.2), we used

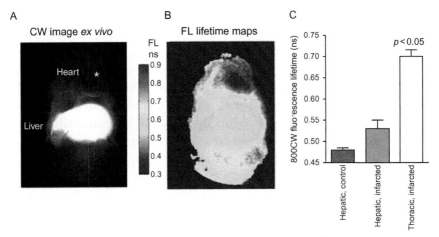

**Figure 9.6** Fluorescence lifetime contrast imaging versus fluorescence intensity changes in experimental myocardial infarction *in situ*. (A) The image taken by using traditional continuous wave fluorescence excitation demonstrates high levels of liver fluorescence that obscure fluorescence in the heart. (B) Lifetime maps showing that longer lifetime fluorescence comes only from infarcted myocardium. (C) The thoracic lifetime increase in mice with myocardial infarction, which is significantly higher than in hepatic areas of control or experimental mice ($p < 0.05$). (See Color Insert.)

the IRDye 800CW dye conjugated to PGC (PGC-800, Table 9.2), as this dye fluorescence can be excited farther in the NIR range, thus decreasing excitation and emission light scattering and nonspecific absorbance of light. The measurements of PGC-800 activation in the area of experimentally induced myocardial ischemia showed that, despite very low changes in fluorescence intensity compared to that in the liver area (see Fig. 9.6A), the differences in lifetimes were clearly detectable using FL maps (Fig. 9.6B). These results demonstrate the benefits of FLI in which the signal difference detected between different tissue compartments (liver and heart) is not sensitive to the local concentration of dye in these organs. As a result, we were able to measure statistically significant changes of FL in the regions affected with myocardial ischemia and the inflammation that followed the infarction (see Fig. 9.6C).

# 10. CONCLUSIONS

Optically "silent" macromolecular imaging probes that undergo enzyme-mediated activation *in vivo* are rapidly becoming essential tools in basic and applied biomedical research, and several of these are now commercially available. These probes can be used for various fluorescence-based detection methods: from whole-body tomography to endoscopic imaging using miniaturized cameras. These "sensing" probes deliver high payloads of fluorophores and have long circulation times *in vivo*, allowing detection of target enzymes over extended periods while keeping the overall dose of the imaging probe low. Their use *in vivo* will include more extensive preclinical applications in investigating responses to experimental therapy and probing the tissue microenvironment. The future will inevitably bring development of more efficient "activatable" probes with optimized enzyme-sensing properties, biocompatibility, and lack of immunogenicity and toxicity. Together with new optical imaging methods such as FL and optoacoustic readout, the ease and precision of differentiating between diseased and normal tissue *in vivo* should lead to a greater understanding of human disease and better methodologies for early diagnosis.

## REFERENCES

1. Cataldo AM, Nixon RA. Enzymatically active lysosomal proteases are associated with amyloid deposits in Alzheimer brain. *Proc Natl Acad Sci USA* 1990;**87**:3861–5.
2. Edwards DR, Murphy G. Cancer. Proteases—invasion and more. *Nature* 1998;**394**:527–8.
3. Fang J, Shing Y, Wiederschain D, Yan L, Butterfield C, Jackson G, et al. Matrix metalloproteinase-2 is required for the switch to the angiogenic phenotype in a tumor model. *Proc Natl Acad Sci USA* 2000;**97**:3884–9.
4. Murray GI, Duncan ME, O'Neil P, Melvin WT, Fothergill JE. Matrix metalloproteinase-1 is associated with poor prognosis in colorectal cancer. *Nat Med* 1996;**2**:461–2.
5. Chambers AF, Matrisian LM. Changing views of the role of matrix metalloproteinases in metastasis. *J Natl Cancer Inst* 1997;**89**:1260–70.
6. Folkman J. Angiogenic zip code. *Nat Biotechnol* 1999;**17**:749.
7. Davidson B, Goldberg I, Kopolovic J, Lerner-Geva L, Gotlieb WH, Ben-Baruch G, et al. MMP-2 and TIMP-2 expression correlates with poor prognosis in cervical carcinoma—a clinicopathologic study using immunohistochemistry and mRNA in situ hybridization. *Gynecol Oncol* 1999;**73**:372–82.
8. Kanayama H, Yokota K, Kurokawa Y, Murakami Y, Nishitani M, Kagawa S. Prognostic values of matrix metalloproteinase-2 and tissue inhibitor of metalloproteinase-2 expression in bladder cancer. *Cancer* 1998;**82**:1359–66.
9. Sakakibara M, Koizumi S, Saikawa Y, Wada H, Ichihara T, Sato H, et al. Membrane-type matrix metalloproteinase-1 expression and activation of gelatinase A as prognostic markers in advanced pediatric neuroblastoma. *Cancer* 1999;**85**:231–9.
10. Shalinsky DR, Brekken J, Zou H, McDermott CD, Forsyth P, Edwards D, et al. Broad antitumor and antiangiogenic activities of AG3340, a potent and selective MMP inhibitor undergoing advanced oncology clinical trials. *Ann N Y Acad Sci* 1999;**878**:236–70.
11. Smith HW, Marshall CJ. Regulation of cell signalling by uPAR. *Nat Rev Mol Cell Biol* 2010;**11**:23–36.
12. Jedeszko C, Sloane BF. Cysteine cathepsins in human cancer. *Biol Chem* 2004;**385**:1017–27.

13. Mohamed MM, Sloane BF. Cysteine cathepsins: multifunctional enzymes in cancer. *Nat Rev Cancer* 2006;**6**:764–75.
14. Keppler D, Sameni M, Moin K, Mikkelsen T, Diglio CA, Sloane BF. Tumor progression and angiogenesis: cathepsin B & Co. *Biochem Cell Biol* 1996;**74**:799–810.
15. Foekens JA, Kos J, Peters HA, Krasovec M, Look MP, Cimerman N, et al. Prognostic significance of cathepsins B and L in primary human breast cancer. *J Clin Oncol* 1998;**16**:1013–21.
16. Harbeck N, Alt U, Berger U, Kruger A, Thomssen C, Janicke F, et al. Prognostic impact of proteolytic factors (urokinase-type plasminogen activator, plasminogen activator inhibitor 1, and cathepsins B, D, and L) in primary breast cancer reflects effects of adjuvant systemic therapy. *Clin Cancer Res* 2001;**7**:2757–64.
17. Jagodic M, Vrhovec I, Borstnar S, Cufer T. Prognostic and predictive value of cathepsins D and L in operable breast cancer patients. *Neoplasma* 2005;**52**:1–9.
18. Garcia M, Platet N, Liaudet E, Laurent V, Derocq D, Brouillet JP, et al. Biological and clinical significance of cathepsin D in breast cancer metastasis. *Stem Cells* 1996;**14**:642–50.
19. Schneider C, Pozzi A. Cyclooxygenases and lipoxygenases in cancer. *Cancer Metastasis Rev* 2011;**30**:277–94.
20. Cortez V, Nair BC, Chakravarty D, Vadlamudi RK. Integrin-linked kinase 1: role in hormonal cancer progression. *Front Biosci (Schol Ed)* 2011;**3**:788–96.
21. Erickson JW, Cerione RA. Glutaminase: a hot spot for regulation of cancer cell metabolism? *Oncotarget* 2010;**1**:734–40.
22. Macphee CH, Nelson JJ, Zalewski A. Lipoprotein-associated phospholipase A2 as a target of therapy. *Curr Opin Lipidol* 2005;**16**:442–6.
23. Fu X, Kassim SY, Parks WC, Heinecke JW. Hypochlorous acid oxygenates the cysteine switch domain of pro-matrilysin (MMP-7). A mechanism for matrix metalloproteinase activation and atherosclerotic plaque rupture by myeloperoxidase. *J Biol Chem* 2001;**276**:41279–87.
24. White H. Editorial: why inhibition of lipoprotein-associated phospholipase A2 has the potential to improve patient outcomes. *Curr Opin Cardiol* 2010;**25**:299–301.
25. Ondetti MA, Rubin B, Cushman DW. Design of specific inhibitors of angiotensin-converting enzyme: new class of orally active antihypertensive agents. *Science* 1977;**196**:441–4.
26. Alam SR, Newby DE, Henriksen PA. Role of the endogenous elastase inhibitor, elafin, in cardiovascular injury: from epithelium to endothelium. *Biochem Pharmacol* 2012;**83**:695–704.
27. Collier AC, Coombs RW, Schoenfeld DA, Bassett RL, Timpone J, Baruch A, et al. Treatment of human immunodeficiency virus infection with saquinavir, zidovudine, and zalcitabine. AIDS Clinical Trials Group. *N Engl J Med* 1996;**334**:1011–7.
28. Orlowski RZ. Bortezomib and its role in the management of patients with multiple myeloma. *Expert Rev Anticancer Ther* 2004;**4**:171–9.
29. McElvaney NG, Nakamura H, Birrer P, Hebert CA, Wong WL, Alphonso M, et al. Modulation of airway inflammation in cystic fibrosis. in vivo suppression of interleukin-8 levels on the respiratory epithelial surface by aerosolization of recombinant secretory leukoprotease inhibitor. *J Clin Invest* 1992;**90**:1296–301.
30. Evans JM, Donnelly LA, Emslie-Smith AM, Alessi DR, Morris AD. Metformin and reduced risk of cancer in diabetic patients. *BMJ* 2005;**330**:1304–5.
31. Devy L, Rabbani SA, Stochl M, Ruskowski M, Mackie I, Naa L, et al. PEGylated DX-1000: pharmacokinetics and antineoplastic activity of a specific plasmin inhibitor. *Neoplasia* 2007;**9**:927–37.
32. Devy L, Huang L, Naa L, Yanamandra N, Pieters H, Frans N, et al. Selective inhibition of matrix metalloproteinase-14 blocks tumor growth, invasion, and angiogenesis. *Cancer Res* 2009;**69**:1517–26.

33. Burden RE, Gormley JA, Jaquin TJ, Small DM, Quinn DJ, Hegarty SM, et al. Antibody-mediated inhibition of cathepsin S blocks colorectal tumor invasion and angiogenesis. *Clin Cancer Res* 2009;**15**:6042–51.

34. Elie BT, Gocheva V, Shree T, Dalrymple SA, Holsinger LJ, Joyce JA. Identification and pre-clinical testing of a reversible cathepsin protease inhibitor reveals anti-tumor efficacy in a pancreatic cancer model. *Biochimie* 2010;**92**:1618–24.

35. Leblond F, Davis SC, Valdes PA, Pogue BW. Pre-clinical whole-body fluorescence imaging: review of instruments, methods and applications. *J Photochem Photobiol B* 2010;**98**:77–94.

36. Sevick-Muraca EM. Translation of near-infrared fluorescence imaging technologies: emerging clinical applications. *Annu Rev Med* 2012;**63**:217–31.

37. Marshall MV, Rasmussen JC, Tan I-C, Aldrich MB, Adams KE, Wang X, et al. Near-infrared fluorescence imaging in humans with indocyanine green: a review and update. *Open Surg Oncol J* 2010;**2**:12–25.

38. Stummer W, Pichlmeier U, Meinel T, Wiestler OD, Zanella F, Reulen HJ. Fluorescence-guided surgery with 5-aminolevulinic acid for resection of malignant glioma: a randomised controlled multicentre phase III trial. *Lancet Oncol* 2006;**7**:392–401.

39. Feigl GC, Ritz R, Moraes M, Klein J, Ramina K, Gharabaghi A, et al. Resection of malignant brain tumors in eloquent cortical areas: a new multimodal approach combining 5-aminolevulinic acid and intraoperative monitoring. *J Neurosurg* 2010;**113**:352–7.

40. Troyan SL, Kianzad V, Gibbs-Strauss SL, Gioux S, Matsui A, Oketokoun R, et al. The FLARE intraoperative near-infrared fluorescence imaging system: a first-in-human clinical trial in breast cancer sentinel lymph node mapping. *Ann Surg Oncol* 2009;**16**:2943–52.

41. Lee BT, Hutteman M, Gioux S, Stockdale A, Lin SJ, Ngo LH, et al. The FLARE intraoperative near-infrared fluorescence imaging system: a first-in-human clinical trial in perforator flap breast reconstruction. *Plast Reconstr Surg* 2010;**126**:1472–81.

42. Ntziachristos V. Fluorescence molecular imaging. *Annu Rev Biomed Eng* 2006;**8**:1–33.

43. Dimitrow E, Riemann I, Ehlers A, Koehler MJ, Norgauer J, Elsner P, et al. Spectral fluorescence lifetime detection and selective melanin imaging by multiphoton laser tomography for melanoma diagnosis. *Exp Dermatol* 2009;**18**:509–15.

44. Poellinger A, Burock S, Grosenick D, Hagen A, Ludemann L, Diekmann F, et al. Breast cancer: early- and late-fluorescence near-infrared imaging with indocyanine green—a preliminary study. *Radiology* 2011;**258**:409–16.

45. Goergen C, Chen H, Bogdanov AJ, Sosnovik D, Kumar A. In vivo fluorescence lifetime detection of an activatable probe in infarcted myocardium. *J Biomed Opt* 2012;**17**. http://dx.doi.org/10.1117/1.JBO.17.5.056001.

46. Funovics M, Weissleder R, Tung CH. Protease sensors for bioimaging. *Anal Bioanal Chem* 2003;**377**:956–63.

47. Pogue BW, Davis SC, Song X, Brooksby BA, Dehghani H, Paulsen KD. Image analysis methods for diffuse optical tomography. *J Biomed Opt* 2006;**11**:33001.

48. Kumar AT, Chung E, Raymond SB, van de Water JA, Shah K, Fukumura D, et al. Feasibility of in vivo imaging of fluorescent proteins using lifetime contrast. *Opt Lett* 2009;**34**:2066–8.

49. Kumar AT, Raymond SB, Boverman G, Boas DA, Bacskai BJ. Time resolved fluorescence tomography of turbid media based on lifetime contrast. *Opt Express* 2006;**14**:12255–70.

50. Han SH, Farshchi-Heydari S, Hall DJ. Analytical method for the fast time-domain reconstruction of fluorescent inclusions in vitro and in vivo. *Biophys J* 2010;**98**:350–7.

51. Hall DJ, Sunar U, Farshchi-Heydari S, Han SH. In vivo simultaneous monitoring of two fluorophores with lifetime contrast using a full-field time domain system. *Appl Opt* 2009;**48**:D74–D78.

52. Mansfield JR, Gossage KW, Hoyt CC, Levenson RM. Autofluorescence removal, multiplexing, and automated analysis methods for in-vivo fluorescence imaging. *J Biomed Opt* 2005;**10**:41207.
53. Davis SC, Dehghani H, Wang J, Jiang S, Pogue BW, Paulsen KD. Image-guided diffuse optical fluorescence tomography implemented with Laplacian-type regularization. *Opt Express* 2007;**15**:4066–82.
54. Sato A, Klaunberg B, Tolwani R. In vivo bioluminescence imaging. *Comp Med* 2004;**54**:631–4.
55. Zrazhevskiy P, Sena M, Gao X. Designing multifunctional quantum dots for bioimaging, detection, and drug delivery. *Chem Soc Rev* 2010;**39**:4326–54.
56. Ntziachristos V, Ripoll J, Wang LV, Weissleder R. Looking and listening to light: the evolution of whole-body photonic imaging. *Nat Biotechnol* 2005;**23**:313–20.
57. Blum G, Weimer RM, Edgington LE, Adams W, Bogyo M. Comparative assessment of substrates and activity based probes as tools for non-invasive optical imaging of cysteine protease activity. *PLoS One* 2009;**4**:e6374.
58. Signore A, Mather SJ, Piaggio G, Malviya G, Dierckx RA. Molecular imaging of inflammation/infection: nuclear medicine and optical imaging agents and methods. *Chem Rev* 2010;**110**:3112–45.
59. Turk B. Targeting proteases: successes, failures and future prospects. *Nat Rev Drug Discov* 2006;**5**:785–99.
60. Chudakov DM, Matz MV, Lukyanov S, Lukyanov KA. Fluorescent proteins and their applications in imaging living cells and tissues. *Physiol Rev* 2010;**90**:1103–63.
61. Welser K, Adsley R, Moore BM, Chan WC, Aylott JW. Protease sensing with nanoparticle based platforms. *Analyst* 2011;**136**:29–41.
62. Gao W, Xing B, Tsien RY, Rao J. Novel fluorogenic substrates for imaging betalactamase gene expression. *J Am Chem Soc* 2003;**125**:11146–7.
63. Maeda H, Ishida N, Kawauchi H, Tsujimura K. Reaction of fluorescein-isothiocyanate with proteins and amino acids. I. Covalent and non-covalent binding of fluorescein-isothiocyanate and fluorescein to proteins. *J Biochem* 1969;**65**:777–83.
64. French T, So PT, Weaver Jr. DJ, Coelho-Sampaio T, Gratton E, Voss Jr. EW, et al. Two-photon fluorescence lifetime imaging microscopy of macrophage-mediated antigen processing. *J Microsc* 1997;**185**:339–53.
65. Horino K, Kindezelskii AL, Elner VM, Hughes BA, Petty HR. Tumor cell invasion of model 3-dimensional matrices: demonstration of migratory pathways, collagen disruption, and intercellular cooperation. *FASEB J* 2001;**15**:932–9.
66. Bogdanov AJ, Mazzanti M, Castillo G, Bolotin E. Protected graft copolymer (PGC) in imaging and therapy: a platform for the delivery of covalently and non-covalently bound drugs. *Theranostics* 2012;**2**:553–76. http://dx.doi.org/10.7150/thno.4070.
67. Weissleder R, Tung CH, Mahmood U, Bogdanov Jr. A. In vivo imaging of tumors with protease-activated near-infrared fluorescent probes. *Nat Biotechnol* 1999;**17**:375–8.
68. Mahmood U, Tung CH, Bogdanov Jr. A, Weissleder R. Near-infrared optical imaging of protease activity for tumor detection. *Radiology* 1999;**213**:866–70.
69. Tung CH, Mahmood U, Bredow S, Weissleder R. In vivo imaging of proteolytic enzyme activity using a novel molecular reporter. *Cancer Res* 2000;**60**:4953–8.
70. Bremer C, Tung CH, Weissleder R. In vivo molecular target assessment of matrix metalloproteinase inhibition. *Nat Med* 2001;**7**:743–8.
71. Chen J, Tung CH, Mahmood U, Ntziachristos V, Gyurko R, Fishman MC, et al. In vivo imaging of proteolytic activity in atherosclerosis. *Circulation* 2002;**105**:2766–71.
72. Ntziachristos V, Tung CH, Bremer C, Weissleder R. Fluorescence molecular tomography resolves protease activity in vivo. *Nat Med* 2002;**8**:757–60.
73. Jaffer FA, Tung CH, Gerszten RE, Weissleder R. In vivo imaging of thrombin activity in experimental thrombi with thrombin-sensitive near-infrared molecular probe. *Arterioscler Thromb Vasc Biol* 2002;**22**:1929–35.

74. Marten K, Bremer C, Khazaie K, Sameni M, Sloane B, Tung CH, et al. Detection of dysplastic intestinal adenomas using enzyme-sensing molecular beacons in mice. *Gastroenterology* 2002;**122**:406–14.
75. Bremer C, Tung CH, Bogdanov Jr. A, Weissleder R. Imaging of differential protease expression in breast cancers for detection of aggressive tumor phenotypes. *Radiology* 2002;**222**:814–8.
76. Wunderbaldinger P, Turetschek K, Bremer C. Near-infrared fluorescence imaging of lymph nodes using a new enzyme sensing activatable macromolecular optical probe. *Eur Radiol* 2003;**13**:2206–11.
77. Shah K, Tung CH, Chang CH, Slootweg E, O'Loughlin T, Breakefield XO, et al. In vivo imaging of HIV protease activity in amplicon vector-transduced gliomas. *Cancer Res* 2004;**64**:273–8.
78. Wunder A, Tung CH, Muller-Ladner U, Weissleder R, Mahmood U. In vivo imaging of protease activity in arthritis: a novel approach for monitoring treatment response. *Arthritis Rheum* 2004;**50**:2459–65.
79. Messerli SM, Prabhakar S, Tang Y, Shah K, Cortes ML, Murthy V, et al. A novel method for imaging apoptosis using a caspase-1 near-infrared fluorescent probe. *Neoplasia* 2004;**6**:95–105.
80. Lai WF, Chang CH, Tang Y, Bronson R, Tung CH. Early diagnosis of osteoarthritis using cathepsin B sensitive near-infrared fluorescent probes. *Osteoarthritis Cartilage* 2004;**12**:239–44.
81. Lamfers ML, Gianni D, Tung CH, Idema S, Schagen FH, Carette JE, et al. Tissue inhibitor of metalloproteinase-3 expression from an oncolytic adenovirus inhibits matrix metalloproteinase activity in vivo without affecting antitumor efficacy in malignant glioma. *Cancer Res* 2005;**65**:9398–405.
82. Chen J, Tung CH, Allport JR, Chen S, Weissleder R, Huang PL. Near-infrared fluorescent imaging of matrix metalloproteinase activity after myocardial infarction. *Circulation* 2005;**111**:1800–5.
83. Grimm J, Kirsch DG, Windsor SD, Kim CF, Santiago PM, Ntziachristos V, et al. Use of gene expression profiling to direct in vivo molecular imaging of lung cancer. *Proc Natl Acad Sci USA* 2005;**102**:14404–9.
84. Hsiao JK, Law B, Weissleder R, Tung CH. In-vivo imaging of tumor associated urokinase-type plasminogen activator activity. *J Biomed Opt* 2006;**11**:34013.
85. Deguchi JO, Aikawa M, Tung CH, Aikawa E, Kim DE, Ntziachristos V, et al. Inflammation in atherosclerosis: visualizing matrix metalloproteinase action in macrophages in vivo. *Circulation* 2006;**114**:55–62.
86. Nahrendorf M, Sosnovik DE, Waterman P, Swirski FK, Pande AN, Aikawa E, et al. Dual channel optical tomographic imaging of leukocyte recruitment and protease activity in the healing myocardial infarct. *Circ Res* 2007;**100**:1218–25.
87. Jaffer FA, Kim DE, Quinti L, Tung CH, Aikawa E, Pande AN, et al. Optical visualization of cathepsin K activity in atherosclerosis with a novel, protease-activatable fluorescence sensor. *Circulation* 2007;**115**:2292–8.
88. Alencar H, Funovics MA, Figueiredo J, Sawaya H, Weissleder R, Mahmood U. Colonic adenocarcinomas: near-infrared microcatheter imaging of smart probes for early detection—study in mice. *Radiology* 2007;**244**:232–8.
89. Hama Y, Urano Y, Koyama Y, Kamiya M, Bernardo M, Paik RS, et al. A target cell-specific activatable fluorescence probe for in vivo molecular imaging of cancer based on a self-quenched avidin-rhodamine conjugate. *Cancer Res* 2007;**67**:2791–9.
90. Hama Y, Urano Y, Koyama Y, Gunn AJ, Choyke PL, Kobayashi H. A self-quenched galactosamine-serum albumin-rhodamineX conjugate: a "smart" fluorescent molecular imaging probe synthesized with clinically applicable material for detecting peritoneal ovarian cancer metastases. *Clin Cancer Res* 2007;**13**:6335–43.

91. Haller J, Hyde D, Deliolanis N, de Kleine R, Niedre M, Ntziachristos V. Visualization of pulmonary inflammation using noninvasive fluorescence molecular imaging. *J Appl Physiol* 2008;**104**:795–802.

92. von Burstin J, Eser S, Seidler B, Meining A, Bajbouj M, Mages J, et al. Highly sensitive detection of early-stage pancreatic cancer by multimodal near-infrared molecular imaging in living mice. *Int J Cancer* 2008;**123**:2138–47.

93. Scherer RL, VanSaun MN, McIntyre JO, Matrisian LM. Optical imaging of matrix metalloproteinase-7 activity in vivo using a proteolytic nanobeacon. *Mol Imaging* 2008;**7**.118–31.

94. Sheth RA, Upadhyay R, Stangenberg L, Sheth R, Weissleder R, Mahmood U. Improved detection of ovarian cancer metastases by intraoperative quantitative fluorescence protease imaging in a pre-clinical model. *Gynecol Oncol* 2009;**112**:616–22.

95. Klohs J, Baeva N, Steinbrink J, Bourayou R, Boettcher C, Royl G, et al. In vivo near-infrared fluorescence imaging of matrix metalloproteinase activity after cerebral ischemia. *J Cereb Blood Flow Metab* 2009;**29**:1284–92.

96. Ignat M, Aprahamian M, Lindner V, Altmeyer A, Perretta S, Dallemagne B, et al. Feasibility and reliability of pancreatic cancer staging using fiberoptic confocal fluorescence microscopy in rats. *Gastroenterology* 2009;**137**: 1584-92 e1.

97. Nahrendorf M, Waterman P, Thurber G, Groves K, Rajopadhye M, Panizzi P, et al. Hybrid in vivo FMT-CT imaging of protease activity in atherosclerosis with customized nanosensors. *Arterioscler Thromb Vasc Biol* 2009;**29**:1444–51.

98. Habibollahi P, Figueiredo JL, Heidari P, Dulak AM, Imamura Y, Bass AJ, et al. Optical Imaging with a Cathepsin B Activated Probe for the Enhanced Detection of Esophageal Adenocarcinoma by Dual Channel Fluorescent Upper GI Endoscopy. *Theranostics* 2012;**2**:227–34.

99. Tung CH, Bredow S, Mahmood U, Weissleder R. Preparation of a cathepsin D sensitive near-infrared fluorescence probe for imaging. *Bioconjug Chem* 1999;**10**:892–6.

100. Sloane BF, Yan S, Podgorski I, Linebaugh BE, Cher ML, Mai J, et al. Cathepsin B and tumor proteolysis: contribution of the tumor microenvironment. *Semin Cancer Biol* 2005;**15**:149–57.

101. Zhang R, Brennan ML, Fu X, Aviles RJ, Pearce GL, Penn MS, et al. Association between myeloperoxidase levels and risk of coronary artery disease. *JAMA* 2001;**286**:2136–42.

102. Kedem O, Katchalsky A. Thermodynamic analysis of the permeability of biological membranes to non-electrolytes. *Biochim Biophys Acta* 1958;**27**:229–46.

103. Dreher MR, Liu W, Michelich CR, Dewhirst MW, Yuan F, Chilkoti A. Tumor vascular permeability, accumulation, and penetration of macromolecular drug carriers. *J Natl Cancer Inst* 2006;**98**:335–44.

104. Qin S, Seo JW, Zhang H, Qi J, Curry FR, Ferrara KW. An imaging-driven model for liposomal stability and circulation. *Mol Pharm* 2010;**7**:12–21.

105. Kim J, Yu W, Kovalski K, Ossowski L. Requirement for specific proteases in cancer cell intravasation as revealed by a novel semiquantitative PCR-based assay. *Cell* 1998;**94**:353–62.

106. McCawley LJ, Matrisian LM. Matrix metalloproteinases: multifunctional contributors to tumor progression. *Mol Med Today* 2000;**6**:149–56.

107. Gherardi E, Birchmeier W, Birchmeier C, Vande Woude G. Targeting MET in cancer: rationale and progress. *Nat Rev Cancer* 2012;**12**:89–103.

108. Stetler-Stevenson WG, Aznavoorian S, Liotta LA. Tumor cell interactions with the extracellular matrix during invasion and metastasis. *Annu Rev Cell Biol* 1993;**9**:541–73.

109. Bogdanov Jr. AA, Lin CP, Simonova M, Matuszewski L, Weissleder R. Cellular activation of the self-quenched fluorescent reporter probe in tumor microenvironment. *Neoplasia* 2002;**4**:228–36.

110. Pham W, Choi Y, Weissleder R, Tung CH. Developing a peptide-based near-infrared molecular probe for protease sensing. *Bioconjug Chem* 2004;**15**:1403–7.
111. Law B, Curino A, Bugge TH, Weissleder R, Tung CH. Design, synthesis, and characterization of urokinase plasminogen-activator-sensitive near-infrared reporter. *Chem Biol* 2004;**11**:99–106.
112. Tung CH, Gerszten RE, Jaffer FA, Weissleder R. A novel near-infrared fluorescence sensor for detection of thrombin activation in blood. *Chembiochem* 2002;**3**:207–11.
113. Penna FJ, Freilich DA, Alvarenga C, Nguyen HT. Improving lymph node yield in retroperitoneal lymph node dissection using fluorescent molecular imaging: a novel method of localizing lymph nodes in Guinea pig model. *Urology* 2011;**78**(232): e15–e18.
114. Jiang T, Olson ES, Nguyen QT, Roy M, Jennings PA, Tsien RY. Tumor imaging by means of proteolytic activation of cell-penetrating peptides. *Proc Natl Acad Sci USA* 2004;**101**:17867–72.
115. Olson ES, Aguilera TA, Jiang T, Ellies LG, Nguyen QT, Wong EH, et al. In vivo characterization of activatable cell penetrating peptides for targeting protease activity in cancer. *Integr Biol (Camb)* 2009;**1**:382–93.
116. Kato D, Boatright KM, Berger AB, Nazif T, Blum G, Ryan C, et al. Activity-based probes that target diverse cysteine protease families. *Nat Chem Biol* 2005;**1**:33–8.
117. Blum G, Mullins SR, Keren K, Fonovic M, Jedeszko C, Rice MJ, et al. Dynamic imaging of protease activity with fluorescently quenched activity-based probes. *Nat Chem Biol* 2005;**1**:203–9.
118. Blum G, von Degenfeld G, Merchant MJ, Blau HM, Bogyo M. Noninvasive optical imaging of cysteine protease activity using fluorescently quenched activity-based probes. *Nat Chem Biol* 2007;**3**:668–77.
119. Kobayashi H, Ogawa M, Alford R, Choyke PL, Urano Y. New strategies for fluorescent probe design in medical diagnostic imaging. *Chem Rev* 2010;**110**:2620–40.
120. Jaffer FA, Libby P, Weissleder R. Molecular imaging of cardiovascular disease. *Circulation* 2007;**116**:1052–61.
121. Malle E, Waeg G, Schreiber R, Grone EF, Sattler W, Grone HJ. Immunohistochemical evidence for the myeloperoxidase/H2O2/halide system in human atherosclerotic lesions: colocalization of myeloperoxidase and hypochlorite-modified proteins. *Eur J Biochem* 2000;**267**:4495–503.
122. Baldus S, Heeschen C, Meinertz T, Zeiher AM, Eiserich JP, Munzel T, et al. Myeloperoxidase serum levels predict risk in patients with acute coronary syndromes. *Circulation* 2003;**108**:1440–5.
123. Brennan ML, Penn MS, Van Lente F, Nambi V, Shishehbor MH, Aviles RJ, et al. Prognostic value of myeloperoxidase in patients with chest pain. *N Engl J Med* 2003;**349**:1595–604.
124. Exner M, Minar E, Mlekusch W, Sabeti S, Amighi J, Lalouschek W, et al. Myeloperoxidase predicts progression of carotid stenosis in states of low high-density lipoprotein cholesterol. *J Am Coll Cardiol* 2006;**47**:2212–8.
125. Rosell A, Ortega-Aznar A, Alvarez-Sabin J, Fernandez-Cadenas I, Ribo M, Molina CA, et al. Increased brain expression of matrix metalloproteinase-9 after ischemic and hemorrhagic human stroke. *Stroke* 2006;**37**:1399–406.
126. Montaner J, Molina CA, Monasterio J, Abilleira S, Arenillas JF, Ribo M, et al. Matrix metalloproteinase-9 pretreatment level predicts intracranial hemorrhagic complications after thrombolysis in human stroke. *Circulation* 2003;**107**:598–603.
127. Rosell A, Alvarez-Sabin J, Arenillas JF, Rovira A, Delgado P, Fernandez-Cadenas I, et al. A matrix metalloproteinase protein array reveals a strong relation between MMP-9 and MMP-13 with diffusion-weighted image lesion increase in human stroke. *Stroke* 2005;**36**:1415–20.

128. Montaner J, Rovira A, Molina CA, Arenillas JF, Ribo M, Chacon P, et al. Plasmatic level of neuroinflammatory markers predict the extent of diffusion-weighted image lesions in hyperacute stroke. *J Cereb Blood Flow Metab* 2003;**23**:1403–7.
129. Shepherd J, Hilderbrand SA, Waterman P, Heinecke JW, Weissleder R, Libby P. A fluorescent probe for the detection of myeloperoxidase activity in atherosclerosis-associated macrophages. *Chem Biol* 2007;**14**:1221–31.
130. Sosnovik DE, Nahrendorf M, Deliolanis N, Novikov M, Aikawa E, Josephson L, et al. Fluorescence tomography and magnetic resonance imaging of myocardial macrophage infiltration in infarcted myocardium in vivo. *Circulation* 2007;**115**:1384–91.
131. Nahrendorf M, Zhang H, Hembrador S, Panizzi P, Sosnovik DE, Aikawa E, et al. Nanoparticle PET-CT imaging of macrophages in inflammatory atherosclerosis. *Circulation* 2008;**117**:379–87.
132. Kozloff KM, Quinti L, Patntirapong S, Hauschka PV, Tung CH, Weissleder R, et al. Non-invasive optical detection of cathepsin K-mediated fluorescence reveals osteoclast activity in vitro and in vivo. *Bone* 2009;**44**:190–8.

CHAPTER TEN

# Fluorescent Proteins as Visible *In Vivo* Sensors

## Robert M. Hoffman*,†

*AntiCancer, Inc., San Diego, California, USA
†Department of Surgery, University of California, San Diego, California, USA

## Contents

*Progress in Molecular Biology and Translational Science*, Volume 113
ISSN 1877-1173
http://dx.doi.org/10.1016/B978-0-12-386932-6.00010-7

## Abstract

Fluorescent proteins have enabled a whole new technology of visible *in vivo* genetic sensors. Fluorescent proteins have revolutionized biology by enabling what was formerly invisible to be seen clearly. These proteins have allowed us to visualize, in real time, important aspects of cancer in living animals, including tumor cell mobility, invasion, metastasis, and angiogenesis. These multicolored proteins have allowed the color coding of cancer cells growing *in vivo* and enabled the distinction of host from tumor with single-cell resolution. Whole-body imaging with fluorescent proteins has been shown to be a powerful technology to noninvasively follow the dynamics of metastatic cancer. Whole-body imaging of cancer cells expressing fluorescent proteins has enabled the facile determination of efficacy of candidate antitumor and antimetastatic agents in mouse models. The use of fluorescent proteins to differentially label cancer cells in the nucleus and cytoplasm and high-powered imaging technology have enabled the visualization of the nuclear–cytoplasmic dynamics of cancer cells *in vivo*, including noninvasive techniques. Fluorescent proteins thus enable both macro- and microimaging technology and thereby provide the basis for the new field of *in vivo* cell biology.

## 1. INTRODUCTION

Our laboratory has pioneered the field of *in vivo* visible genetic sensors using fluorescent proteins.[1–6] This development was enabled by the discovery of the green fluorescent protein (GFP) by Shimomura,[7] the discovery and purification of the red fluorescent protein (RFP) by Labas and Savitsky,[8] and the cloning of the GFP gene by Prasher.[9]

## 2. NONINVASIVE IMAGING

There have been many misleading papers that GFP is not usable for *in vivo* imaging[10–12] because of interference by skin autofluorescence. Apparently, these authors were not aware that very simple equipment with a narrow-band excitation filter and a bandpass emission filter can be used to image the whole body of mice implanted with cells expressing fluorescent proteins, without interference from skin autofluorescence.[13] These same authors have propagated the misconceptions in the literature suggesting fluorescent proteins are inferior to luciferase for imaging.[14–16] The results described here should greatly clarify this subject.

This chapter will demonstrate the following: (i) very strong signals are emitted from GFP- and RFP-expressing tumors inside the animal; (ii) the

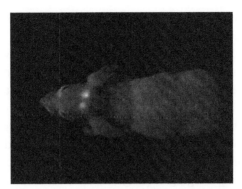

**Figure 10.1** Non-invasive image of GFP and RFP human tumors in the brain in a nude mouse. GFP- and RFP-expressing U87 human glioma cells were implanted in the brain in a single nude mouse. The excitation light was produced with a simple blue-LED flashlight equipped with an excitation filter with a central peak of 470 nm. The image was acquired with a Hamamatsu charge-coupled device (CCD) camera.[13] (See Color Insert.)

images are readily quantifiable; (iii) there is negligible interference from autofluorescence; and (iv) very simple and low-cost instruments can be used for GFP and RFP whole-body macroimaging, including LED (light-emitting diode) flashlights with proper filters (Fig. 10.1).[13]

A novel, RFP-expressing pancreatic cancer model was orthotopically established in nude mice.[17] The MIA-PaCa-2 human pancreatic cancer cell line was transduced with RFP and grown subcutaneously. Fluorescent tumor fragments were then surgically transplanted onto the nude mouse pancreas. Groups treated with intraperitoneal gemcitabine or intravenous irinotecan were sequentially imaged to compare, in real time, the anti–metastatic and anti–tumor efficacy of these agents compared with untreated controls. The anti–metastatic efficacy of each drug was followed noninvasively in real time by imaging the RFP-expressing tumor and metastases and was confirmed by fluorescence open imaging of autopsy specimens. This highly metastatic model reliably simulates the aggressive course of human pancreatic cancer. Noninvasive, sequential imaging permits quantification of tumor growth and dissemination and, thereby, real-time evaluation of therapeutic efficacy. These features make this model an ideal pre-clinical system with which to study novel therapeutics for pancreatic cancer.

## 2.1. Noninvasive cellular and subcellular imaging *in vivo*

Our laboratory has developed dual-color fluorescent cells with one color in the nucleus and the other in the cytoplasm, which enable real-time nuclear–cytoplasmic dynamics to be visualized in living cells *in vivo* as well as *in vitro*.[18] To obtain the dual-color cells, RFP was expressed in the cytoplasm of cancer cells, and GFP linked to histone H2B was expressed in the nucleus. Nuclear GFP expression enabled visualization of nuclear dynamics, whereas simultaneous cytoplasmic RFP expression enabled visualization of nuclear–cytoplasmic ratios as well as simultaneous cell and nuclear shape changes. The cell-cycle position of individual living cells was readily visualized by the nuclear–cytoplasmic ratio and nuclear morphology. Real-time induction of apoptosis was observed by nuclear size changes and progressive nuclear fragmentation. Mitotic cells were visualized by whole-body imaging after injection into the mouse ear. Common carotid artery injection of dual-color cells and a reversible skin flap enabled the external visualization of the dual-color cells in microvessels in the mouse brain where extreme elongation of the cell body as well as the nucleus occurred.

## 2.2. Noninvasive cellular and subcellular imaging of cancer cell–stromal cell interaction

To noninvasively image cancer cell/stromal cell interaction in the tumor microenvironment (TME) and drug response at the cellular level in live animals in real time, we developed a new imageable three-color animal model.[19] The model consists of GFP-expressing mice transplanted with dual-color cancer cells labeled with GFP in the nucleus and RFP in the cytoplasm. An Olympus IV100 laser scanning microscope, with ultra-narrow microscope objectives (stick objectives), was used for three-color whole-body imaging of the two-color cancer cells interacting with the GFP-expressing stromal cells. Drug response of both cancer and stromal cells in the intact live animal was also imaged in real time. Various *in vivo* phenomena of tumor–host interaction and cellular dynamics were noninvasively imaged, including mitotic and apoptotic tumor cells, stromal cells interacting with cancer cells, tumor vasculature, and tumor blood flow (Fig. 10.2).

## 2.3. Noninvasive cellular imaging in graft versus host disease

Noninvasive *in vivo* fluorescence imaging of the ear pinna enabled visualization of GFP donor cells at the single-cell level in real time after allogeneic hematopoietic stem-cell transplantation (HSCT).[20] Movement of donor cells was increased by treatment with croton oil as an inflammatory reagent. Treatment with dexamethasone, as an anti-inflammatory reagent, suppressed donor cell infiltration.

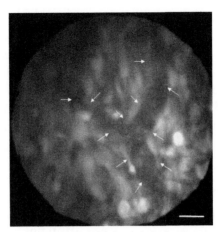

**Figure 10.2** Non-invasive, subcellular imaging of dual-color mouse mammary cancer cells and GFP stromal cells in the live GFP-nude mouse. Dual-color mouse mammary tumor (MMT) cells were injected in the footpad of GFP transgenic nude mice. Numerous dual-color spindle-shaped MMT cells interacted with GFP-expressing host cells. Well-developed tumor blood vessels and real-time blood flow were visualized by noninvasive imaging (arrows). Bar: 20 μm.[19] (See Color Insert.)

The ability to image nuclear–cytoplasmic dynamics *in vivo* is a major advance in our ability to understand the proliferation, quiescence, dormancy, trafficking, and death of cancer cells in the living animal. With this powerful technology, we will be able to visualize *in vivo* the most fundamental properties of cancer, including the reversible transition between cancer cell proliferation and quiescence, how prolonged quiescence may lead to dormancy, the dynamics of cell death, and the nuclear–cytoplasmic dynamics of cancer cell spread. Most importantly, we will have an opportunity to visualize, in real time, in the live animal the activity of novel drugs on these processes as well as how drugs induce cell death at the subcellular level. With this technology, we can expect to discover new classes of drugs for cancer.

## 3. LIGHTING UP THE TUMOR STROMA WITH FLUORESCENT PROTEINS

### 3.1. Imaging the recruitment of cancer-associated fibroblasts by liver-metastatic colon cancer

The TME is critical for tumor growth and progression. In order to image the TME, a GFP-expressing mouse was used as the host that expressed GFP in all organs but not the parenchymal cells of the liver. Noncolored HCT-116 human colon cancer cells were injected into the spleen of GFP-nude mice, which led to the formation of experimental liver metastasis.

TME formation resulting from the liver metastasis was observed using an Olympus OV100 small animal fluorescence imaging system. HCT-116 cells formed tumor colonies in the liver 28 days after cell transplantation to the spleen. GFP-expressing host cells were recruited by the metastatic tumors as visualized by fluorescence imaging. A desmin-positive area increased around and within the liver metastasis over time, suggesting that cancer-associated fibroblasts (CAFs) were recruited by the liver metastasis which have a role in tumor progression.[21]

## 3.2. Multicolor palette of the TME

Six different implantation models were used to image the TME using multiple colors of fluorescent proteins: (I) RFP- or GFP-expressing HCT-116 human colon cancer cells were implanted subcutaneously in the cyan fluorescent protein (CFP)-expressing nude mice. CFP stromal elements from the subcutaneous TME were visualized interacting with the RFP- or GFP-expressing tumors. (II) RFP-expressing HCT-116 cells were transplanted into the spleen of CFP-nude mice, and experimental metastases were then formed in the liver. CFP stromal elements from the liver TME were visualized interacting with the RFP-expressing tumor. (III) RFP-expressing HCT-116 cancer cells were transplanted in the tail vein of CFP-expressing nude mice, forming experimental metastases in the lung. CFP stromal elements from the lung were visualized interacting with the RFP-expressing tumor. (IV) In order to visualize two different tumors in the TME, GFP-expressing and RFP-expressing HCT-116 cancer cells were co-implanted subcutaneously in CFP-expressing nude mice. A three-color TME was formed subcutaneously in the CFP mouse, and CFP stromal elements were visualized interacting with the RFP- and GFP-expressing tumors. (V) In order to have two different colors of stromal elements, GFP-expressing HCT-116 cells were initially injected subcutaneously in RFP-expressing nude mice. After 14 days, the tumor, which consisted of GFP cancer cells and RFP stromal cells derived from the RFP-nude mouse, was harvested and transplanted into the CFP-nude mouse. CFP stromal cells invaded the growing transplanted tumor containing GFP cancer cells and RFP stroma. (VI) Mouse mammary tumor (MMT) cells expressing GFP in the nucleus and RFP in the cytoplasm were implanted in the spleen of a CFP-nude mouse. Cancer cells were imaged in the liver 3 days after cell injection. The dual-color dividing MMT cells and CFP hepatocytes, as well as CFP nonparenchymal cells of the liver, were imaged interacting with the two–color cancer cells. CFP-expressing host CAFs were predominantly observed in the TME models developed in the CFP-nude mouse (Fig. 10.3).[22]

## 4. STROMA CELLS ARE REQUIRED FOR CANCER METASTASIS

After splenic injection of colon cancer cells, splenocytes co-trafficked with the tumor cells to the liver and facilitated metastatic colony formation. Extensive clasmocytosis (destruction of the cytoplasm) of the cancer cells occurred within 6 h after portal vein (PV) injection, and essentially all the cancer cells died. In contrast, splenic injection of these tumor cells resulted in the formation of liver and distant metastasis. GFP spleen cells were found in the liver metastases, which resulted from intrasplenic injection of the cancer cells in transgenic nude mice ubiquitously expressing GFP. When GFP spleen cells and the RFP cancer cells were coinjected

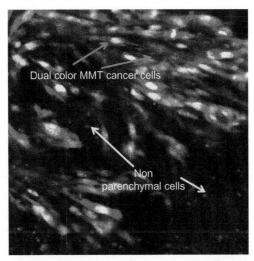

**Figure 10.3** MMT cells with GFP in the cytoplasm and RFP in the nucleus, growing in the liver of a CFP-nude mouse. Dual-color MMT cells formed tumors in the liver of a CFP mouse 28 days after splenic injection. Hepatocytes, non-parenchymal liver cells (yellow arrows), and dual-color MMT cancer cells (red arrows) were visualized simultaneously. The image was taken with an FV1000 confocal microscope. Bar: 50 μm.[22] (See Color Insert.)

in the PV, liver metastasis resulted that contained GFP spleen cells. These results suggest that liver metastasis requires the presence of stromal cells.[23]

## 5. FLUORESCENT TUMORGRAFTS MADE FROM HUMAN CANCER PATIENTS

### 5.1. Multicolor palette of fluorescent proteins for lighting up patent tumors in mouse models

Tumor specimens from patients with from pancreatic cancer were initially established subcutaneously in SCID-NOD mice immediately after surgery. The patient tumors were then harvested from SCID-NOD mice and passaged orthotopically in transgenic nude mice ubiquitously expressing RFP. The primary patient tumors acquired RFP-expressing stroma. The RFP-expressing stroma included CAFs and tumor-associated macrophages (TAMs). Further passage to transgenic nude mice ubiquitously expressing GFP resulted in tumors and metastasis that acquired GFP stroma in addition to their RFP stroma, including CAFs and TAMs and blood vessels. The RFP stroma persisted in the tumors growing in the GFP mouse. Further passage to transgenic nude mice ubiquitously expressing CFP resulted in tumors and metastasis acquiring CFP stroma in addition to persisting RFP and GFP stroma, including RFP- and GFP-expressing CAFs and TAMs and blood vessels (Fig. 10.4).[24]

### 5.2. Lighting up metastasis from patient tumors in mouse models

Primary patient tumors acquired GFP-expressing stroma. Subsequent liver metastases and disseminated peritoneal metastases maintained the stroma from the primary tumor and possibly recruited additional GFP-expressing stroma, resulting in their very bright fluorescence. The GFP-expressing stroma included CAFs and TAMs in both primary and metastatic tumors. This imageable model of metastasis from a patient tumor is an important advance over patient "tumorgraft" models currently in use, which are implanted subcutaneously, do not metastasize, and are not imageable.[25]

### 5.3. Noninvasive fluorescence imaging of patient tumors in mouse models

The tumors from pancreatic cancer patients, with very bright GFP and RFP stroma, as described above, were orthotopically passaged to noncolored nude mice. The brightly fluorescent patient tumors could be non-invasively

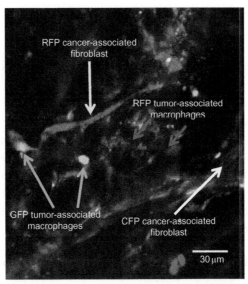

**Figure 10.4** Stroma of patient pancreatic cancer after passage to RFP-, GFP-, and CFP-nude mice. RFP cancer-associated fibroblasts (CAFs) (yellow arrow) and GFP tumor-associated macrophages (TAMs) (green arrows) in the tumor from pancreatic cancer patients grew in the CFP host tumor. White arrow indicates CFP CAFs. Image was taken with the Olympus FV1000 confocal microscope. Bar: 30 mm.[24] (See Color Insert.)

imaged longitudinally as they progressed in the noncolored nude mice. This non-invasively imageable tumorgraft model will be valuable for screening for effective treatment options for individual pancreatic cancer patients, as well as for the discovery of improved agents for this treatment-resistant disease.[26]

## 6. REAL-TIME IMAGING OF TRAFFICKING CANCER CELLS

### 6.1. Real-time imaging of deforming cancer cells in capillaries

The migration velocity of cancer cells in capillaries was measured by capturing images of the dual-color fluorescent cells over time. The cells and nuclei in the capillaries elongated to fit the width of these vessels. The average length of the major axis of the cancer cells in the capillaries increased to approximately four times their normal length. The nuclei increased their length 1.6 times in the capillaries. Cancer cells in capillaries over 8 μm in diameter could migrate up to 48.3 μm/h. These data suggest that the minimum diameter of capillaries where cancer cells are able to migrate is approximately 8 μm.[27]

## 6.2. Real-time imaging of trafficking cancer cells in blood vessels

Dual-color cancer cells were injected by a vascular route in an abdominal skin flap in nude mice. Nuclear and cytoplasmic behavior of cancer cells were observed in real time in blood vessels as dual-colored cancer cells trafficking by various means or adhered to the vessel inner surface in the abdominal skin flap. During extravasation, real-time dual-color imaging showed that cytoplasmic processes of the cancer cells exited the vessels first, with nuclei following along the cytoplasmic projections. Both cytoplasm and nuclei underwent deformation during extravasation.[28]

## 6.3. Imaging the trafficking of cancer cells in lymphatic vessels

Cancer cells labeled with both GFP in the nucleus and RFP in the cytoplasm were injected into the inguinal lymph node of nude mice. The labeled cancer cells trafficked through lymphatic vessels, where they were imaged via a skin flap in real time at the cellular level until they entered the axillary lymph node. Using this imaging technology, we investigated the role of pressure on tumor cell shedding into lymphatic vessels. Pressure was generated by placing 25- and 250-g weights for 10 s on the bottom surface of a tumor-bearing footpad. Increasing pressure on the tumor increased the numbers of shed cancer cells, and emboli. Pressure also deformed the shed emboli, increasing their maximum major axis.[29]

## 6.4. The role of the intravascular microenvironment in spontaneous metastasis development

Real-time imaging of cancer cell–endothelium interactions during spontaneous metastatic colonization of the liver and lung in live mice was carried out. We observed that prior to the detection of extravascular metastases, GFP-expressing PC-3 cancer cells resided initially inside the blood vessels of the liver and the lung, where they proliferated and expressed Ki-67 and exhibited matrix metalloproteinase (MMP) activity. Thus, the intravascular cancer cells produced their own microenvironment, where they could continue to proliferate. Extravasation occurred earlier in the lung than in the liver. These results demonstrate that the intravascular microenvironment is a critical staging area for the development of metastasis that later can invade the parenchyma. Intravascular tumor cells may represent a therapeutic target to inhibit the development of extravascular metastases.[30]

## 6.5. Imaging of nuclear–cytoplasmic dynamics, proliferation, and cell death of cancer cells in the portal vein area

Human HCT-116 colon cancer and MMT cells were injected in the PV of nude mice. Both cell lines were labeled with GFP in the nucleus and RFP in the cytoplasm. The cells were observed intravitally in the liver at the single-cell level using the Olympus OV100 system. Most HCT-116–GFP–RFP cells remained in sinusoids near peripheral PVs. Only a small fraction of the cancer cells invaded the lobular area. Extensive clasmocytosis of the HCT-116–GFP–RFP cells occurred within 6 h. The number of apoptotic cells rapidly increased within the PV within 12 h of injection. Apoptosis was readily visualized in the dual-color cells by separation of nucleus and cytoplasm. The data suggest rapid death of HCT-116–GFP–RFP cells in the PV. In contrast, dual-color MMT–GFP–RFP cells injected into the PV mostly survived in the liver of nude mice 24 h after injection. Many surviving MMT–GFP–RFP cells showed invasive figures with cytoplasmic protrusions. The cells grew aggressively and formed colonies in the liver. However, when the host mice were pretreated with cyclophosphamide, the HCT-116–GFP–RFP cells also survived and formed colonies in the liver after PV injection. These results suggest that a cyclophosphamide-sensitive host cellular system attacked the HCT-116–GFP–RFP cells but could not effectively kill the MMT–GFP–RFP cells.[31]

## 6.6. Imaging the effect of cyclophosphamide on cancer cells quiescence, proliferation, and death in blood vessels

Intravascular proliferation, extravasation, and colony formation by cancer cells, which are critical steps of metastasis, can be enhanced by pretreatment of host mice with cyclophosphamide. In contrast, in the un-pretreated mice, most cancer cells remained quiescent in vessels without extravasation. HT1080 human fibrosarcoma cells, labeled in the nucleus with GFP and in the cytoplasm with RFP, were injected into the epigastric cranialis vein of nude mice. Twenty-four hours before cancer cell injection, cyclophosphamide was administered i.p. Double-labeled cancer cells were imaged at the cellular level in live mice with the Olympus OV100 system. Cyclophosphamide appeared to interfere with a host process that inhibited intravascular proliferation, extravasation, and extravascular colony formation. Cyclophosphamide did not directly affect the cancer cells, as cyclophosphamide had been cleared by the time the cancer cells were injected. These results demonstrate an important, unexpected "opposite effect" of chemotherapy that enhances critical steps in malignancy rather than inhibiting them, suggesting that certain current approaches to cancer chemotherapy should be modified.[32]

## 7. METHOD OF CHOICE FOR WHOLE-BODY IMAGING

The features of fluorescent-protein-based imaging, such as a very strong and stable signal, enable noninvasive whole-body imaging down to the subcellular level,[19] especially with red-shifted proteins, and make it far superior to luciferase-based imaging. Luciferase-based imaging, with its very weak signal,[12] precluding image acquisition and allowing only photon counting with pseudocolor-generated images, has very limited applications.[6] For example, cellular imaging *in vivo* is not possible with luciferase. The dependence on circulating luciferin makes the signal from luciferase imaging unstable.[6] The one possible advantage of luciferase-based imaging is that no excitation light is necessary. However, far-red absorbing proteins such as Katushka greatly reduce problems with excitation, even in deep tissues, as shown by Shcherbo *et al.*[33]

## 8. CONCLUSIONS

With the imaging technologies described here, essentially any *in vivo* process can be imaged, enabling the new field of *in vivo* cell biology using fluorescent proteins that act as visible *in vivo* genetic sensors.[34] Recent applications of the technology described here include linking fluorescent proteins with cell-cycle-specific proteins such that the cells change color from red to green as they transit from G1 to S phases.[35] Another recent application is the combinatorial expression of a series of four different color fluorescent proteins, resulting in at least 90 different colors of cells in the brain such that the lineage of each can be traced. The technique has been called "Brainbow."[36] The possibilities seem limitless for sensing any phenomena *in vivo* with visible genetic reports.

## REFERENCES

1. Chishima T, Miyagi Y, Wang X, Yamaoka H, Shimada H, Moossa AR, et al. Cancer invasion and micrometastasis visualized in live tissue by green fluorescent protein expression. *Cancer Res* 1997;**57**:2042–7.
2. Yang M, Baranov E, Jiang P, Sun F-X, Li X-M, Li L, et al. Whole-body optical imaging of green fluorescent protein-expressing tumors and metastases. *Proc Natl Acad Sci USA* 2000;**97**:1206–11.

3. Hoffman RM. The multiple uses of fluorescent proteins to visualize cancer in vivo. *Nat Rev Cancer* 2005;**5**:796–806.
4. Hoffman RM, Yang M. Subcellular imaging in the live mouse. *Nat Protoc* 2006;**1**:775–82.
5. Hoffman RM, Yang M. Color-coded fluorescence imaging of tumor-host interactions. *Nat Protoc* 2006;**1**:928–35.
6. Hoffman RM, Yang M. Whole-body imaging with fluorescent proteins. *Nat Protoc* 2006;**1**:1429–38.
7. Shimomura O. Discovery of Green Fluorescent Protein (GFP) (Nobel Lecture). *Angew Chem Int Ed* 2009;**48**:5590–602.
8. Matz MV, Fradkov AF, Labas YA, Savitsky AP, Zaraisky AG, Markelov ML, et al. Fluorescent proteins from nonbioluminescent Anthozoa species. *Nat Biotechnol* 1999;**17**:969–73.
9. Prasher DC, Eckenrode VK, Ward WW, Prendergast FG, Cormier MJ. Primary structure of the Aequorea victoria green fluorescent protein. *Gene* 1992;**111**:229–33.
10. Weissleder R, Tung CH, Mahmood U, Bogdanov Jr. A. In vivo imaging of tumors with protease-activated near-infrared fluorescent probes. *Nat Biotechnol* 1999;**17**:375–8.
11. Contag CH, Jenkins D, Contag PR, Negrin RS. Use of reporter genes for optical measurements of neoplastic disease in vivo. *Neoplasia* 2000;**2**:41–52.
12. Ray P, De A, Min J-J, Tsien RY, Gambhir SS. Imaging tri-fusion multimodality reporter gene expression in living subjects. *Cancer Res* 2004;**64**:1323–30.
13. Yang M, Luiken G, Baranov E, Hoffman RM. Facile whole-body imaging of internal fluorescent tumors in mice with an LED flashlight. *Biotechniques* 2005;**39**:170–2.
14. Gross S, Piwnica-Worms D. Spying on cancer: molecular imaging in vivo with genetically encoded reporters. *Cancer Cell* 2005;**7**:5–15.
15. Weissleder R, Ntziachristos V. Shedding light onto live molecular targets. Shedding light onto live molecular targets. *Nat Med* 2003;**9**:123–8.
16. Ntziachristos V, Ripoll J, Wang LV, Weissleder R. Looking and listening to light: the evolution of whole-body photonic imaging. *Nat Biotechnol* 2005;**23**:313–20.
17. Katz M, Takimoto S, Spivac D, Moossa AR, Hoffman RM, Bouvet M. A novel red fluorescent protein orthotopic pancreatic cancer model for the preclinical evaluation of chemotherapeutics. *J Surg Res* 2003;**113**:151–60.
18. Yamamoto N, Jiang P, Yang M, Xu M, Yamauchi K, Tsuchiya H, et al. Cellular dynamics visualized in live cells *in vitro* and *in vivo* by differential dual-color nuclear-cytoplasmic fluorescent-protein expression. *Cancer Res* 2004;**64**:4251–6.
19. Yang M, Jiang P, Hoffman RM. Whole-body subcellular multicolor imaging of tumor-host interaction and drug response in real time. *Cancer Res* 2007;**67**:5195–200.
20. Yamazaki T, Aoki K, Heike Y, Kim S-W, Ochiya T, Wakeda T, et al. Real-time in vivo cellular imaging of graft-versus-host disease and its reaction to immunomodulatory reagents. *Immunol Lett* 2012;**144**:33–40.
21. Suetsugu A, Osawa Y, Nagaki M, Saji S, Moriwaki H, Bouvet M, et al. Imaging the recruitment of cancer-associated fibroblasts by liver-metastatic colon cancer. *J Cell Biochem* 2011;**112**:949–53.
22. Suetsugu A, Hassanein MK, Reynoso J, Osawa Y, Nagaki M, Moriwaki H, et al. The cyan fluorescent protein nude mouse as a host for multicolor-coded imaging models of primary and metastatic tumor microenvironments. *Anticancer Res* 2012;**32**:31–8.
23. Bouvet M, Tsuji K, Yang M, Jiang P, Moossa AR, Hoffman RM. *In vivo* color-coded imaging of the interaction of colon cancer cells and splenocytes in the formation of liver metastases. *Cancer Res* 2006;**66**:11293–7.
24. Suetsugu A, Katz M, Fleming J, Truty M, Thomas R, Moriwaki H, et al. Multi-color palatte of fluorescent proteins for imaging the tumor microenvironment of orthotopic

tumorgraft mouse models of clinical pancreatic cancer specimens. *J Cell Biochem* 2012;**113**:2290–5.

25. Suetsugu A, Katz M, Fleming J, Truty M, Thomas R, Saji S, et al. Imageable fluorescent metastasis resulting in transgenic GFP mice orthotopically implanted with human patient primary pancreatic cancer specimens. *Anticancer Res* 2012;**32**:1175–80.

26. Suetsugu A, Katz M, Fleming J, Truty M, Thomas R, Saji S, et al. Non-invasive fluorescent protein imaging of orthotopic pancreatic-cancer-patient tumorgraft progression in nude mice. *Anticancer Res* 2012;**32**:3063–7.

27. Yamauchi K, Yang M, Jiang P, Yamamoto N, Xu M, Amoh Y, et al. Real-time in vivo dual-color imaging of intracapillary cancer cell and nucleus deformation and migration. *Cancer Res* 2005;**65**:4246–52.

28. Yamauchi K, Yang M, Jiang P, Xu M, Yamamoto N, Tsuchiya H, et al. Development of real-time subcellular dynamic multicolor imaging of cancer cell trafficking in live mice with a variable-magnification whole-mouse imaging system. *Cancer Res* 2006; **66**:4208–14.

29. Hayashi K, Jiang P, Yamauchi K, Yamamoto N, Tsuchiya H, Tomita K, et al. Real-time imaging of tumor-cell shedding and trafficking in lymphatic channels. *Cancer Res* 2007;**67**:8223–8.

30. Zhang Q, Yang M, Shen J, Gerhold LM, Hoffman RM, Xing HR. The role of the intravascular microenvironment in spontaneous metastasis development. *Int J Cancer* 2010;**126**:2534–41.

31. Tsuji K, Yamauchi K, Yang M, Jiang P, Bouvet M, Endo H, et al. Dual-color imaging of nuclear-cytoplasmic dynamics, viability, and proliferation of cancer cells in the portal vein area. *Cancer Res* 2006;**66**:303–6.

32. Yamauchi K, Yang M, Hayashi K, Jiang P, Yamamoto N, Tsuchiya H, et al. Induction of cancer metastasis by cyclophosphamide pretreatment of host mice: an opposite effect of chemotherapy. *Cancer Res* 2008;**68**:516–20.

33. Shcherbo D, Merzlyak EM, Chepurnykh TV, Fradkov AF, Ermakova GV, Solovieva EA, et al. Bright far-red fluorescent protein for whole-body imaging. *Nat Methods* 2007;**4**:741–6.

34. Jiang P, Yamauchi K, Yang M, Tsuji K, Xu M, Maitra A, et al. Tumor cells genetically labeled with GFP in the nucleus and RFP in the cytoplasm for imaging cellular dynamics. *Cell Cycle* 2006;**5**:1198–201.

35. Sakaue-Sawano A, Kurokawa H, Morimura T, Hanyu A, Hama H, Osawa H, et al. Visualizing spatiotemporal dynamics of multicellular cell-cycle progression. *Cell* 2008;**132**:487–98.

36. Livet J, Weissman TA, Kang H, Draft RW, Lu J, Bennis RA, et al. Transgenic strategies for combinatorial expression of fluorescent proteins in the nervous system. *Nature* 2007;**450**:56–62.

# INDEX

Note: Page numbers followed by "*f*" indicate figures, and "*t*" indicate tables.

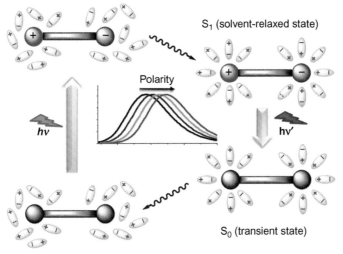

S₁ (Franck-Condon state)

S₁ (solvent-relaxed state)

Polarity

hν

hν′

S₀ (transient state)

S₀ (ground state)

Klymchenko and Mely, Figure 2.2

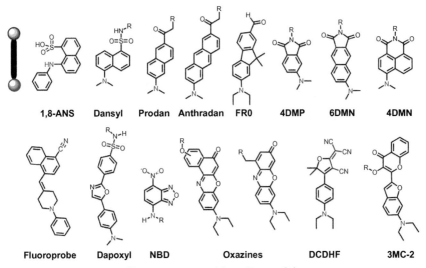

1,8-ANS  Dansyl  Prodan  Anthradan  FR0  4DMP  6DMN  4DMN

Fluoroprobe  Dapoxyl  NBD  Oxazines  DCDHF  3MC-2

Klymchenko and Mely, Figure 2.3

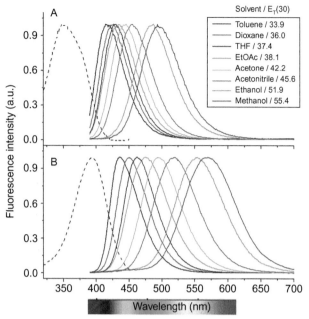

KLYMCHENKO AND MELY, FIGURE 2.4

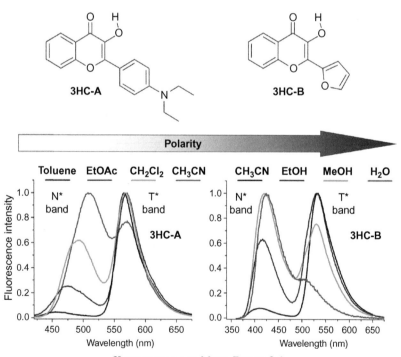

KLYMCHENKO AND MELY, FIGURE 2.6

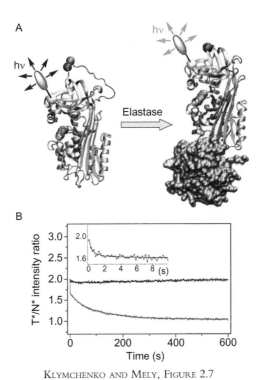

A

hν

hν

Elastase

B

KLYMCHENKO AND MELY, FIGURE 2.7

Time domain—Time-correlated single-photon counting

Simplified Perrin-Jablonski diagram

$S_1$

$\Delta t_1$

$S_0$

$\Delta t_1$

$S_2$

$\Delta t_2$

$S_0$

$\Delta t_2$

$S_2$

$\Delta t_2$

$S_0$

$\Delta t_n$

Fluorescence lifetime

Laser pulse    Emitted photon    Previously emitted photon    Fit curve

Frequency domain—Phase and modulation lifetime

Modulated excitation
Fluorescence emission
Fitting for each pixel

φ1
φ2

Phase shift (φ)

m1
m2

Demodulation (m)

Intensity

reference

Elapsed time

Phase lifetime ($\tau_t$)    4 ns

Modulation lifetime ($\tau_m$)    1 ns

SIPIETER AND VANDAME ET AL., FIGURE 5.13

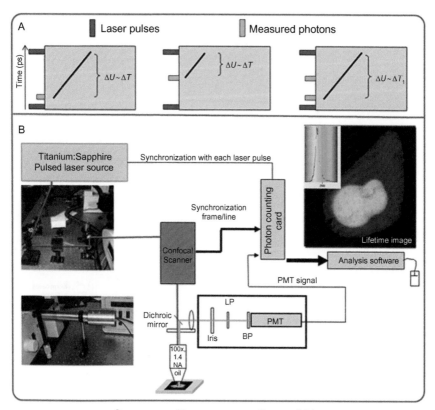

SIPIETER AND VANDAME *ET AL.*, FIGURE 5.14

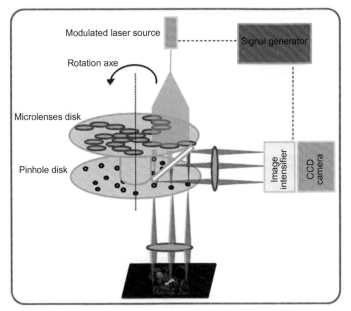

SIPIETER AND VANDAME *ET AL.*, FIGURE 5.15

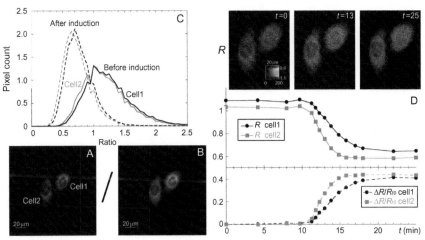

SIPIETER AND VANDAME *ET AL.*, FIGURE 5.16

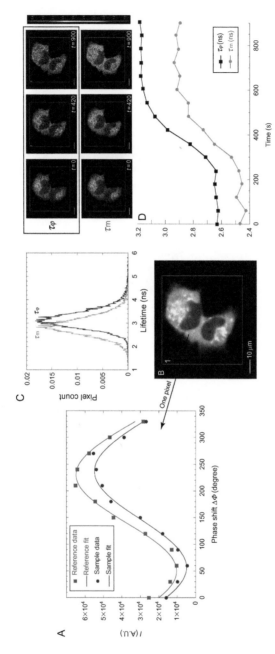

SIPIETER AND VANDAME *ET AL.*, FIGURE 5.17

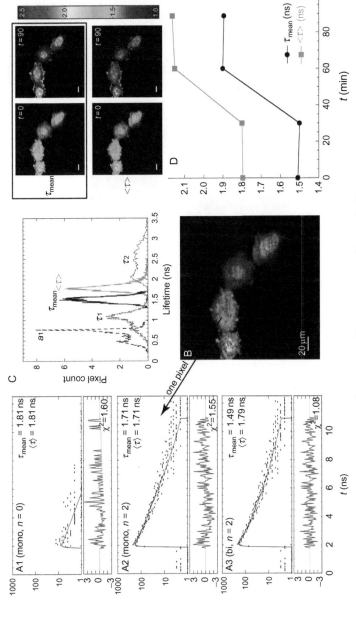

SIPIETER AND VANDAME *ET AL.*, FIGURE 5.18

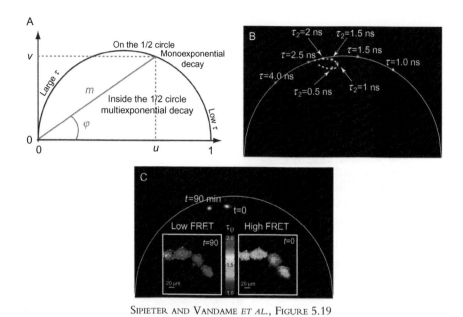

SIPIETER AND VANDAME *ET AL.*, FIGURE 5.19

SIPIETER AND VANDAME *ET AL.*, FIGURE 5.21

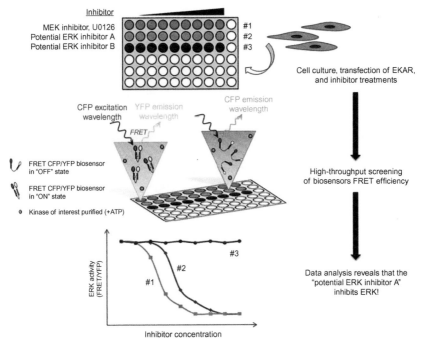

SIPIETER AND VANDAME *ET AL.*, FIGURE 5.23

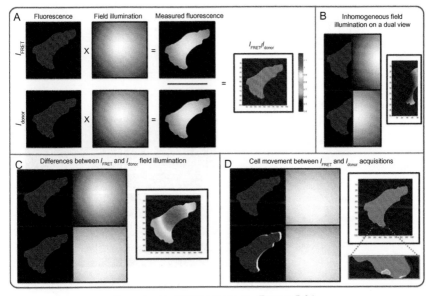

SIPIETER AND VANDAME *ET AL.*, FIGURE 5.24

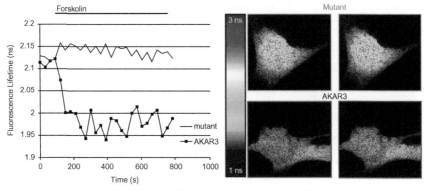

SIPIETER AND VANDAME *ET AL.*, FIGURE 5.25

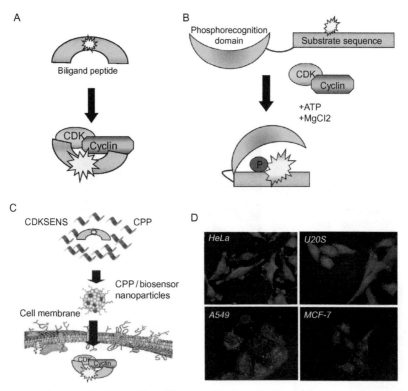

THI NHU NGOC VAN AND MORRIS, FIGURE 6.14

## A  Temporal selectivity

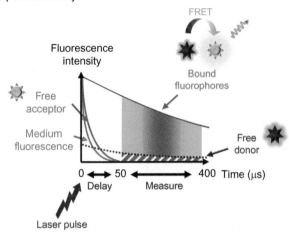

## B  Spectral selectivity

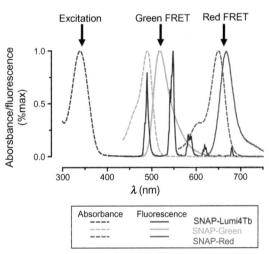

SCHOLLER *ET AL.*, FIGURE 7.1

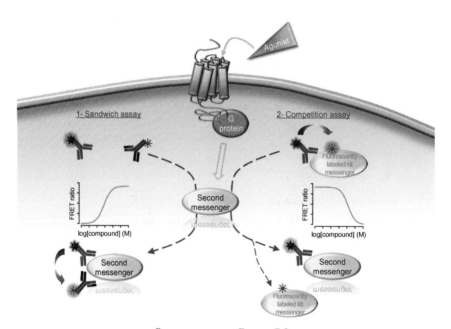

SCHOLLER *ET AL.*, FIGURE 7.2

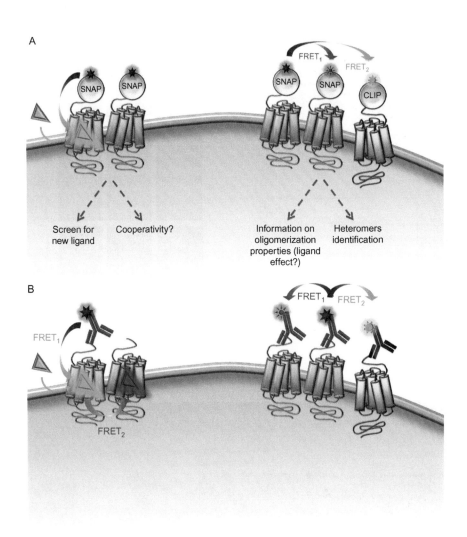

A

SNAP  SNAP

Screen for     Cooperativity?
new ligand

FRET$_1$
FRET$_2$

SNAP  SNAP  CLIP

Information on    Heteromers
oligomerization   identification
properties (ligand
effect?)

B

FRET$_1$

FRET$_2$

FRET$_1$  FRET$_2$

SCHOLLER *ET AL.*, FIGURE 7.3

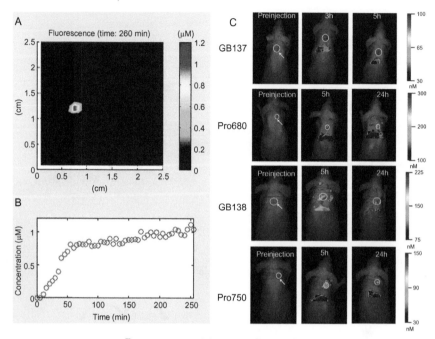

BOGDANOV AND MAZZANTI, FIGURE 9.1

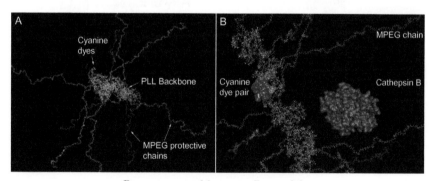

BOGDANOV AND MAZZANTI, FIGURE 9.3

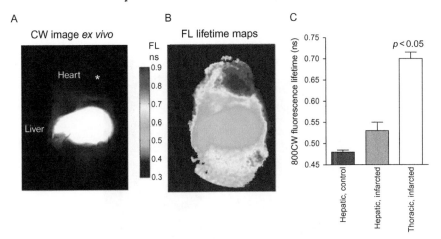

A
CW image *ex vivo*

Heart     *

Liver

FL
ns
0.9
0.8
0.7
0.6
0.5
0.4
0.3

B
FL lifetime maps

C
$p < 0.05$

BOGDANOV AND MAZZANTI, FIGURE 9.6

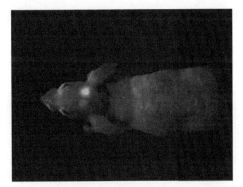

HOFFMAN, FIGURE 10.1

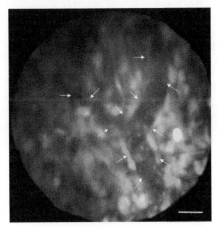

HOFFMAN, FIGURE 10.2

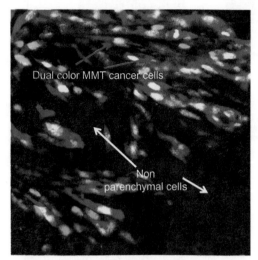

HOFFMAN, FIGURE 10.3

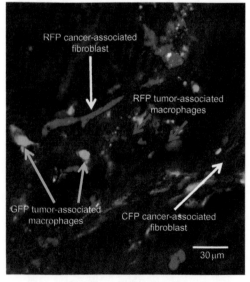

HOFFMAN, FIGURE 10.4

Printed and bound by CPI Group (UK) Ltd, Croydon, CR0 4YY

08/05/2025

01864953-0004